AF598138

Methods in Molecular Biology

Series Editor
John M. Walker
School of Life Sciences
University of Hertfordshire
Hatfield, Hertfordshire, AL10 9AB, UK

For further volumes:
http://www.springer.com/series/7651

Cancer Cell Signaling

Methods and Protocols

Second Edition

Edited by

Martha Robles-Flores

Universidad Nacional Autónoma de México, México D.F., México

Editor
Martha Robles-Flores
Universidad Nacional Autónoma de México
México D.F., México

ISSN 1064-3745 ISSN 1940-6029 (electronic)
ISBN 978-1-4939-0855-4 ISBN 978-1-4939-0856-1 (eBook)
DOI 10.1007/978-1-4939-0856-1
Springer New York Heidelberg Dordrecht London

Library of Congress Control Number: 2014937488

Printed on acid-free paper

Humana Press is a brand of Springer
Springer is part of Springer Science+Business Media (www.springer.com)

Preface

The aim of this book, "Cancer Cell Signaling: Methods and Protocols Second Edition," is to bring together the most recent advances in cancer cell signaling knowledge with the recent progress in the development of novel strategies and experimental cell models to study human cancer. This area has exploded over the last years with the identification of tumor-initiating cancer stem cells, with the identification of intensive interactions of oncogenic signaling pathways with tumor cell metabolism, and with the identification of epigenetic changes associated with tumor initiation and progression. More recently, other discoveries, such as the property of cancer cells to communicate between themselves via extracellular vesicles, or that long noncoding RNAs play key regulatory roles in cancer biology, have also contributed to this explosion. The accumulated experimental evidence in all these topics has begun to show significant effects on approaches to study cancer: new technologies are being developed to exploit this knowledge that surely will have a profound impact in cancer prevention and treatment.

The book is a suite of reviews and protocols written by a team of acknowledged researchers and covers some of these hot topics in cancer cell signaling. It is organized into four parts: Part I is concerned with targeting cancer cell metabolism as an anticancer strategy. Part II contains an overview of the interdependency between genetic and epigenetic regulatory effects in cancer and several protocols to study epigenetic control of cancer. In Part III, the reader will be introduced to the theme of metastasis promotion with a review of microvesicles as mediators of intercellular communication in cancer, followed by a set of protocols to study cancer metastasis. Finally, in Part IV, two new techniques for cancer study are described.

I would like to acknowledge and thank all authors for their valuable contributions, particularly for sharing with the readers all hints, tips, and observations that one learns from using a method regularly. I also thank Dr. John Walker for his technical guidance and support.

México D.F., México *Martha Robles-Flores, Ph.D.*

Contents

Contributors

NARIMAN ANSARI • *Physical Biology Group, Buchmann Institute for Molecular Life Sciences (BMLS), Goethe Universität Frankfurt am Main, Frankfurt am Main, Germany*

PATRICK ANTONIETTI • *Experimental Neurosurgery, Neuroscience Center, Goethe University Hospital, Frankfurt am Main, Germany*

MARC A. ANTONYAK • *Department of Molecular Medicine, Cornell University, Ithaca, NY, USA*

ERANDI AYALA-ORTEGA • *Departamento de Genética Molecular, Instituto de Fisiología Celular, Universidad Nacional Autónoma de México, México, DF, México*

FÁTIMA BALTAZAR • *Surgical Sciences Domain, Life and Health Sciences Research Institute (ICVS), School of Health Sciences, University of Minho, Braga/Guimarães, Portugal; ICVS/3B's-PT Government Associate Laboratory, Braga/Guimarães, Portugal*

IGNACIO CAMACHO-ARROYO • *Departamento de Biología, Facultad de Química, Universidad Nacional Autónoma de México, México, DF, México*

RICHARD A. CERIONE • *Department of Molecular Medicine, C3-155 Veterinary Medical Center, Cornell University, Ithaca, NY, USA*

COLLEEN M. CONNELLY • *Department of Chemistry, North Carolina State University, Raleigh, NC, USA*

ALEXANDER DEITERS • *Department of Chemistry, University of Pittsburgh, Chevron Science Center, Pittsburgh, PA, USA*

ALFONSO DUEÑAS-GONZÁLEZ • *Instituto de Investigaciones Biomédicas, Universidad Nacional Autónoma de México UNAM/Instituto Nacional de Cancerología, México City, México*

YI FENG • *Abramson Family Cancer Research Institute, School of Medicine, University of Pennsylvania, Philadelphia, PA, USA*

JONATHAN GARCÍA-ROMÁN • *Unidad de Bioquímica, Instituto Nacional de Ciencias Médicas y Nutrición "Salvador Zubirán" SSA, México, DF, México*

PILAR GARCÍA-TOBILLA • *Instituto Nacional de Medicina Genómica, México, DF, México*

EDGAR GONZÁLEZ-BUENDÍA • *Departamento de Genética Molecular, Instituto de Fisiología Celular, Universidad Nacional Autónoma de México, México, DF, México*

CLAUDIA GONZÁLEZ-ESPINOSA • *Departamento de Farmacobiología, Centro de Investigación y de Estudios Avanzados (Cinvestav) Sede Sur., México, DF, México*

SARA GRANJA • *Surgical Sciences Domain, Life and Health Sciences Research Institute (ICVS), School of Health Sciences, University of Minho, Braga/Guimarães, Portugal; ICVS/3B's-PT Government Associate Laboratory, Braga/Guimarães, Portugal*

GEORGINA GUERRERO • *Departamento de Genética Molecular, Instituto de Fisiología Celular, Universidad Nacional Autónoma de México, México, DF, México*

CÉSAR ALEJANDRO GUZMÁN-PÉREZ • *Departamento de Medicina Genómica y Toxicología Ambiental, Instituto de Investigaciones Biomédicas, UNAM, México, DF, México; Unidad de Bioquímica, Instituto Nacional de Ciencias Médicas y Nutrición "Salvador Zubirán" SSA, México DF, México*

VALERIA HANSBERG-PASTOR • *Departamento de Biología, Facultad de Química, Universidad Nacional Autónoma de México, México, DF, México*

STEFANIE HARDUNG • *Physical Biology Group, Buchmann Institute for Molecular Life Sciences (BMLS), Goethe Universität Frankfurt am Main, Frankfurt am Main, Germany*

EMILIANO HISAKI-ITAYA • *Unidad de Desarrollo e Investigación en Bioprocesos (UDIBI), Escuela Nacional de Ciencias Biológicas, IPN, México, DF, México*

KATHARINA HÖTTE • *Physical Biology Group, Buchmann Institute for Molecular Life Sciences (BMLS), Goethe Universität Frankfurt am Main, Frankfurt am Main, Germany*

XIAOWEN HU • *Ovarian Cancer Research Center, School of Medicine, University of Pennsylvania, Philadelphia, PA, USA*

ZHUO-WEI HU • *Molecular Immunology and Cancer Pharmacology Group, State Key Laboratory of Bioactive Substance and Function of Natural Medicines, Institute of Materia Medica, Chinese Academy of Medical Sciences and Peking Union Medical College, Beijing, China*

FANG HUA • *Molecular Immunology and Cancer Pharmacology Group, State Key Laboratory of Bioactive Substance and Function of Natural Medicines, Institute of Materia Medica, Chinese Academy of Medical Sciences and Peking Union Medical College, Beijing, China*

ALFREDO IBARRA-SÁNCHEZ • *Departamento de Farmacobiología, Centro de Investigación y de Estudios Avanzados (Cinvestav) Sede Sur., México, DF, México*

DONAT KÖGEL • *Experimental Neurosurgery, Neuroscience Center, Goethe University Hospital, Frankfurt am Main, Germany*

CHUNSHENG LI • *Ovarian Cancer Research Center, School of Medicine, University of Pennsylvania, Philadelphia, PA, USA*

QI LV • *Molecular Immunology and Cancer Pharmacology Group, State Key Laboratory of Bioactive Substance and Function of Natural Medicines, Institute of Materia Medica, Chinese Academy of Medical Sciences and Peking Union Medical College, Beijing, China*

FERMÍN MAR-AGUILAR • *Laboratorio de Inmunología y Virología, Facultad de Ciencias Biológicas, Universidad Autónoma de Nuevo León (UANL), Nuevo León, México*

IBTISSAM MARCHIQ • *Institute for Research on Cancer and Aging, University of Nice, CNRS, INSERM, Nice, France*

YUHEI NISHIMURA • *Department of Systems Pharmacology, Mie University Graduate School of Medicine, Tsu, Mie, Japan*

FRANCESCO PAMPALONI • *Physical Biology Group, Buchmann Institute for Molecular Life Sciences (BMLS), Goethe Universität Frankfurt am Main, Frankfurt am Main, Germany*

ENRIQUE PÉREZ-CÁRDENAS • *Division of Basic Research, Instituto Nacional de Cancerología, México City, México*

ROSARIO PÉREZ-MOLINA • *Departamento de Genética Molecular, Instituto de Fisiología Celular, Universidad Nacional Autónoma de México, México, DF, México*

JACQUES POUYSSÉGUR • *Institute for Research on Cancer and Aging, University of Nice, CNRS, INSERM, Nice, France*

STEFANIE RAKEL • *Experimental Neurosurgery, Neuroscience Center, Goethe University Hospital, Frankfurt am Main, Germany*

FÉLIX RECILLAS-TARGA • *Departamento de Genética Molecular, Instituto de Fisiología Celular, Universidad Nacional Autónoma de México, México, DF, México*

DIANA RESÉNDEZ-PÉREZ • *Laboratorio de Inmunología y Virología, Facultad de Ciencias Biológicas, Universidad Autónoma de Nuevo León (UANL), Nuevo León, Mexico*

MAURICIO RODRÍGUEZ-DORANTES • *Instituto Nacional de Medicina Genómica, México, DF, México*

CRISTINA RODRÍGUEZ-PADILLA • *Laboratorio de Inmunología y Virología, Facultad de Ciencias Biológicas, Universidad Autónoma de Nuevo León (UANL), Nuevo León, México*

A. IVAN SALIDO-GUADARRAMA • *Instituto Nacional de Medicina Genómica, México, DF, México*

NOHEMÍ SALINAS-JAZMÍN • *Unidad de Desarrollo e Investigación en Bioprocesos (UDIBI), Escuela Nacional de Ciencias Biológicas, IPN, México, DF, México*

VLAD C. SANDULACHE • *Bobby R. Alford Department of Otolaryngology – Head and Neck Surgery, Baylor College of Medicine, Houston, TX, USA*

YASUHITO SHIMADA • *Department of Systems Pharmacology, Mie University Graduate School of Medicine, Tsu, Mie, Japan*

HEATH D. SKINNER • *Department of Radiation Oncology, The University of Texas M. D. Anderson Cancer Center, Houston, TX, USA*

ERNST H.K. STELZER • *Physical Biology Group, Buchmann Institute for Molecular Life Sciences (BMLS), Goethe Universität Frankfurt am Main, Frankfurt am Main, Germany*

TOSHIO TANAKA • *Mie University Life Science Research Center/Mie University Industrial Technology Innovation Institute, Tsu, Mie, Japan*

CATALINA TREJO-BECERRIL • *Division of Basic Research, Instituto Nacional de Cancerología, México City, México*

MARCO A. VELASCO-VELÁZQUEZ • *Departamento de Farmacología, Facultad de Medicina, Universidad Nacional Autónoma de México, México, DF, México*

JOSÉ LUIS VENTURA-GALLEGOS • *Departamento de Medicina Genómica y Toxicología Ambiental, Instituto de Investigaciones Biomédicas, UNAM, México DF, México; Unidad de Bioquímica, Instituto Nacional de Ciencias Médicas y Nutrición "Salvador Zubirán" SSA, México DF, México*

SANG HYEOK WOO • *Department of Head and Neck Surgery, The University of Texas M. D. Anderson Cancer Center, Houston, TX, USA*

LIANGPENG YANG • *Department of Radiation Oncology, The University of Texas M. D. Anderson Cancer Center, Houston, TX, USA*

ALEJANDRO ZENTELLA-DEHESA • *Instituto de Investigaciones Biomédicas, UNAM, México, DF, Mexico; Unidad de Bioquímica, Instituto Nacional de Ciencias Médicas y Nutrición "Salvador Zubirán" SSA, México, DF, México*

DONGMEI ZHANG • *Ovarian Cancer Research Center, School of Medicine, University of Pennsylvania, Philadelphia, PA, USA*

LIN ZHANG • *Smilow Center for Translational Research, School of Medicine, University of Pennsylvania, Philadelphia, PA, USA*

YOUYOU ZHANG • *Ovarian Cancer Research Center, School of Medicine, University of Pennsylvania, Philadelphia, PA, USA*

Part I

Targeting Cancer Cell Metabolism as Anticancer Therapy

Chapter 1

Use of Biguanides to Improve Response to Chemotherapy

Vlad C. Sandulache, Liangpeng Yang, and Heath D. Skinner

Abstract

Metformin is a commonly utilized antidiabetic agent, which has been associated with improved clinical outcomes in cancer patients. The precise mechanism of action remains unclear, but preclinical evidence suggests that metformin can sensitize tumor cells to the effects to conventional chemotherapeutic agents and ionizing radiation (IR). In this chapter we discuss the general background of an approach to evaluate the effects of metformin on conventional chemotherapeutic agent toxicity in a preclinical model.

Key words Metformin, Chemotherapy, Cancer, AMPK, Cytotoxicity, Apoptosis, Senescence, Reactive oxygen species (ROS), Metabolism, Clonogenic survival

1 Introduction

Metformin, a biguanide, is a commonly utilized antidiabetic agent [1]. Metformin was initially synthesized in the 1920s and was introduced into clinical use in the USA in 1995. Currently it is one of the most commonly utilized antidiabetic drugs in the USA and is considered a first line agent for type II diabetes mellitus due to its salutatory effects on cardiovascular mortality and circulating triglyceride levels [2]. Unlike other antidiabetic agents, metformin is associated with a low risk of hypoglycemia, primarily due to its mechanism of action: inhibition of gluconeogenesis [2]. Metformin use is associated with a theoretical risk of lactic acidosis, although a 2006 Cochrane review failed to demonstrate this association in the context of reviewed clinical trials [3].

Over the last decade, multiple studies have demonstrated a beneficial effect associated with metformin use in the context of malignant tumors. Across a wide array of histologies, cancer patients taking metformin demonstrate better treatment response and improved survival [4–8]. This effect is particularly impressive considering that cancer patients who are diabetic generally have significantly poorer clinical outcomes compared to their nondiabetic counterparts [8, 9]. Our preliminary studies have demonstrated

Martha Robles-Flores (ed.), *Cancer Cell Signaling: Methods and Protocols*, Methods in Molecular Biology, vol. 1165, DOI 10.1007/978-1-4939-0856-1_1,

that patients with head and neck squamous cell carcinoma (HNSCC) taking metformin during their cancer treatment have decreased rates of local-regional recurrence compared to matched controls resulting in improved disease free survival [5].

Although no prospective clinical studies have been conducted yet to determine the definitive mechanism by which metformin improves clinical outcomes of cancer patients, preclinical data point to several possibilities. Metformin has been shown to activate AMPK, one of master regulators of tumor cell metabolism [10]. As such, it is thought to impair aspects of tumor cell metabolism, which support the massive biogenesis required for rapid proliferation and tumor growth [11]. In a preclinical model of HNSCC, metformin was shown to inhibit mitochondrial respiration, forcing tumor cells to rely primarily on glycolysis as a source of energy [5, 10]. Inhibition of mitochondrial activity was further linked to increased levels of reactive oxygen species (ROS) resulting in sensitization of tumor cells to the effects of ionizing radiation (IR), [5].

Metformin has been combined with multiple metabolically based, as well as conventional chemotherapeutic agents in the preclinical setting [10, 12–14]. Both in vitro and in vivo experiments indicate that metformin can potentiate the effects of conventional agents as well as IR [5, 10, 12]. Although the clinical viability of this approach remains unproven, it does represent an intriguing possibility given the long track record of metformin as an antidiabetic agent, and its generally safe profile in the single agent setting.

Combination of metformin with conventional chemotherapeutic agents is not in and of itself a complex methodological undertaking. However, drawing appropriate conclusions from the experiments performed requires a robust understanding of metformin effects in the tumor cell system chosen for experimentation. As a result, we strongly suggest a step-wise approach to investigation of metformin combinations with any chosen chemotherapeutic agent. This will assist in appropriate timing of drug administration and improved understanding of results.

In their recent article, Smith and Houghton raise an important issue regarding preclinical drug testing: dose selection [15]. They correctly point out that investigators will often utilize in vitro drug concentrations which are either above their intended target-dose point (based on specific molecular marker profiles) or above achievable plasma concentrations. In the case of metformin the authors point out that the millimolar concentrations often used in preclinical cancer studies are well above the circulating plasma concentration of low double-digit micromolar ones [16, 17]. This discrepancy raises the possibility that in vitro effects may not appropriately reflect clinical reality and may be misleading. Although the point is well taken, it is important to note that circulating plasma concentrations are primarily a function of absorption and excretion. Tumor, or rather tumor cell functional concentrations are often not reflective of plasma concentrations; unfortunately, there exists no current data regarding the concentration of metformin in tumors using a clinically relevant dosing schedule. It is known that some tissues (i.e., liver) and various cell types can take up and concentrate metformin through the OC (organic cation)-1 transporter resulting in increased mitochondrial concentrations which can yield 400-fold greater concentrations compared to plasma levels [18–21]. As such, the appropriate concentration for in vitro experiments cannot be stated with any reasonable degree of certitude, although, we do subscribe to the notion that lower is better [14]. Because of this, we recommend a broad range of tested concentrations, starting at the single micromolar dose, extending into the single millimolar dose range.

2 Metformin Sensitivity

2.1 Metabolic Effects

Appropriate testing of cellular sensitivity to metformin in the single agent setting should precede any combinatorial testing. To date, two general categories of metformin metabolic effects have been described. The first is activation of metabolic regulatory pathways such as those regulated by AMPK. Metformin is an agonist for AMPK, resulting in its phosphorylation and activation [5, 10–13, 22]. As such, the cells under investigation should be evaluated for the presence of AMPK and for its potential for activation as previously described [5, 10, 13]. The second described mechanism of metformin activity is direct inhibition of mitochondrial respiration, likely through direct interference with components of the electron transport chain (ETC), [10, 23–25]. This results in primary shunting of metabolic activity into glycolytic pathways as well as altered ROS generation [10]. The time frame for this direct metabolic inhibition is on the order of minutes to hours and can be characterized using either high throughput commercially available kits or real-time measurements of oxygen consumption and lactate generation as previously described [10, 26].

We believe that a preliminary investigation of metformin metabolic effects on the chosen tumor cells is an essential prerequisite to subsequent chemotherapeutic investigation. The effects of metformin can vary dramatically based on the metabolic profile of the chosen cell type [10, 13]. Cellular genomic background can also affect metformin effects, by driving relative metabolic flexibility. For example, we have found that metformin effects, metabolic and therapeutic, are more pronounced in tumor cells that exhibit mutations in *TP53*, secondary to loss of metabolic flexibility [5, 10]. Comparisons between multiple cell lines/types must take into account potentially relevant genetic or epigenetic variability.

2.2 ROS Generation

Metformin has been shown to alter intracellular ROS levels [5, 10, 25]. Some studies have shown that metformin can decrease production of ROS by normal cells, in part through suppression of NAD(P)H oxidase activity [27, 28]. Our group has found that metformin can increase tumor cell ROS levels, primarily through perturbations in ETC activity [5, 10]. This can itself result in a mild single agent cytotoxic effect [5]. We suggest measuring metformin effects on cellular ROS prior to combinatorial approaches. This can be accomplished using a variety of methods including measurements of total ROS using 5-(and-6)-carboxy-2′,7′-dichlorofluorescein (CM-H_2DCFDA) [29].

2.3 Mechanism of Cell Death

Because of its complex impact on cell metabolism, meformin can exert measurable effects on cellular proliferation and death. These effects need to first be established in the single agent setting prior to combinatorial experiments with specific chemotherapeutic agents. They also need to be considered carefully when designing combinatorial strategies as they may impact relative effectiveness.

2.3.1 Arrest

Metformin has been shown to induce cell cycle arrest [13, 20]. Since arrest can be a primary driver of subsequent cell death through multiple mechanisms, metformin effects in the single agent setting on cell cycle progression should be assayed. We recommend the commonly utilized propidium iodide (PI) method.

2.3.2 Apoptosis

Metformin has been reported to cause apoptosis through caspase dependent and caspase independent mechanisms [30, 31]. Since apoptosis is a common mechanism of death following exposure to chemotherapeutic agents, quantitative assessment of the apoptotic cell fraction following administration of metformin is likely to be instructive. This can be assayed by measuring Annexin V staining.

2.3.3 Senescence

Although less studied than apoptosis, cell senescence has been shown to play an important role in tumor cell death following exposure to ionizing radiation or agents, which induce ROS [5]. Metformin has been shown to increase intracellular ROS levels in multiple tumor cell types, leading to an increased fraction of senescent cells [5]. Senescence can be routinely assayed using the β-galactosidase technique as previously described, using both qualitative and semiquantitative derivatives [5].

3 Metformin Combination with Chemotherapy

3.1 Timing

Combinatorial strategies generally must take into account timing of drug administration. To a large degree, timing should be driven by the expected mechanism of drug synergy. For example, potentiation of tumor cell ROS levels prior to IR exposure can significantly increase radiosensitivity [5]. Conversely, priming tumor cells with ROS scavenging compounds such as *N*-acetyl cysteine can significantly blunt the effectiveness of IR or other ROS dependent cytotoxic agents [5]. In our view, metformin does not represent a conventional chemotherapeutic agent in and of itself. As such, it should be viewed primarily as a metabolic modulator, which can prime tumor cell metabolism in order to maximize the toxicity of specific traditional or targeted chemotherapeutic agents. We therefore recommend priming tumor cells with metformin prior to drug administration. The concentration range should be broad for the reasons described earlier. Metabolic effects (i.e., ROS perturbations) can be detected within minutes of administration [5, 10].

3.2 Dosing

We recommend a broad range of initial dose testing, ranging from 5 μM to 5 mM. To streamline experimental protocols, we would suggest first performing metabolic and toxicity testing in the single agent setting using the assays described above, prior to proceeding to combinatorial experiments. This approach can significantly reduce the number of tested combinatorial concentrations.

3.3 Output

Drug toxicity can be assayed using a wide variety of techniques. In general, more specific techniques, which evaluate individual cell death mechanisms (apoptosis, senescence), require higher drug concentrations to achieve detectable effects due to the compressed timeline for most of these protocols (12–48 h). Other common assays such as the routinely employed 3-(4,5-Dimethylthiazol-2-yl) -2,5-diphenyltetrazolium bromide (MTT) and CellTiterGlo type assays can provide increased sensitivity due to their longer time frame (5–7 days depending on growth rate). We would like to note, however, that the MTT or CellTiterGlo assays are ***not an appropriate*** method of analyzing drug effects on cell viability in the case of metformin, or any other metabolic agent. The MTT assay relies on cellular reducing potential, which is measurably impacted by metformin exposure [28]. The CellTiteGlo assay measures cellular ATP levels which are profoundly affected by metformin exposure not only by direct inhibition of mitochondrial respiration but also through secondary effects via AMPK activation [22, 32].

In our experience, clonogenic survival assays are the most effective means of testing drug effects on tumor cell death irrespective of mechanism. Because of the extended time frame of this assay (14–21 days), clonogenic survival experiments can often detect drug effects in much lower ranges compared to other assays. In addition, they are the gold standard for measuring radiation toxicity, and as such are perfectly suited for combinatorial experiments with IR [5, 10].

References

1. Dunn CJ, Peters DH (1995) Metformin. A review of its pharmacological properties and therapeutic use in non-insulin-dependent diabetes mellitus. Drugs 49:721–749
2. Scheen AJ, Paquot N (2013) Metformin revisited: a critical review of the benefit-risk balance in at-risk patients with type 2 diabetes. Diabetes Metab 39:179–190
3. Salpeter S, Greyber E, Pasternak G, Salpeter E (2006) Risk of fatal and nonfatal lactic acidosis with metformin use in type 2 diabetes mellitus. Cochrane Database Syst Rev 1, CD002967
4. Skinner HD, McCurdy MR, Echeverria AE et al (2013) Metformin use and improved response to therapy in esophageal adenocarcinoma. Acta Oncol 52:1002–1009
5. Skinner HD, Sandulache VC, Ow TJ et al (2012) TP53 disruptive mutations lead to head and neck cancer treatment failure through inhibition of radiation-induced senescence. Clin Cancer Res 18:290–300
6. Jiralerspong S, Palla SL, Giordano SH et al (2009) Metformin and pathologic complete responses to neoadjuvant chemotherapy in diabetic patients with breast cancer. J Clin Oncol 27:3297–3302
7. Noto H, Goto A, Tsujimoto T, Noda M (2012) Cancer risk in diabetic patients treated with metformin: a systematic review and meta-analysis. PLoS One 7:e33411
8. Lee MS, Hsu CC, Wahlqvist ML, Tsai HN, Chang YH, Huang YC (2011) Type 2 diabetes increases and metformin reduces total, colorectal, liver and pancreatic cancer incidences in Taiwanese: a representative population prospective cohort study of 800,000 individuals. BMC Cancer 11:20
9. Bosetti C, Rosato V, Polesel J et al (2012) Diabetes mellitus and cancer risk in a network of case-control studies. Nutr Cancer 64:643–651
10. Sandulache VC, Skinner HD, Ow TJ et al (2012) Individualizing antimetabolic treatment strategies for head and neck squamous cell carcinoma based on TP53 mutational status. Cancer 118:711–721
11. Gallagher EJ, LeRoith D (2011) Diabetes, cancer, and metformin: connections of metabolism and cell proliferation. Ann N Y Acad Sci 1243:54–68
12. Rocha GZ, Dias MM, Ropelle ER et al (2011) Metformin amplifies chemotherapy-induced AMPK activation and antitumoral growth. Clin Cancer Res 17:3993–4005
13. Ben Sahra I, Laurent K, Giuliano S et al (2010) Targeting cancer cell metabolism: the combination of metformin and 2-deoxyglucose induces p53-dependent apoptosis in prostate cancer cells. Cancer Res 70:2465–2475
14. Erices R, Bravo ML, Gonzalez P et al (2013) Metformin, at concentrations corresponding to the treatment of diabetes, potentiates the cytotoxic effects of carboplatin in cultures of ovarian cancer cells. Reprod Sci 20(1433):1446
15. Smith MA, Houghton P (2013) A proposal regarding reporting of in vitro testing results. Clin Cancer Res 19:2828–2833
16. Bardin C, Nobecourt E, Larger E, Chast F, Treluyer JM, Urien S (2012) Population pharmacokinetics of metformin in obese and non-obese patients with type 2 diabetes mellitus. Eur J Clin Pharmacol 68:961–968
17. Charles B, Norris R, Xiao X, Hague W (2006) Population pharmacokinetics of metformin in late pregnancy. Ther Drug Monit 28:67–72
18. Wang DS, Kusuhara H, Kato Y, Jonker JW, Schinkel AH, Sugiyama Y (2003) Involvement of organic cation transporter 1 in the lactic acidosis caused by metformin. Mol Pharmacol 63:844–848
19. Davidoff F (1968) Effects of guanidine derivatives on mitochondrial function. II. Reversal of guanidine-derivative inhibiton by free fatty acids. J Clin Invest 47:2344–2358
20. Davidoff F (1968) Effects of guanidine derivatives on mitochondrial function. I. Phenethylbiguanide inhibition of respiration in mitochondria from guinea pig and rat tissues. J Clin Invest 47:2331–2343
21. Davidoff F (1971) Effects of guanidine derivatives on mitochondrial function. 3. The mechanism of phenethylbiguanide accumulation and its relationship to in vitro respiratory inhibition. J Biol Chem 246:4017–4027
22. Zhang L, He H, Balschi JA (2007) Metformin and phenformin activate AMP-activated protein kinase in the heart by increasing cytosolic AMP concentration. Am J Physiol Heart Circ Physiol 293:H457–H466
23. Ota S, Horigome K, Ishii T et al (2009) Metformin suppresses glucose-6-phosphatase expression by a complex I inhibition and AMPK activation-independent mechanism. Biochem Biophys Res Commun 388:311–316
24. Guigas B, Detaille D, Chauvin C et al (2004) Metformin inhibits mitochondrial permeability transition and cell death: a pharmacological in vitro study. Biochem J 382:877–884
25. Batandier C, Guigas B, Detaille D et al (2006) The ROS production induced by a reverse-electron flux at respiratory-chain complex 1 is hampered by metformin. J Bioenerg Biomembr 38:33–42

26. Beeson CC, Beeson GC, Schnellmann RG (2010) A high-throughput respirometric assay for mitochondrial biogenesis and toxicity. Anal Biochem 404:75–81
27. Ouslimani N, Peynet J, Bonnefont-Rousselot D, Therond P, Legrand A, Beaudeux JL (2005) Metformin decreases intracellular production of reactive oxygen species in aortic endothelial cells. Metabolism 54:829–834
28. Piwkowska A, Rogacka D, Jankowski M, Dominiczak MH, Stepinski JK, Angielski S (2010) Metformin induces suppression of NAD(P)H oxidase activity in podocytes. Biochem Biophys Res Commun 39:268–273
29. Eruslanov E, Kusmartsev S (2010) Identification of ROS using oxidized DCFDA and flow-cytometry. Methods Mol Biol 594:57–72
30. Storozhuk Y, Hopmans SN, Sanli T et al (2013) Metformin inhibits growth and enhances radiation response of non-small cell lung cancer (NSCLC) through ATM and AMPK. Br J Cancer 108:2021–2032
31. An D, Kewalramani G, Chan JK et al (2006) Metformin influences cardiomyocyte cell death by pathways that are dependent and independent of caspase-3. Diabetologia 49:2174–2184
32. Silva FM, da Silva MH, Bracht A, Eller GJ, Constantin RP, Yamamoto NS (2010) Effects of metformin on glucose metabolism of perfused rat livers. Mol Cell Biochem 340:283–289

Chapter 2

Evaluating Response to Metformin/Cisplatin Combination in Cancer Cells via Metabolic Measurement and Clonogenic Survival

Sang Hyeok Woo, Vlad C. Sandulache, Liangpeng Yang, and Heath D. Skinner

Abstract

Metformin is a commonly utilized antidiabetic agent, which has been associated with improved clinical outcomes in cancer patients. The precise mechanism of action remains unclear, but preclinical evidence suggests that metformin can sensitize tumor cells to the effects to conventional chemotherapeutic agents and ionizing radiation (IR). In this chapter we describe two assays to investigate the effects of combination of metformin and a chemotherapeutic agent (in this case cisplatin) in head and neck cancer squamous cell carcinoma (HNSCC) cell lines.

Key words Metformin, Head and neck cancer, Cisplatin, Cytotoxicity, Apoptosis, Metabolism, Reactive oxygen species (ROS), Clonogenic survival

1 Introduction

In the previous chapter, we discussed the general background regarding the investigation of metformin as a novel therapeutic in cancer and some general methods of assaying this effect. We will now describe examining the effects of metformin in combination with cisplatin on the metabolic functioning of the cell. We will also describe a standard clonogenic assay using metformin in combination with cisplatin. In the current series of experiments we will be describing the use of the HN 31 HNSCC cell line. This cell line exhibits rapid growth in vitro and is suitable for the experiments described. Although many HNSCC cell lines can be used for the experiments described here, in Subheading 4, we mention several cell lines that, in our hands, do not form assayable colonies on clonogenic assay.

Martha Robles-Flores (ed.), *Cancer Cell Signaling: Methods and Protocols*, Methods in Molecular Biology, vol. 1165, DOI 10.1007/978-1-4939-0856-1_2, © Springer Science+Business Media New York 2014

2 Materials

All working solutions should be prepared with purified deionized water and analytical grade reagents and be buffered to a neutral pH to minimize unintended cytotoxicity. Reagents must be disposed of in a manner compliant with institutional regulations. Although the reagents and kits listed below have proven effective in our hands, we recognize that many alternatives are available, and we do not strongly endorse any single product.

2.1 Cisplatin and Metformin Solutions

We utilize cisplatin in an aqueous solution provided by the institutional inpatient pharmacy in order to insure that we are utilizing the agent in a manner consistent with clinical practice. Cisplatin is also available from multiple suppliers including Sigma-Aldrich (St. Louis, MO). We generate stock solutions in sodium chloride with a concentration range of 1–5 mM, which are stored in the dark at 2–8 °C. Further dilutions are accomplished in cell media immediately prior to administration (*see* **Note 1**). Metformin is available from Sigma-Aldrich (St. Louis, MO) as 1,1-dimethylbiguanide hydrochloride, which is soluble in deionized purified water and can be prepared to a stock concentration of 1 M (*see* **Note 2**).

2.2 Metabolic Alteration Due to Metformin and Cisplatin Treatment

Metformin has significant activity against components of the mitochondrial electron transport chain (ETC). By inhibiting flow through the ETC, metformin forces tumor cells to rely on glycolytic activity as the primary source of energy. In addition, inhibition of ETC flow results in increased ROS generation, possibly through direct increases in the natural ROS leak encountered during mitochondrial respiration. Cisplatin itself can act as a free radical both within tumor cells and normal tissue. As such, it is subject to the same intracellular forces, which govern the balance of reducing equivalents and ROS moieties. We suggest two separate experiments designed to investigate the combinatorial effects of metformin and cisplatin on tumor cell metabolism. These experiments should be paired with the clonogenic assays detailed below to provide a potential mechanistic explanation for the cytotoxic effects of the drug combination.

2.2.1 Experiment 1 (Acute Exposure)

The Seahorse XF Extracellular Flux Analyzer allows for real-time measurements in glycolytic flux and mitochondrial respiration. In our experience, both measurements can be altered by changes in the reducing potential-ROS balance within tumor cells. To test the effects of acute drug exposure, we suggest first establishing a glycolytic and respiratory baseline as detailed below. Following completion of baseline metabolic experiments, cells should be plated and prepared as described below. Using the provided experimental media, HN31 cells should be exposed to cisplatin, metformin, or a combination of both for 4 h. We recommend using the A port of the analyzer platform to deliver the drugs in order to maximize time of exposure. In general we utilize cisplatin concentrations ranging from 1 to 10 μM and metformin concentrations from 1 to 10 mM.

2.2.2 Experiment 2 (Prolonged Exposure)

Cells should be plated at 75 % confluence and allowed to attach overnight. Cells should then be exposed to the chosen drug concentrations for 24 h in regular growth media. Using the Seahorse XF, platform baseline and reserve glycolytic and mitochondrial respiratory measurements should be conducted using the protocol detailed below. This experiment will shed light on drug effects on tumor cell baseline and reserve glycolytic and respiratory capacity, as well as potential shifts in the balance of metabolic fluxes inside tumor cells following exposure to the individual drugs or the drug combination.

2.3 Materials and Reagents

2.3.1 Metabolic Measurements

XF Glycolysis Stress Test Kit.

XF Cell Mito Stress Test Kit.

XF microplates.

XF 24- or 96-well 4-port FluxPak.

XF assay medium.

CO_2-incubator set to 37 °C.

CO_2-free incubator set to 37 °C.

XF24 or 96 Extracellular Flux Analyzer.

Experimental media provided by the manufacturer containing 25 mM glucose, 1 mM pyruvate, 4 mM glutamine, and 0 % serum.

Experimental media used for baseline measurements containing 1 mM pyruvate, 4 mM glutamine, and 0 % serum.

Glucose 25 mM solution.

Oligomycin 1 μg/ml.

2-deoxy-D-glucose 10 mM solution.

Carbonyl cyanide-p-trifluoromethoxyphenylhydrazone FCCP (1 μM).

Rotenone (1 μM).

Antimycin A (1 μM).

2.3.2 Clonogenic Survival

Crystal violet, 0.5 % solution in formalin.

PBS buffer.

Methanol.

Glacial acetic acid.

Six-well plates.

Hemocytometer or Coulter counter.

Stereomicroscope.

Fixation solution: acetic acid–methanol 1:7 (vol/vol).

3 Methods

3.1 Glycolytic Function

Glycolytic activity converts glucose into pyruvate. In normal cells, pyruvate can either be converted into CO_2 and water in the mitochondria or converted into lactate. Tumor cell metabolism differs qualitatively and quantitatively from that of normal cells as a result of the oncogenic events, which drive tumorigenesis. Measurements of glycolytic and mitochondrial activity can offer insight into tumor cell metabolism at baseline as well as under various stress conditions. One benefit of the Seahorse XF platform is that it allows for real-time metabolic measurements under various test conditions. Results from this platform can always be validated using traditional biochemical assays that allow for quantitative measurements for lactate, pyruvate, glucose, and other metabolic intermediates.

The platform measures proton generation (extracellular acidification rate, ECAR) and oxygen consumption (OCR), surrogates for glycolytic flux and mitochondrial respiration. Briefly, investigators can measure baseline glycolytic activity in regular growth media as well as reserve glycolytic activity, which can be assayed in the presence of oligomycin which inhibits mitochondrial ATP production and forces the cell to primarily utilize glycolysis for energy production. Additional measurements can be obtained in the presence of 2-deoxy-D-glucose, a competitive inhibitor of glucose which can be used to drive down glycolytic activity. After finishing the assay the software embedded in the analyzer will automatically generate four parameters: Glycolysis, Glycolytic Capacity, Glycolytic Reserve, and Non-glycolytic Acidification.

Procedure:

1. Seed cells in each well of an XF cell culture microplate (2×10^4 cells/well for 96-well plates, 4×10^4 for 24-well plates) and place in a 37 °C incubator (5 % CO_2). As the cell density will affect the rate of glycolysis observed, the optimal cell seeding density has to be determined prior to running the test assay (*see* **Notes 3** and **4**).
2. The XF Sensor Cartridge must be hydrated at least for 12 h prior to the assay. Place the XF Sensor Cartridge on top of the utility plate and put in CO_2-free incubator at 37 °C.
3. For acute exposure experiments (Experiment 1), cisplatin, metformin, or the combination should be added in the appropriate concentration to Port A and metabolic measurements

should be carried out for 4 h. For prolonged exposure experiments (Experiment 2), cells should be pretreated with the drugs alone or in combination for 24 h and then a baseline metabolic profile should be carried out as detailed below. All drug exposure metabolic measurements should be carried out in experimental media provided by the manufacturer containing 25 mM glucose, 1 mM pyruvate, 4 mM glutamine, and 0 % serum.

For baseline glycolytic measurements add 25 μl of the prepared reagents into the appropriate injection port: Port A for glucose (25 mM), port B for oligomycin (1 μg/ml), and port C for 2-DG (10 mM); experimental media used for baseline measurements contains 1 mM pyruvate, 4 mM glutamine, and 0 % serum.

4. Remove the growth medium and rinse the cells three times with 300 μl of test assay medium. Add 175 μl of fresh test assay medium to each well.
5. Run the assay with the XF software specified for XF Cell glycolytic stress test.

3.2 Mitochondrial Function

Oxygen consumption rate (OCR) measurements provide the investigator with information regarding mitochondrial activity in tumor cells under baseline and stress conditions. The protocol detailed here has the benefit of providing real-time information about acute metabolic perturbations, as well as allowing for measurements of mitochondrial activity under baseline and stress conditions. The Seahorse XF platform utilizes several compounds to perturb OCR and thus highlight mitochondrial activity and reserve capacity. The first is oligomycin, which inhibits ATP synthesis by blocking mitochondrial complex V activity. It can be used to measure the portion of OCR devoted to ATP synthesis, or coupling Efficiency (difference between basal OCR value and minimum OCR value after injection of oligomycin). The second is carbonyl cyanide-p-trifluoromethoxyphenylhydrazone (FCCP) that acts as an uncoupling agent collapsing the mitochondrial membrane potential. The collapse leads to a rapid consumption of energy and oxygen without the generation of ATP. FCCP treatment generates the highest value of OCR in the assay. Spare or reserve Respiratory Capacity will be calculated as the quantitative difference between the highest OCR and the initial basal OCR. The last reagent is a combination of rotenone, a complex I inhibitor, and antimycin A, a complex III inhibitor. The combination shuts down mitochondrial respiration completely, enabling measurements of non-mitochondrial respiration, defined as the quantitative difference between the initial basal OCR and the lowest OCR value obtained by the combined treatment.

Procedure:

1. Seed cells in each well of an XF cell culture microplate (2×10^4 cells/well for 96-well plates, 4×10^4 for 24-well plates), and place in a 37 °C incubator with the desired level of CO_2. As the cell density will affect mitochondrial activity observed, the optimal cell seeding density has to be determined prior to running the test assay.
2. The XF Sensor Cartridge must be hydrated at least for 12 h prior to the assay. Place the XF Sensor Cartridge on top of the utility plate, and incubate in CO_2-free incubator at 37 °C.
3. For acute exposure experiments (Experiment 1), cisplatin, metformin, or the combination should be added in the appropriate concentration to Port A and metabolic measurements should be carried out for 4 h. For prolonged exposure experiments (Experiment 2), the cells should be pretreated with the drugs alone or in combination for 24 h and then a baseline metabolic profile should be carried out as detailed below. For baseline mitochondrial measurements add 25 μl of the prepared reagents into the appropriate injection port: Port A for oligomycin (1 μg/ml), port B for FCCP (1 μM), and port C for rotenone (1 μM) and antimycin A (1 μM).
4. Remove the growth medium and rinse the cells three times with 300 μl of test assay medium. Add 175 μl of fresh test assay medium to each well.
5. Run the assay with the XF software specified for XF Cell mitochondrial stress test.

3.3 Clonogenic Survival

Clonogenic survival assay enables an assessment of the differences in ability of cells to produce progeny between control untreated cells and cells have been exposed to ionizing radiation or various cytotoxic reagents [1]. Although the assay was initially described for studying the effects of radiation on cells, it is now widely used to determine the effects of agents having potential applications in clinic. The loss of clonogenic ability as a function of dose of chemotherapy agent is described by the dose-survival curve (*see* **Note 5**).

Procedure:

1. Culture the cells in regular growth medium.
2. Remove the medium and then rinse the cells with PBS.
3. Add trypsin to the cells and incubate at 37 °C for 5 min until the cells detached.
4. Add medium with 10 % FBS and resuspend the cells by pipetting.
5. Count the cells using hemocytometer.

6. Plate 500 cells per well on 6-well plate for HN31 cells, at least in duplicate.
7. Incubate the cells for 24 h in a CO_2 incubator at 37 °C, allowing them to attach to the plate.
8. Replace attachment medium with medium containing cisplatin (1, 2 or 3 μM), metformin (1, 3 or 5 mM) or the combination.
9. Incubate the cells for 48 h, remove drug containing medium and replace with regular growth medium.
10. Incubate the cells in a CO_2 incubator at 37 °C for 1–3 weeks until cells in control plates have formed colonies with substantially good size (50 cells per colony is the minimum for scoring).
11. Remove the medium and then rinse the cells with PBS.
12. Remove PBS and add 2 ml of fixation solution and leave the plates at room temperature for 5 min.
13. Remove fixation solution and add 0.5 % crystal violet solution and incubate at room temperature for 30 min.
14. Remove crystal violet carefully and rinse off residual the solution with H_2O.
15. Air-dry the plate at room temperature for up to a day.
16. Count number of colonies and calculate plating efficiency (no. of colonies formed/no. of cells seeded X 100 %) and surviving fraction (no. of colonies formed after treatment/no. of cells seeded X plating efficiency).

4 Notes

1. It is important to note that cisplatin will interact with proteins, and as such care must be used when including serum in cell media. Since this additive is essential for cancer cell proliferation, we suggest that the investigator utilize consistent methods and concentrations from experiment to experiment to insure reproducibility.
2. Additional dilutions can be performed by using water or cell culture media as needed. We recommend that fresh working solutions be prepared for all experiments.
3. In our experience, HN31 cells should be approximately 80–90 % confluent at the time of metabolic measurements. Complete confluency can result in crowding and cell cycle arrest which can profoundly affect metabolic activity.
4. Cell growth. Most currently available cell lines are grown in media containing supraphysiologic concentrations of glucose

(i.e., 25 mM) which far exceed those encountered either in preclinical animal models or the clinical setting. Since metformin effects are expected to be primarily metabolic in nature, the available glucose concentration can play an important role in the relative drug concentrations required to achieve specific cellular activities. Investigators may choose to perform some of the experiments detailed above in lower, more physiologically relevant glucose concentrations. It is likely, however, that if the cell lines in question are conditioned to high glucose concentrations, lowering this nutrient's availability for short periods of time could have a profound effect on cell proliferation, survival and overall metabolic profile. Additionally, HNSCC cells are susceptible to infection with mycoplasma. In order to reduce the risk of mycoplasma contamination, we do not recommend using any cell line continuously for more than 3–4 weeks at a time.

5. Clonogenic survival. Not all tumor cells can form tight colonies. As such, this assay will be limited by the morphologic characteristics of the chosen cell type. If clonogenic assays are not feasible, an alternative method will need to be employed, however, we do once again strongly recommend against any metabolically based assays. In addition, cell lines may exhibit slow growth rate, in which case, the colony formation time frame may have to be extended significantly. We recommend counting colonies which are at least 50 cells in size.

Reference

1. Franken NA, Rodermond HM, Stap J, Haveman J, van Bree C (2006) Clonogenic assay of cells in vitro. Nat Protoc 1:2315–2319

Chapter 3

Quantifying the Autophagy-Triggering Effects of Drugs in Cell Spheroids with Live Fluorescence Microscopy

Nariman Ansari, Stefanie Hardung, Katharina Hötte, Stefanie Rakel, Patrick Antonietti, Donat Kögel, Ernst H.K. Stelzer, and Francesco Pampaloni

Abstract

We present a 3D assay for the quantification of the autophagic flux in live cell spheroids by using the fluorescent reporter mRFP-GFP-LC3. The protocol describes the formation of the spheroids from the astrocytoma cell line U343, live long-term 3D fluorescence imaging of drug-treated spheroids, and the image processing workflow required to extract quantitative data on the autophagic flux.

Key words Three-dimensional cell cultures, Live-cell assay, Glioma tumor spheroids, Autophagic flux, Drug development, Live cell imaging, Automated fluorescence microscopy, Live confocal microscopy

1 Introduction

Autophagy is an essential cellular function required for protein quality control and energy metabolism, and dysfunction of this pathway is implicated in a variety of pathologies including cancer and neurodegenerative diseases [1]. According to the homeostasis model proposed by Levine, both a deficiency and an excess of autophagy are detrimental to cellular function [2]. In many studies, induction of autophagy is exclusively analyzed by conversion of LC3-I to LC3-II, formation of autophagosomes and subcellular translocation of GFP-LC3 [3]. However, these techniques are restricted to the early stage of autophagy and do not allow to measure the autophagic flux, i.e., the fusion of autophagosomes with lysosomes for the degradation of cargo by lysosomal proteases. In light of the potential pathological relevance of impaired and overactivated autophagy, robust cell-based assays that allow to quantify the autophagic flux are therefore urgently requested. Three-dimensional cellular spheroids are more realistic models of

Martha Robles-Flores (ed.), *Cancer Cell Signaling: Methods and Protocols*, Methods in Molecular Biology, vol. 1165, DOI 10.1007/978-1-4939-0856-1_3, © Springer Science+Business Media New York 2014

tumors and healthy tissues compared to standard two-dimensional cultures [4]. Employing spheroids improves the reliability and the physiological significance of cell-based assays [5–7]. We describe a detailed cell-based assay that quantifies the autophagic flux in live cellular spheroids. We employ automated live confocal laser-scanning microscopy to investigate the drug-induced autophagy in the spheroids over several days. We describe the spheroid preparation, manipulation, live fluorescence imaging, as well as data processing. As an example, we quantify the autophagy-triggering effects of the drugs (-)-gossypol and rapamycin [8] in glioma cell spheroids. The formation of the autophagosomes and the fusion of the autophagosomes with lysosomes in the treated spheroids are monitored over time and space with an mRFP-GFP-LC3 tandem fluorescent protein [9]. Recently, a high-content screen of autophagy drug-inducers has been performed by flow-cytometry [10]. However, cell-based fluorescence microscopy delivers more cell-specific information, including data on morphology and organelles distribution over time and space. This is the reason why the combination of live fluorescence microscopy with motorized and computer-controlled microscopes and protein-specific fluorescent probes is increasingly employed in both in vitro and in vivo preclinical studies.

2 Materials

2.1 Chemical and Reagents

- *Cell growth medium*: Add 50 ml FBS (100 % Fetal Bovine Serum), 5 ml Penicillin–Streptomycin, 5 ml 200 mM L-Glutamine, and 0.5 ml Geneticin selective antibiotic to 445 ml DMEM (1× Dulbecco's Modified Eagle Medium phenol-free). Mix and prepare 50 ml aliquots. Store at 4 °C.
- *Cell dissociation*: StemPro Accutase Cell Dissociation Reagent (Life technologies).
- *Spheroid embedding medium*: Add 0.5 g of low-melting agarose to 50 ml 1× PBS. Mix and prepare 2 ml aliquots. Store at 4 °C.

2.2 Drug Solutions

The two drugs gossypol and rapamycin, which are well-known inducers of autophagy, are employed for the assay.

Drug	Final concentration (μM)	Stock concentration (mM)	
Gossypol	15	15	(ENZO Life Science)
Rapamycin	1	1	(ENZO Life Science)

The drugs are dissolved in 100 % DMSO, aliquoted in 50 μl aliquots, and stored at −20 °C. For the drug treatment of the cells the stock solutions are diluted in growth medium to the indicated final concentrations (*see* also **Note 1**).

2.3 Cell Culture Equipment

- HydroCell Surface™ 96 microWell plates (Nunc/Thermo Scientific).
- 8-canal multi-pipette.
- Neubauer hemocytometer

2.4 Cell Line and Transfection

Human glioblastoma cell line U343 [8] was stably transfected by using FuGene® 6 (Promega) according to the manufacturer's protocol. Stable transformants were selected in complete growth medium containing 1 mg/ml G418. After 10 days with frequent changes of medium individual colonies were trypsinized and transferred to flasks for further propagation.

2.5 Autophagy Flux Fluorescence Marker

The autophagy flux, i.e., the autophagosome maturation process, is monitored over time by imaging the reporter protein mRFP-GFP-LC3, a tandem fluorescent-tagged LC3 (microtubule-associated protein 1A/B light chain 3) stably transfected in the cells [8]. Autophagosomes marked by this marker portein showed both mRFP and GFP signals. After fusion with lysosomes, GFP signals were attenuated, and only mRFP signals were observed. Using this protein it is possible to discriminate between the single steps of the autophagy flux.

2.6 Imaging

- Routine cell culture inverted wide-field epifluorescence microscope.
- Automated laser-scanning confocal microscope equipped with a motorized stage and a carrier for multiwell plates, as well as an environmental control system able to maintain a constant temperature of 37 °C and a 5 % CO_2 concentration on the specimen.
- Long working distance objective lens, e.g., Nikon CFI Plan Fluor 10×, NA 0.30, WD 16 mm or Nikon CFI S Plan Fluor ELWD 20×, NA 0.45, WD 8.2–6.9 mm.

2.7 Software for Image Acquisition and Image Processing

- Image acquisition software of the automated confocal microscope.
- Image processing software (e.g., Fiji—Fiji Is Just ImageJ, http://fiji.sc/).
- Spreadsheet software.

3 Methods

3.1 Formation of Cellular Spheroids in HydroCell Surface™ 96 Multiwell Plates

3.1.1 Thaw Frozen Cells

Thaw the frozen U343 cells rapidly (within 5 min) in a 37 °C water bath (*see* **Note 2**). Dilute the thawed cells in 5 ml pre-warmed growth medium. Centrifuge the cells for 4 min at 300×*g*. Resuspend the pellet in 5 ml pre-warmed cell growth medium. Plate the cells at high density in a T25 cell culture flasks to speed up recovery.

3.1.2 Propagate Cells

Change medium after 24 h. Propagate cells at 90–100 % confluence. Detach cells with 500 ml StemPro Accutase Cell Dissociation Reagent and resuspend cells in 4.5 ml cell growth medium.

3.1.3 Cell Counting

Determine the concentration of cells in the medium with the hemocytometer (e.g., Neubauer). Clean the counting chamber and the coverslip with water and dried with tissue paper (e.g., Kimwipes) before use. The coverslip is placed over the counting chamber and filled with approx. 10 μl cell suspension. The volume under the coverslip fills by capillary action. The counting chamber is then placed on the microscope and the counting grid is focused. Cells residing in the four large squares are counted and averaged. The value is multiplied by the factor 10^4 to obtain the number of cells in 1 ml (*see* also **Notes 3** and **4**).

3.1.4 Spheroid Formation by the Liquid Overly Method

Employ 1,000 cells for spheroid formation in HydroCell Surface™ 96-well plates. 500–10,000 cells/spheroid are suitable for drug screening assays. The original cell suspension is diluted respectively in growth medium and 100 μl are transferred in each well by employing an 8-channel pipette and a pipetting trough (Fig. 1). In order to reduce the formation time and obtain more compact spheroids, a centrifugation step of 3 min at $300\times g$ is recommended. The HydroCell Surface™ 96 multiwell plates are finally incubated under cell culture conditions. Spheroid formation is completed after 2–6 days (Fig. 2).

The U343 spheroids (Fig. 3a) closely reproduce several morphological features of real glioblastoma (Fig. 3b).

3.2 Live Cell Imaging of Drug-Induced Autophagy in mRFP-GFP-LC3 U343 Cellular Spheroids by Confocal Laser-Scanning Microscopy

Figure 4 shows snapshots of gossypol- and rapamycin-treated U343 spheroids. An increase of mRFP intensity with respect to GFP intensity is observed over time. The increasing of the mRFP intensity over GFP signalizes the ongoing autophagosome maturation, which culminates in the fusion of autophagosomes and lysosomes to an autolysosome. The increase of the mRFP signal is due to the instability of GFP in the acidic environment of the autolysosome. This causes the fading of the GFP signal. Since mRFP is not pH-sensitive, the mRFP signal remains constant in the autolysosome [8].

3.2.1 Drug Treatment of Cellular Spheroids

Start the drug treatment of the cellular spheroids by pipetting the desired amounts of the drug directly into the medium in the well. In the present protocol, the autophagy-inducing drugs gossypol and rapamycin are employed as an example. At least three spheroids are treated with each drug, in order to achieve statistical significance. For the investigation of the autophagic flux, use growth medium without antibiotics. In fact, antibiotics could influence the autophagic flux.

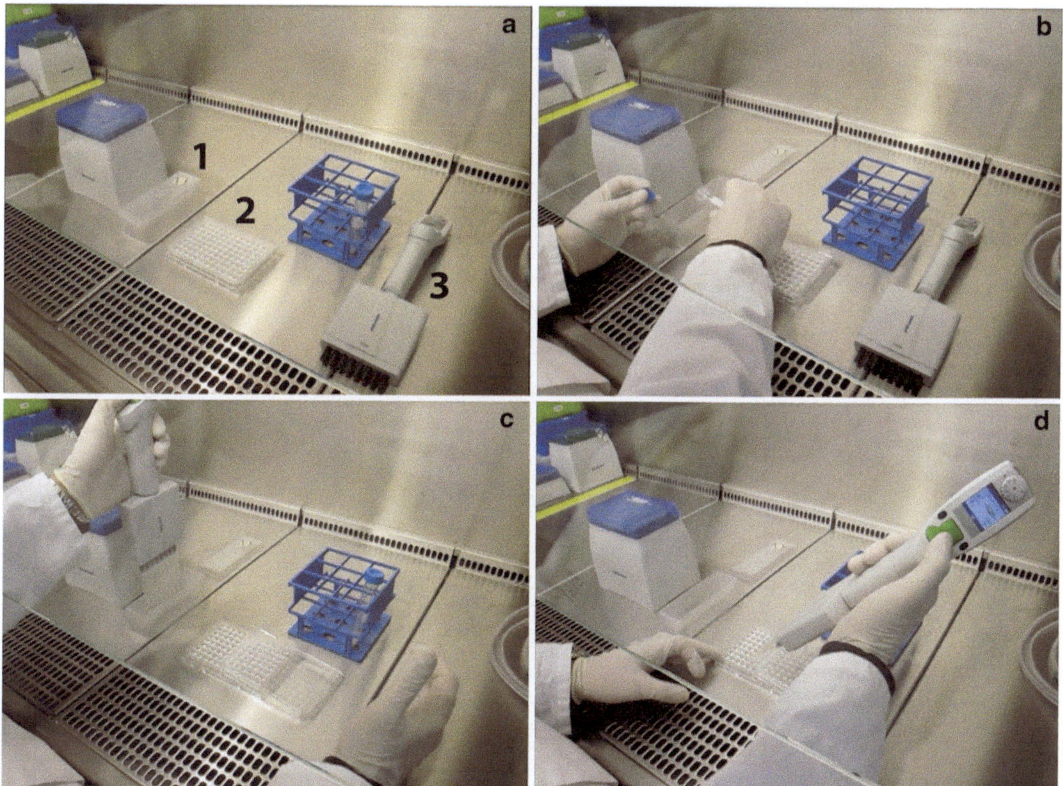

Fig. 1 Pipetting the U343 cells in a 96-well plate for spheroid formation. (**a**) The required equipment is (*1*) a multichannel pipetting trough, (*2*) a 96-well U-bottom plate, and (*3*) an 8-channel pipette. The 10,000 cells/ml cell suspension is contained in a 10 ml tube. (**b**) The cell suspension is transferred to the pipetting trough. (**c**) 100 μl of the cell suspension are loaded in the pipette and (**d**) transferred in the 96-well plate

3.2.2 Live Cell Imaging Using Confocal Laser Scanning Microscopy (CLSM)

Image over time the drug effects in cellular spheroids with an automated motorized confocal microscope. Employ a long working-distance objective lens with high numerical aperture and an intermediate magnification (e.g., 20×), in order to obtain a large enough field of view (Fig. 5). This is important, since spheroids move across the well during time-lapse imaging (*see* **Note 5**). Each substance should be tested at least in triplicate, and at least one control triplicate should be conducted in order to achieve a statistical significance. In the case that a motorized automated confocal microscope is not available, the spheroids can be cultured in a CO_2 incubator for the duration of the experiment, and images can be recorded at fixed intervals (*see* **Note 6**). Define the confocal *z*-stack by choosing a relatively small number of planes (e.g., 10) across the whole spheroids (outer superficial region, intermediate inner region, spheroid's core). The number of planes should be limited in order to minimize photobleaching and phototoxicity (*see* **Note 7**).

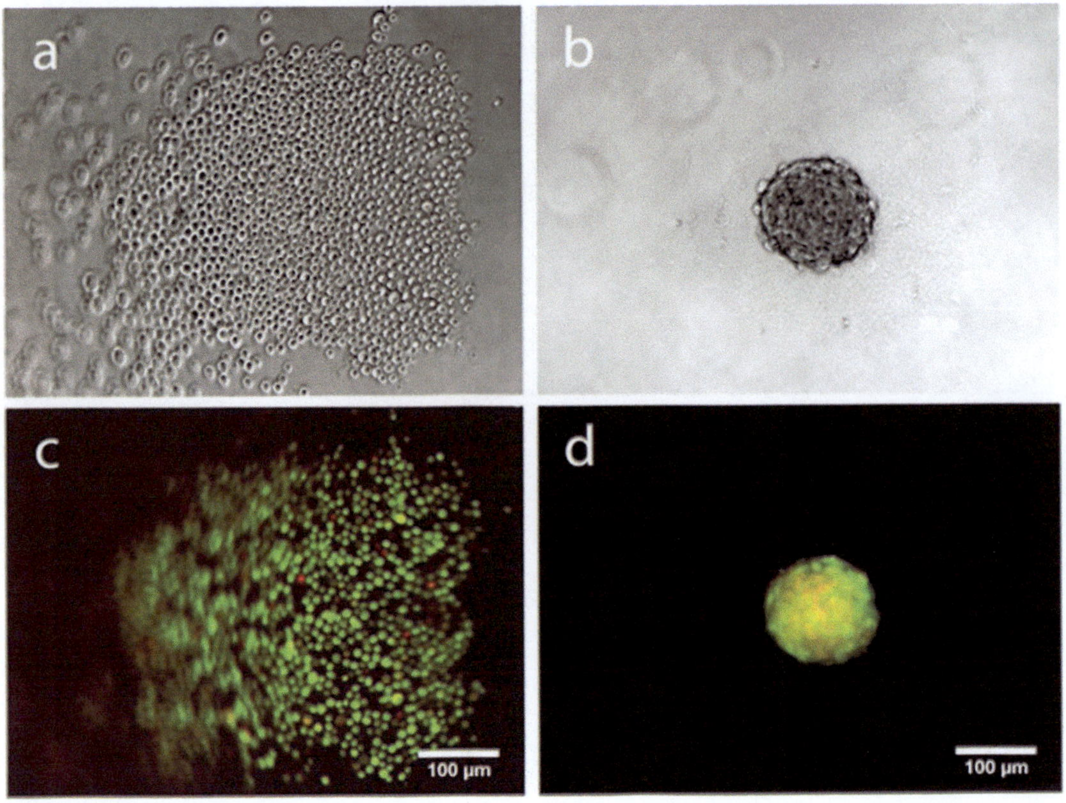

Fig. 2 Formation of mRFP-GFP-LC3 U343 tumor spheroids. (**a**) 1,000 U343 cells are initially seeded in each U-well. (**b**) After 3 days in culture at 37 °C and 5 % CO_2 the cells form a compact spheroid.(**c, d**) the same cells imaged with the epifluorescence microscope, by employing 525 ± 50 nm and 605 ± 70 nm emission filters to detect the GFP and mRFP fluorescence emission from the autophagy marker mRFP-GFP-LC3, respectively

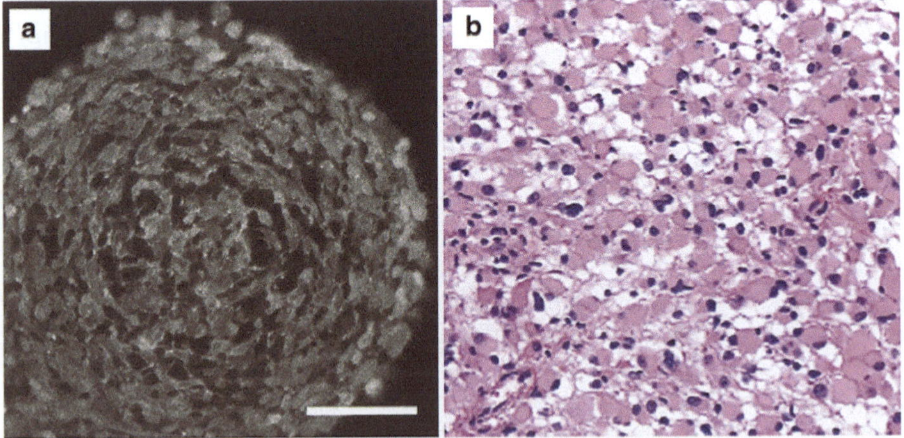

Fig. 3 Comparison between a U343 tumor spheroid formed accordingly to the protocol and a real neuronal glioblastoma. (**a**) confocal microscopy image of an U343 cell spheroid. The image was recorded approximately in the middle of the spheroid. ex./em. 543 nm/620 nm, marker LC3-GFP. (**b**) H and E histological specimen of a patient's glioblastoma (From Wesseling P et al. (2011) Diagnostic histopathology. 17:11)

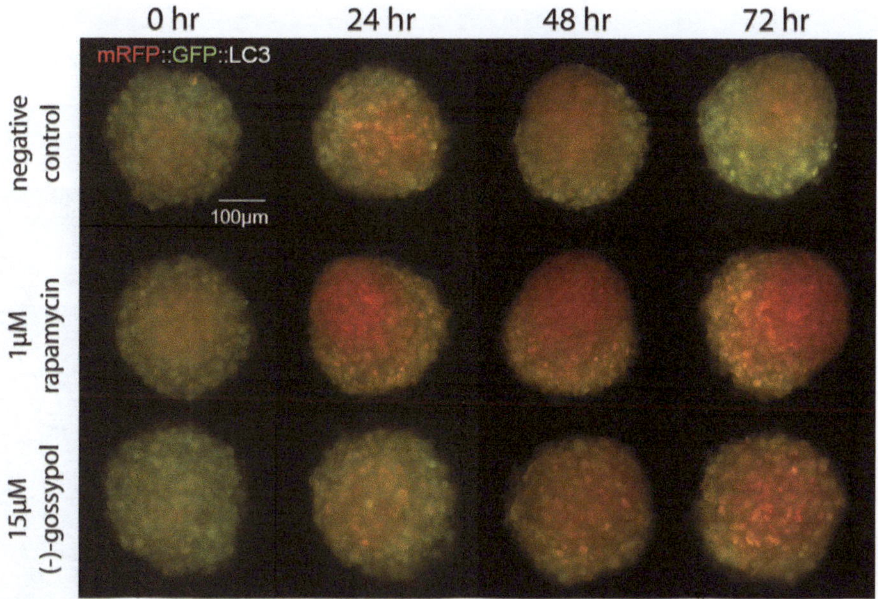

Fig. 4 The relative increase of the intensity of mRFP over GFP signalizes the formation of autolysosomes and thus the progression of autophagy. The U-343 spheroids were treated with autophagy-inducing drugs and the GFP and mRFP fluorescence signal measured over time with live imaging. 1,000 cells/spheroid were seeded. The spheroids were formed by liquid overlay over 48 h. The rapamycin and gossypol-treated spheroids show a clear increase in the mRFP intensity. Imaging parameters: Zeiss Cell Observer Microscope, objective lens: CZ 10×/NA 0.25, ex./em. GFP: 470/525 ± 50 nm, ex./em. mRFP: 550/605 ± 70 nm

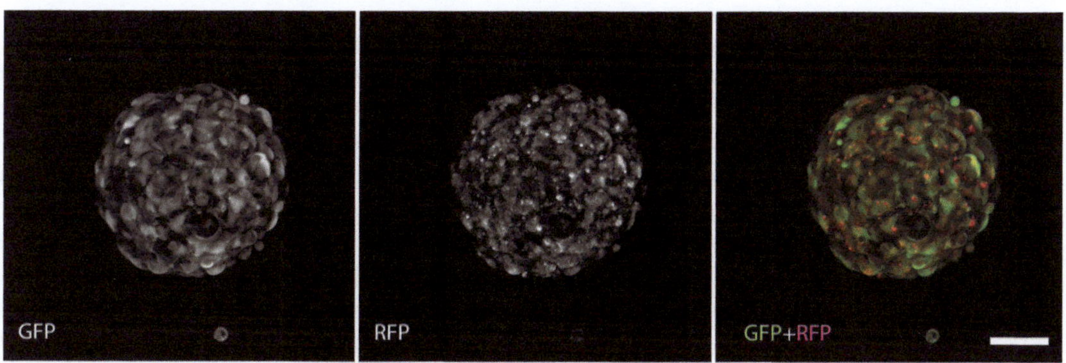

Fig. 5 mRFP-GFP-LC3 U-343 spheroid imaged with the confocal microscope. Maximum projection of 13 planes, spacing 10 μm. Objective lenses Nikon CFI Plan Fluor 10×, NA 0.30, WD 16 mm. Scale bar 50 μm

3.3 Quantitative Measurement of the Autophagic Flux

The fluorescence intensity in both the mRFP and GFP channel is measured by calculating the *corrected total cell fluorescence* (CTCF) [10].

The CTFC is calculated with Fiji. The quantitative measurement of the autophagic flux over time is obtained by calculating the mRFP–GFP intensity ratio at each time point of the time-lapse. Employ the following workflow:

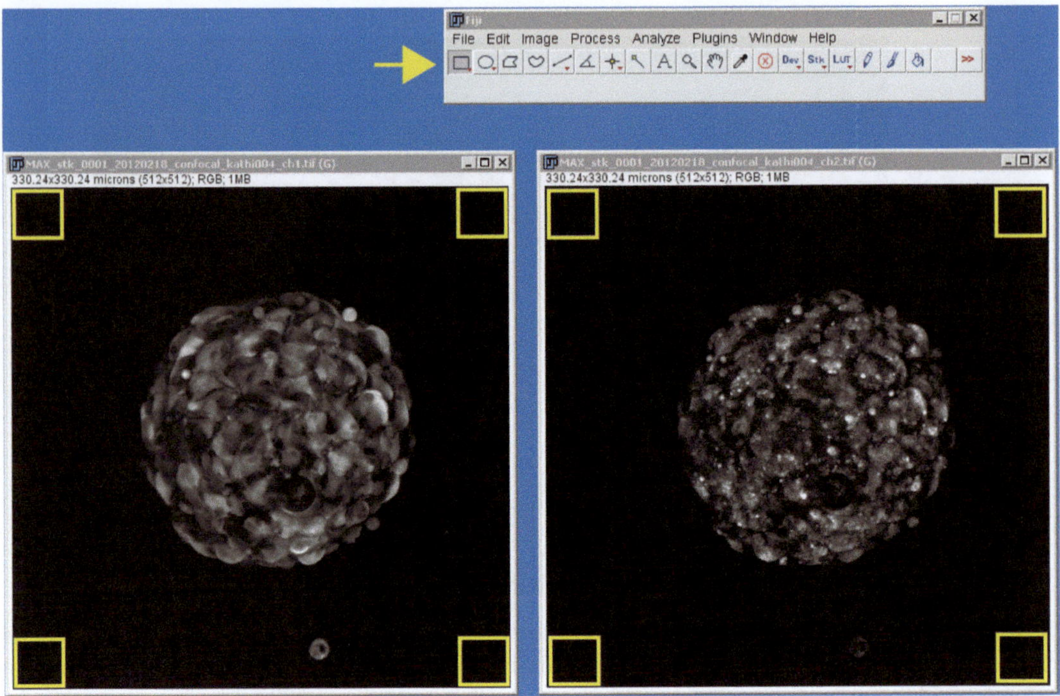

Fig. 6 Measure of the background with Fiji. Four square ROIs are defined for the mRFP and GFP channels in the *upper and bottom corners*. The mean background gray value for each channel is then calculated by using the analyze tools in Fiji. In this picture the maximum projections of a 3D stack composed by 13 single confocal planes for each channel are shown. Objective lens: Nikon CFI Plan Fluor 10×, NA 0.30, WD 16 mm

3.3.1 Preprocessing

In many cases, the raw data from the confocal microscope need to be preprocessed. Preprocessing can involve conversion of the microscope raw data into formats that can be read by the image processing software. Fiji can import many different proprietary formats by the LOCI (Laboratory for Optical and Computational Instrumentation, University of Winsconsin-Madison, USA) plugin. Further preprocessing can involve the sorting specific single planes of the stack. In the examples presented in this protocol, preprocessing includes the conversion of Nikon confocal raw data (.nd2) to .tiff with the Nikon NIS Viewer (www.nis-elements.com/resources-downloads.html). Afterwards, import the .tiff stacks in Fiji and calculating the maximum projections of each stack at each time point for both the mRFP and GFP channels. The calculation of the maximum intensity projection selects preferentially the fluorescence emitted by "hot-spots" such as the autophagosomes and autolysosomes.

3.3.2 Background Subtraction

With the square selection tool define four small squares region of interest (e.g., 30 × 30 pixels ROI) at the four corners of the image (Fig. 6), where only background noise should be present. Note that only one ROI at a time can be selected in Fiji.

Select "*Set measurement*" from the "*Analyze*" menu and mark the check boxes "*Mean gray value*". Select "*Measure*" from the "*Analyze*" menu or hit STRG+M. The window "*Results*" appears showing the mean values for the selected ROI. Copy and paste the data from the *Results* window into a spreadsheet (e.g., Microsoft Excel). Repeat the procedure for each square ROI and finally calculate the mean background over the four ROIs.

3.3.3 Integrated Density

Select the whole field of view with *Edit/Selection/Select all*. Select "*Set measurement*" from the "*Analyze*" menu and mark the check box "*Integrated density*". Select "*Measure*" from the "*Analyze*" menu or hit STRG+M. The window "*Results*" appears showing the integrated density of the image. Copy and paste the data from the *Results* window into the spreadsheet. Repeat this procedure for each channel and each time point.

3.3.4 Corrected Total Cell Fluorescence (CTCF)

Apply the following formula to each channel and each time point:

$$\text{CTCF} = \text{integrated density of the image} - (\text{image area} \times \text{mean fluorescence of background readings})$$

3.3.5 Data Plotting and Fitting

Calculate the ratios of mRFP and GFP integrated density values. The ratios can be normalized for a better comparison between the different drugs. Plot the mRFP–GFP ratio over time with your preferred software (e.g., Microsoft Excel, Mathworks Matlab, OriginLab Origin, Wolfram Mathematica). The plots represent the time evolution of the autophagic flux in the spheroids (Fig. 7). The plots can be fitted with the equations modeling the autophagic flux, e.g., an exponential law (Fig. 7).

4 Notes

1. If possible, avoid dissolving drugs in DMSO, EtOH, or other solvents following not-standardized protocols. Many substances can be supplied or prepared by hospital pharmacies following the same formulations/protocols employed for patients (e.g., solution in physiological saline, stable emulsions).
2. Thawing is a stressful procedure to frozen cells. Do not allow the cells to remain in the thawed freezing media more than few seconds.
3. 0.2 % Trypan blue (solved in PBS) should be used to count only viable cells.
4. The number of cells can also be determined by an automated cell counting device (e.g., Nexcelom Cellometer™ Auto T4).
5. Check the position list several times during the recording if an automated microscope is employed. The spheroids can drift

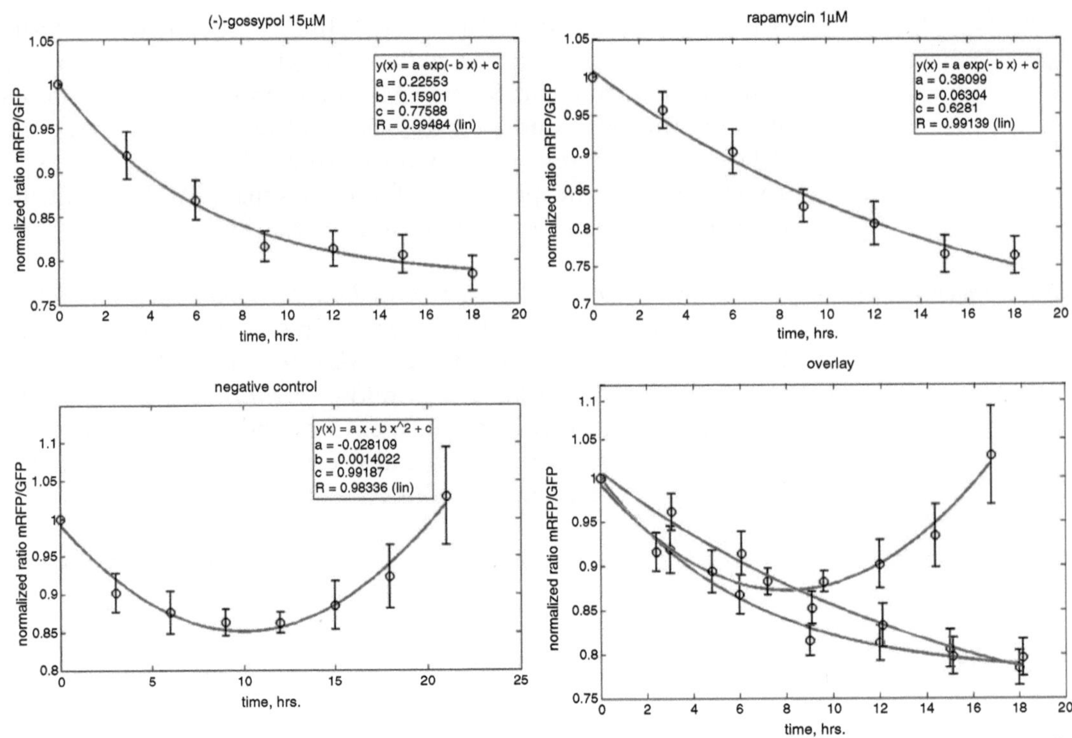

Fig. 7 Plots and fitting curves of the autophagy flux over time in U343 spheroids. The normalized mRFP–GFP ratio is plotted over time. The ratios were calculated as described in Subheadings 3.3.1–3.3.5. The plots quantify the autophagy flux of the two drugs (-)-gossypol 15 μM and rapamycin 1 μM. All the plots were modeled with an exponential law

out of the field of view. In this case it is necessary to redefine the positions.

6. Avoid too short recording intervals with the automated microscope. For this setup we suggest an interval of at least 20 min. Avoid using UV-light to excite fluorescence. This could lead to too high phototoxicity.
7. Keep a negative control during live cell imaging, which is not imaged. This is important to evaluate artifacts due to phototoxicity effects.

Acknowledgements

The authors thank the German Federal Ministry of Education and Research (BMBF) for financial support (project ProMEBS) and Dr. Tamotsu Yoshimori (National Institute of Genetics, Shizuoka, Japan) for providing the mRFP-GFP-LC3 plasmid.

References

1. Levine B, Kroemer G (2008) Autophagy in the pathogenesis of disease. Cell 132:27–42
2. Levine B (2007) Cell biology: autophagy and cancer. Nature 446:745–747
3. Klionsky DJ, Abdalla FC, Abeliovich H et al (2012) Guidelines for the use and interpretation of assays for monitoring autophagy. Autophagy 8:445–544
4. Pampaloni F, Reynaud EG, Stelzer EH (2007) The third dimension bridges the gap between cell culture and live tissue. Nat Rev Mol Cell Biol 8:839–845
5. Friedrich J, Seidel C, Ebner R, Kunz-Schughart LA (2009) Spheroid-based drug screen: considerations and practical approach. Nat Protoc 4:309–324
6. Pampaloni F, Ansari N, Stelzer EH (2013) High-resolution deep imaging of live cellular spheroids with light-sheet-based fluorescence microscopy. Cell Tissue Res 352:161–177
7. Pampaloni F, Stelzer EH, Masotti A (2009) Three-dimensional tissue models for drug discovery and toxicology. Recent Pat Biotechnol 3:103–117
8. Voss V, Senft C, Lang V, Ronellenfitsch MW, Steinbach JP, Seifert V, Kogel D (2010) The pan-Bcl-2 inhibitor (-)-gossypol triggers autophagic cell death in malignant glioma. Mol Cancer Res 8:1002–1016
9. Kimura S, Noda T, Yoshimori T (2007) Dissection of the autophagosome maturation process by a novel reporter protein, tandem fluorescent-tagged LC3. Autophagy 3:452–460
10. Hundeshagen P, Hamacher-Brady A, Eils R, Brady NR (2011) Concurrent detection of autolysosome formation and lysosomal degradation by flow cytometry in a high-content screen for inducers of autophagy. BMC Biol 9:38

Part II

Epigenetic Control of Cancer

Chapter 4

Interdependency Between Genetic and Epigenetic Regulatory Defects in Cancer

Félix Recillas-Targa

Abstract

Epigenetic regulation is understood as heritable changes in gene expression and genome function that can occur without affecting the DNA sequence. In its in vivo context DNA is coupled to a group of small basic proteins that together with the DNA form the chromatin. The organization and regulation of the chromatin alliance with multiple nuclear functions are inconceivable without genetic information. With the advance on the understanding of the chromatin organization of the eukaryotic genome, it has been clear that not only genetics but also epigenetics influence both normal human biology and diseases. As a consequence, the basic concepts and mechanisms of cancer need to be readdressed and viewed not only locally but also at the whole genome scale or even, in the three-dimensional context of the cell nucleus space. Such a vision has a larger impact than has been previously predicted, since phenomena like aging, senescence, the entail of nutrition, stem cell biology, and cancer are orchestrated by epigenetic and genetic processes. Here I describe the relevance and central role of genetic and epigenetic defects in cancer.

Key words Epigenetics, Genetics, Chromatin, Chromatin remodeling, DNA methylation, Cancer, Long noncoding RNA, Nuclear dynamics, Epigenome

1 Introduction

It is generally accepted that carcinogenesis takes place through a complicated and interdependent events consisting of three main phases: initiation, promotion, and progression. More recently, several attributes or "hallmarks" have been identified that are universal to all malignancies [1]. These include insensitivity to anti-growth signals, self-sufficiency in growth signals, evasion of apoptosis, limitless potential for replication, and acquisition of the mechanisms required for tissue invasion, metastasis, and angiogenesis.

The gene-centered dogma upholds the belief that all steps of the process stem ultimately from the acquisition of changes in the DNA sequence due to different types of mutations, i.e., genetic defects. There is indeed a wealth of information that supports this view. Mutations, deletions, rearrangements, translocations, and

Martha Robles-Flores (ed.), *Cancer Cell Signaling: Methods and Protocols*, Methods in Molecular Biology, vol. 1165, DOI 10.1007/978-1-4939-0856-1_4, © Springer Science+Business Media New York 2014

gene amplification are the most common changes. The genetic model appears to explain cancer initiation, since for most common tumor mutations have been identified as early events [2]. On the other hand, mutations in some genes have been shown to be responsible for cancer progression. The genetic model also offers great promise in screening for the presence of early malignancies, and cancer in its familiar form [2].

Furthermore, the discovery of promoter sequences (oncogenes) and suppressor genes offers a structural framework in which mutations in these type of genes act to induce continuous cell replication and tumor growth. These genocentric concepts of carcinogenesis seemed sufficient to explain both cancer family syndromes and sporadic cancers. However, in the last two decades it has become apparent that other mechanisms apart from DNA mutations could be involved in the generation of cancers [3, 4].

It is now firmly established that the complement of genes remains constant during ontogeny and cell differentiation, confirming Conrad Waddington hypothesis [5, 6]. Genes are switched on and off differently to make the more than 100 various types of cells in the human body. In other words, patterns of gene expression, not gene themselves, define each cell type. The mechanisms that are responsible for the fine control of gene expression constitute the emerging field known as epigenetics. The general epigenetics concepts comprise modifications of the genome that can be transmitted during cell division that do not involve a modification of the DNA sequence. Therefore, cell differentiation and other cellular processes depend mainly on epigenetic mechanisms (Fig. 1) [7, 8]. Since neoplastic transformation can be viewed as a de-differentiation process, it would be logical to assume that the epigenetic control of gene expression responsible for terminal cell differentiation would be deranged in malignancies.

1.1 The Eukaryotic Genome is Organized into Chromatin

It the last 15 years experimental evidence has documented unequivocally that epigenetic mechanisms are operative in malignant transformation [4]. The central structure whereby these mechanisms take place is the nuclear chromatin where DNA and protein are distributed in an orderly manner [9]. The fundamental packaging unit of chromatin is the nucleosome organized into 147 base pairs of DNA wound around an octamer core of histones (two each of the H2A, H2B, H3, and H4 histones). The position and density of histone octamers along the DNA are mediated by ATP-dependent chromatin remodeling complexes that use the energy from ATP hydrolysis to move or slide the histone octamers among and along DNA molecules. The interplay between histone post-translational modifications, DNA, and chromatin-associated proteins (readers) is controlled by acetylation, methylation, phosphorylation, poly(ADP-ribosyl)ation, and ubiquitination among others. These modifications and the positioning of histones organize

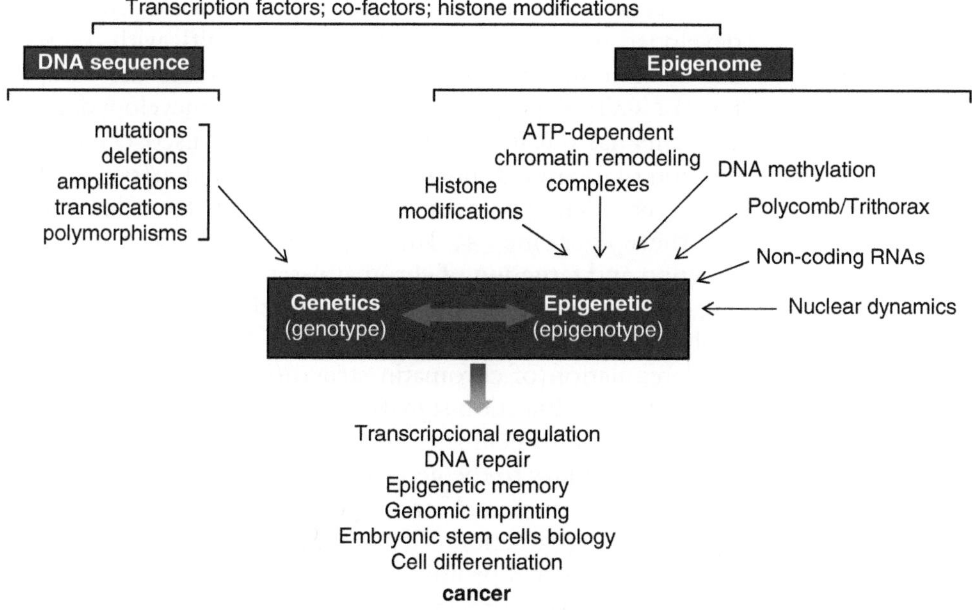

Fig. 1 Genetic and epigenetic interplay in the control of gene expression and in other cellular processes. Genotype is mainly controlled at the level of the nucleotidic sequence. Instead, the epigenotype is governed through the modulation, at distinct levels, of chromatin structure. The organization of the eukaryotic genome into chromatin not only allows gene activation but also controls gene silencing through DNA methylation, Polycomb members, and/or noncoding RNAs among others. It is important to recall that the great majority of the chromatin remodelers are not capable to "read" genetic information and they need to be recruited to their target sites by transcription factors, cofactors (co-activators or co-repressors), or specific histone modifications (histone methylations). Together, the genotype and epigenotype have an impact on a large variety of key biological processes that certainly can be influenced by the environment, which in turn have implications on the human physiology and diseases

the genome into either open (euchromatin) or compacted chromatin (heterochromatin) and thus regulate the accessibility of DNA for many processes including transcription, DNA recombination, replication, DNA repair, and chromosome segregation. The positioning and modification of the histones form a histone code that generates specific signals that converge in regulatory and physiological processes. In addition, one appealing and scientifically attractive aspect of the field is the so-called epigenetic "memory" that is passed from mother cell to daughter cells [7, 9]. Today, abnormal DNA methylation, covalent histone modification, and ATP-dependent chromatin remodeling complexes comprise only one of the facets of epigenetic regulation of gene expression involved in cancer; noncoding RNAs, members of the repressive and activating complexes Polycomb/Trithorax, and, more recently, the cell nucleus enlarge the epigenetic landscape that needs to be taken into consideration to understand the origin and progression of cancer (Fig. 1) [7, 8].

1.2 Genome-Wide Description of the Epigenome

The recent outcome of The Cancer Genome Atlas (TCGA) developed by the National Institutes of Health with the sequencing of thousands of human cancer genomes and the complementary ENCODE project that generated the encyclopedia of DNA elements has put into light the unexpected discovery that a large fraction of identified mutations falls into genes associated to chromatin remodeling components and/or regulatory elements affecting the epigenome [4, 10, 11]. Thus, mutations affecting the function and targeting of chromatin-remodeling complexes generally cause cancers or multisystem developmental disorders. The multisystem nature of these single-gene disorders can be explained by deregulation of chromatin structure at many loci in a sort of malignant amplification signals [4].

According to Feinberg, the epigenetic model of cancer, which is complementary to the genetic model [8], can help to explain the action of environmental carcinogens, as well as tumor progression. Epigenetic deregulation offers another mechanism to explain activation of genes that promote tumor invasion and metastasis as well as activation of latent viral oncogenes and silencing of tumor suppressor genes. This can be further complemented by the varied action of microRNAs targeting chromatin-modifying enzymes, and more recently, long noncoding RNAs (lncRNAs). Epigenetic alterations might also increase the probability of cancer progression after genetic change or vice versa. Due to the fact that epigenetic mechanisms are dynamic and, importantly, reversible as opposed to genetic mutations, the oncogenic deregulation of these processes is amenable, promising therapeutic manipulation.

In this chapter I describe some of the recent advances in cancer epigenetics research that attest the great promise for cancer diagnosis and treatment.

2 Epigenetic Silencing in Cancer by DNA Methylation

DNA methylation is recognized as one of the most important epigenetic processes maybe because it is one of the most studied [12, 13]. DNA methylation is synonymous of gene expression silencing and repressive chromatin structure and until recently, an intense debate was established concerning the reversibility of DNA methylation (see below). It has to do with the incorporation of a methyl group to the cytosine only when the cytosine is preceded by a guanosine in the DNA sequences (known as the CpG dinucleotide) [12, 13]. The incorporation of the methyl group to the CpG dinucleotide is responsibility of a growing family of DNA methyltransferases (DNMTs). Among them, the de novo DNMT3a and DNMT3b are active in early developmental stages and the replication associated DNMT1 is known as the maintenance DNMT [14, 15]. Interestingly, such a classification does not seem to be definitive, since in tumor cell lines, knockdown of one of the DNMTs does not cause a drastic decrement in DNA methylation levels, suggesting a compensation or redundancy on their function, in particular by the DNMT1 and DNMT3b [16]. At this point, it is important to define the normal role of DNA methylation. It has been demonstrated that DNA methylation is crucial during development, specifically through heritable gene expression silencing but also for genomic imprinting, X-chromosome inactivation, regulation of gene expression and maintenance of epigenetic memory [13, 17]. Furthermore, there is evidence that support its role in maintaining genomic stability through keeping chromatin structure of the genomic mobile elements like transposons and many other repetitive sequences in a close and stable conformation [13, 18].

There are two main mechanisms through which DNA methylation negatively influence gene expression. The first one and more direct is the DNA methylation interference of transcription factor recognition for their DNA binding sequences in a promoter or regulatory element. Alternatively, and what seems to be the most widely spread strategy is that DNA methylation is recognized by a family of nuclear proteins, known as CpG-binding proteins (MeCPs) or methyl-binding domain (MBDs) proteins [19, 20]. Furthermore, their repressive effect is mediated through the action of co-repressors, like mSin3A, by the recruitment of histone deacetylases containing complexes [21, 22]. This complex interaction resulted in a more general mechanism in which DNA methylation repressive effect is directly mediated through the induction of a close and compacted chromatin structure [12, 13]. The relevance of the methyl-CpG-binding protein in cancer has been constantly demonstrated since the first demonstration showing that binding of MeCP2 to the CpG-island of *BRCA1* gene promoter produces gene expression silencing [23]. The binding of MBD2 was also shown on the silent p14/p16 locus in different tumor cell lines [24]. Interestingly, in this work, when the inhibitors of histone deacetylases and DNA methylation were used independently, only a partial reactivation of $p14^{ARF}$ and $p16^{INK4a}$ was seen. It is only when both inhibitors were incorporated at the same time to the cell culture, when the highest levels of reactivation were reached supporting the model of the repressive complex formed through the action of DNA methylation and histone deacetylases. Interestingly, in a survey of methyl-CpG-binding protein at the hole genome scale in human breast cancer cells and primary tumors, it was found that for some genes there was an association of multiple MBD proteins, while for others there was only a single MBD, with also a methylation-dependent and independent binding [25, 26]. This observation can be interpreted as the need to establish qualitative and strength differences of methyl-CpG-binding proteins action on gene silencing in different types of human cancers.

One of the main targets for the action of abnormal epigenetic events leading to cancer is the CpG-islands, in particular, those corresponding to promoter elements of tumor suppressor genes [27–29]. CpG-islands are clusters of CpG dinucleotides distributed all over the genome and around 60 % of the CpG-islands are associated with human genes, preferentially to their promoter and first intron regions [17, 27]. In relation to cancer, the great majority of tumor suppressor genes promoter corresponds to CpG-islands [28, 29]. Paradoxically, and contrary to what happens to CpG dinucleotides in the rest of the genome, in their normal context CpG-islands are unmethylated and in consequence, allowing proper regulation of gene expression. In cancer, abnormal patterns of DNA methylation have been identified now for more than two decades. Then, abnormal hypermethylation of tumor suppressor gene promoters turns to be a key event in epigenetic cancer development (see below).

In an attractive review, Issa proposed the "CpG island methylator phenotype" to try to explain and understand the specific weight of DNA methylation in cancer [30]. The "methylator phenotype" established a central concept in which DNA methylation, as epigenetic silencing mechanism, should converge allowing some degree of classification according to the degree of DNA methylation of the affected genes. He proposes that the spreading of DNA methylation could initiate at "methylation centers" in noncoding regions encompassing large genomic areas including repetitive sequences and in certain cases regulatory elements like promoters. Furthermore, boundary elements may exist delimitating the CpG-islands, in particular tumor suppressor gene promoters, protecting from abnormal DNA methylation [30, 31].

With the advent of the genomic era, three novel aspects around the definition of CpG-islands have emerged. The first came on the basis of a genome-wide analysis of the distribution of CpGs along the entire human genome [28, 29]. In such a survey there is a redefinition of the CpG-islands in three different categories: the low-CpG content, the intermediate, and the high CpG-islands. Interestingly, DNA methylation at low CpG-island promoter does not interfere with their function, in contrast to what happens at the intermediate and high CpG content promoters [28]. The other aspect has to do with the new concept of "CpG-island shore" that states that the DNA methylation is not necessarily incorporated where the highest density of CpG is located, but instead, it can occur at distances of 2 kilo bases (kb) from the defined CpG-island [32]. This observation renders the study of DNA methylation patterns more difficult since it is now difficult to predict where abnormal gain of DNA methylation occurs. Finally, another genome-wide study performed in embryonic stem cells in which, non-CpG DNA methylation, CHH or CHG sites (where H: A, C, or T) have been identified in around 25 % of the genome [33]. The function of such non-CpG sequences remains unclear, but from the cancer perspective we cannot discard that in certain tumoral cells abnormal gain of non-CpG DNA methylation can be involved in the pathology.

3 Epigenetic Regulation Through DNA Demethylation

For many years it has been considered that the DNA methylation reversibility was limited or exclusive to earliest embryonic stages, such as the zygote and developing primordial germ cells [34]. Several genome-wide analyses lead to the proposal of a more active DNA demethylation, in particular in embryonic stem cells and in the brain [35]. In 2009, the enzyme ten eleven translocation 1 (TET1; one of the three enzymes of the TET family) was shown to possess enzymatic activity capable to oxidize the 5-methylcytosine (5 mC) to generate 5-hydroxymethylcytosine (5 hmC) [36]. This 5 hmC can serve as substrate for mammalian DNA glycosylases or deaminases. The most studied pathway involves the 5 hmC deamination by the AID family of cytidine deaminase to produce 5-hydroxymethyluracil (5 hmU) that creates a 5hmU:G mismatch which is repaired by the action of DNA glycosylases (such as TDG and SMUG1) and the BER pathway, ending the reaction by a non-methylated cytosine [36]. Therefore, it is evident that this novel epigenetic pathway, guided by the TET family of proteins, represents an opportunity, in addition to the DNA demethylation role in embryonic stem cells and brain, to revert in a regulated and targeted manner the specific and local patterns of DNA methylation in response to diverse stimuli. Consequently, if we can better understand this process of DNA demethylation, we can be optimistic in the development of more effective therapeutic strategies against genomic regions that gained abnormal DNA methylation as in cancer.

4 Other Epigenetic Chromatin Remodeling Activities

Even though researchers had placed DNA methylation on a prime line of abnormal epigenetic regulation in human diseases, there is a growing amount of experimental evidence of epigenetic processes responsible for distinct neoplasias [4, 37]. Furthermore, the order of events, as well as the combinatorial effects caused by genetic and epigenetic abnormalities, remains currently undetermined [3, 4]. In any case, there is little doubt that several epigenetic mechanisms, in addition to DNA methylation, participate in the abnormal gene silencing involved in cancer (Fig. 1).

One of the principal chromatin remodeling activities is represented by the large variety and combinatorial covalent histone modifications. As suggested by Allis and collaborators, those post-transcriptional modifications are known as the "histone code" which includes acetylation, deacetylation, methylation, phosphorylation and more recently, ubiquitination, sumoylation, ADP-ribosylation, and O-GlncNAcylation of amino- and core-residues of the histones [38]. To illustrate such changes, it has been demonstrated that an apparent feature of human cancer is the loss of acetylation at lysine 16 and trimethylation at lysine 20 of the histone H4 [38, 39]. But histone modifications are not the sole epigenetic source of regulation based on chromatin structure. Moreover, having distinct degrees of interdependency, histone modification, ATP-dependent remodeling complexes, Polycomb and Trithorax family members, noncoding RNAs, and even the nuclear structure and dynamics participate in epigenetic regulation and the processes of malignant transformation (Fig. 1) [38–42].

4.1 Epigenetics and Imprinting Associated Diseases

Epigenetics is involved in the control of genomic imprinting and its deregulation has not only to do with DNA methylation but also to do with interference of nuclear factors like CTCF that contribute to the determination of mono-allelic expression [43]. Defects on epigenetic control of genomic imprinting are involved in several syndromes like Angelman's, Prader–Willi, Beckwith–Wiedemann, ATRX-related syndrome, and others [43–45].

In summary, to understand carcinogenesis it is necessary to take into consideration not only genetic defects (acquired and heritable) but also the complex interplay of epigenetic factors in order to have a more holistic vision of a multifactorial disease like cancer.

4.2 DNA Methylation, CpG-Island, and Epigenetic Silencing in Cancer

In cancer one of the most growing fields of research has to do with abnormal methylation of CpG-islands over gene promoters and their correlation with epigenetic silencing of gene expression with the critical inactivation of tumor suppressor genes [29, 46]. There is also a growing list of genes that can be affected by epigenetic mechanisms, in particular, DNA repair genes like BRCA1, the DNA mismatch repair (hMLH1), and O^6-methylguanine DNA methyltransferase (MGMT), in which DNA methylation plays a central role causing a series of genomic instabilities. This is the case of hMLH1 promoter hypermethylation that seems to be responsible of sporadic cases of colorectal, endometrial, and gastric cancer [47].

As mentioned before, there is an increasing list of genes, including tumor suppressor genes, that are silenced through their CpG-island hypermethylation correlating with human cancer. Genes that are affected are involved in DNA repair like MLH1, cell cycle control like $p14^{ARF}$, $p16^{INK4a}$, Rb, and several others, transcription factors, apoptotic and pro-apoptotic genes, and others [48]. Therefore, the epigenetic defects on human diseases can occur at distinct levels.

Epigenetic alterations in cancer include hypomethylation (the first observed), in which there has been a resurgence of interest, for example MAGE in melanoma, cyclin D2 and maspin in gastric cancer, CA9 in renal cell cancer, and demethylation of latent papilloma virus in cervical cancer progression [49–51]; and hypermethylation associated with tumor suppressor gene silencing, for example RB, p16, and VHL in lung and other cancers [52–54]. Hypomethylation facilitates transcription, mainly of oncogenes that promote cellular proliferation, and hypermethylation of promoter regions halts transcription of suppressor sequences.

Other affected genes are those involved in the control of the cell cycle, like $p16^{INK4a}$, $p15^{INK4b}$ and $p14^{ARF}$. The majority of these genes have been studied for their aberrant DNA methylation patterns, but they have also been shown to be epigenetically silenced by alternative and frequently redundant mechanisms. A clear example is given by $p16^{INK4a}$ which can be epigenetically silenced through the action of HDACs or members of the Polycomb repressive family of proteins [24, 37]. Interestingly, in our research group we have been studying the silencing epigenetic mechanisms of two of the most known tumor suppressor genes, Rb and p53. We have found, in accordance to the "methylator phenotype" and the spreading of chromatin modifications, that the multifunctional CTCF nuclear factor is shielding the promoter regions of these two genes against abnormal epigenetic silencing [31, 55, 56]. In addition, in the absence of CTCF binding, the Rb promoter is silenced by DNA methylation. In contrast, p53 do not gain DNA methylation instead there is an atypical enrichment on histone H3K27me3 and H4K20me3 repressive marks [55, 56]. Further investigations are needed to understand if there is a direct relationship between CTCF and silencing, and the mechanisms associated to the loose of CTCF in those promoters.

As part of the combination of genes or group of genes linked to particular pathways, not all the genes susceptible to be inactivated epigenetically are equally affected in cancer. Two genes involved in mismatch repair and responsible for very similar malignant phenotypes, hereditary nonpolyposis colon cancer, MLH1 and MSH2 (mutant S homologue 2) genes, are differentially affected in their expression. Despite the fact that both posses a CpG-island in their promoters, MLH1 is frequently inactivated by epigenetic mechanisms, in particular DNA methylation. In contrast, MSH2 is never methylated in cancer [30, 57].

4.3 Both Genetic and Epigenetic Alterations Contribute to Cancer

A more recent demonstration showing that inactivation of tumor suppressor genes not only occurs via epigenetic mechanisms but also with the contribution of genetic defects is the p53 pathway [58]. In a mouse model, Baylin and collaborators have found that the hypermethylated in cancer 1 (*HIC1*) gene is very frequently epigenetically inactivated, but not mutated in human cancer ([58] and references therein). But when germ line mutation on one allele of p53 occurs, Hic1 is epigenetically silenced inducing the appearance in mice of breast, ovarian carcinomas and metastatic osteosarcomas [58]. Thus, there is no doubt that we need to take into account both genetic and epigenetic alterations. Moreover, we need to take into consideration upstream and/or downstream components of a specific tumor suppressor pathway since they can influence to different extent the transformant phenotype.

What remains unclear is the time course order of events leading to epigenetic silencing in cancer. There are more and more evidences that suggest that DNA methylation is one of the latest, if not the last, epigenetic mark of silencing. Thus, DNA methylation is preceded by distinct and probably inter-connected histone modifications that frequently are the signal for recruitment of other remodeling activities [4]. To exemplify this vision, I would like to present two recent examples. The first has to deal with the loss of histone covalent modifications as a hallmark of human cancer. Using biochemical and molecular approaches, Esteller and collaborators determined the early appearance, during oncogenesis in a controlled mouse model of multistage skin carcinogenesis the loss of acetylation on histone H4 lysine 16 and trimethylation of lysine 20 [39]. What is surprising is the fact that different histones had a large variety of distinct modifications (acetylation, methylation, phosphorylation among others). Then, we need to point out that it is going to be very difficult to define specific patterns of histone posttranslational modifications in cancer.

The second example has to deal with the epigenetic role of Rb-pathways and Rb-family members in the formation of heterochromatin. In senescence Rb-pathway contributes through epigenetic mechanisms to genome stability [59]. It has recently been demonstrated that the integrity of the Rb-pathway is critical for the senescent phenotype of human fibroblasts and for the formation of the so-called "senescence-associated heterochromatin foci" or SAHF, that restricts cell proliferation, favoring cell senescence and consequently keeps genomic stability [59]. Such novel contribution of the Rb-pathway appears to be of extreme importance in cancer since, once the mechanisms are understood, the control of senescence through limiting the proliferation potential of a transformed cell could represent a strategy for cancer control [59]. Another member of the Rb-family, the p130 protein, which is predominantly present in neurons, is able to epigenetically repress pro-apoptotic genes, supporting neuron survival through the action of histone deacetylases and histone methyltransferases and E2F4 [60].

4.4 The Role of Polycomb-Repressive Group of Proteins in Cancer

Polycomb group of repressive proteins (PcG) contributes to the pathogenesis and progression of cancer and other complex diseases. Polycomb proteins are involved in critical aspects of the cell's life like cell differentiation and the higher-order genome organization inside the nucleus. Furthermore, PcG also participates in the biology of the stem cells through the regulation of embryonic stem cells self-renewal capacity and cell differentiation [61]. Interestingly, there is a sort of parallelism between PcG and DNA methylation in terms of their epigenetic influence in normal and cancer cells. Recent evidences support a role of PcG in the segregation post-mitotically of epigenetic traits [62]. In cancer, there are evidences that members of the PcG complex, in particular EZH2, the histone methyltransferase that incorporates the H3K27me3 repressive mark [61], is over-expressed in prostate and breast tumors among other types of cancers [29, 63–65]. Issa and collaborators have shown that in different types of cancers (breast, prostate, and epithelial tumor cell lines) the abnormal epigenetic silencing can depend on DNA methylation, silencing by PcG of proteins, or both mechanisms [29]. Results from our research group are in agreement with this vision, since epigenetic silencing of Rb and p53 tumor suppressor genes occurs by DNA methylation and PcG, respectively [55, 56]. From these results we have found a potential relationship between the abundance and type of CpG-island and the mechanism of epigenetic silencing [31]. The Rb gene promoter, considered as a high- content CpG-island, is silenced by abnormal gain of DNA methylation in glioblastoma transformed cells [54, 66], in contrast, for p53 gene promoter who is a low CpG-island, it is silenced by the gain of H3K27me3 repressive histone mark, which is incorporated by PcG with no gain of DNA methylation.

A novel aspect not only for the PcG group of proteins, but also for the Trithorax proteins, is the link between these histone modifications and lncRNAs (see below). LncRNAs are viewed as a mean to target the PcG and TrxG to specific genomic locations, in particular promoter regions. Importantly, abnormal over up- or down-regulation of lncRNAs can be involved in cancer development as documented by the lncRNA HOTAIR (see below).

Therefore, the progress in recent years in the diverse functions of PcG or TrxG group of proteins show the relevance of these epigenetic complexes and the complexity of their mode of action, including their relationship with lncRNA. The relevance of there epigenetic remodelers is further accentuated with their role in development, cancer biology and complex diseases.

4.5 Genetic Defects with Epigenetic Consequences: Chromosomal Translocations

Interestingly, and as an example of the interdependency of the genetic and epigenetic phenotypes, chromosomal translocations, which are by definition genetic abnormalities, provoke defects on target genes through epigenetic mechanisms [67, 68]. The best example is the oncogenic fusion protein PML-RARα, composed by the promyelocytic leukemia factor and the co-repressor retinoic acid receptor α. Such protein derived from the translocation t(15; 17) participates in the development of acute promyelocytic leukemia [68]. The abnormal epigenetic mode of action of this fusion protein is mediated through the recruitment of the NCoR co-repressor and a histone deacetylase (HDAC) complex inducing a repressive chromatin configuration of the retinoic receptor target genes. Furthermore, it has also been demonstrated, in particular for the promoter of the RARβ2, that the PML-RARα is also capable to attract DNMTs, leading to the abnormal DNA methylation of the RARβ2 CpG-island promoter [67]. Several other translocations have been evident through the translocation t(8; 21) causing the fusion of AML1 to ETO, which in turn associates with epigenetic gene repressors such as HDAC's and Sin3. Such fusion protein abnormally silences the AML1 target genes, interrupting normal myeloid differentiation and favoring cellular transformation [69]. In consequence, a genetic defect in its origins causes sequential epigenetic events leading to the silencing of multiple target genes affecting gene differentiation and proliferation.

5 Noncoding RNAs and Their Role in Cancer

It is surprising how in recent years the different types and functions of noncoding RNAs have increased [70]. Small and lncRNAs possess a sophisticated and frequently unknown regulation, but they can act directly, in *cis* or *trans*, through posttranscriptional mechanisms, like microRNAs, or through epigenetic pathways [70, 71].

Strictly, microRNAs are not considered direct epigenetic regulators since they belong to the class of small noncoding RNAs that act posttranscriptionally degrading mRNAs or interfering with translation [72]. The relationship between microRNAs and epigenetics in cancer is basically at two levels: microRNA targeting epigenetic modifiers and deregulation of microRNAs expression through epigenetic mechanisms. For example, miR-101 and miR-29 target specifically the Polycomb-associated histone methyltransferase EZH2 [73, 74], and DNMTab/b [73, 75]. Thus, abnormal regulation of miR-101 and miR-29 can induce the abnormal de-repression of genes and induce genomic instabilities.

Until recently, the identification of regulatory regions, in particular promoters, represented a difficult task because their intra- or intergenic location and the definition of the microRNA transcriptional unit. Recently, the Encyclopedia of DNA elements (ENCODE) project and many genome-wide studies has allowed the identification of a large number of microRNA promoters [10, 76]. The relevance of those findings comes when there are evidences that DNA methylation and other chromatin modifications can aberrantly silence the expression of microRNAs. A seminal example is the miR-124a, which undergoes transcriptional silencing by the DNA hypermethylation of its promoter region in different types of cancer cells [77]. This inactivation is associated in tumor cells with the activation of the oncogenic cyclin D kinase 6 and phosphorylation of the retinoblastoma tumor suppressor gene [77].

In conclusion, the aberrant regulation of microRNAs through epigenetic mechanisms can have an amplification effect over many gene products in a tissue-specific manner. In addition, we cannot discard genetic defects in microRNAs with consequences in the chromatin remodeling targets, that in their turn, can modify their activity and genome conformation in cancer.

LncRNAs (>200 nucleotides in length) are a novel subgroup of noncoding RNAs linked to the epigenetic regulation of the human genome [71]. LncRNAs are mainly transcribed by the RNA polymerase II, they have a 5′ terminal methylguanosine cap and are often spliced and polyadenylated. One of the most recent genome-wide studies and summary of all the published data in the field showed the existence of 14,880 lncRNAs, with 9,518 classified as intergenic (known as lincRNAs) and 5,362 intragenic RNAs [78]. Several mechanisms for this action have emerged including the decoy model, the scaffold model, the guide activity which directs enzymatic and chromatin remodeling activities to specific target genes, and the enhancer model (for details see ref. 70). Until now, the regulatory role of lncRNAs has been mainly focused in their role as epigenetic modulators.

Historically, the first lncRNA to be characterized was Xist, which is transcribed from the X-inactivation center (Xic) and represents the initial step for a coordinated cascade of repressive chromatin remodeling processes that culminate with the inactivation of one of the two X-chromosomes [79]. More recently, one of the most studied lncRNA is the *Hox Antisense Intergenic RNA* or HOTAIR which is a noncoding RNA of around 2.2 kb that is antisense to the human *HOXC* genes located on chromosome 12 [80]. Strikingly, HOTAIR acts in *trans* repressing HOXD locus on chromosome 2 through the specific recruitment of the Polycomb Repressive Complex 2 (PRC2) and the local incorporation of the histone repressive mark H3K27me3 [80]. Therefore, the lncRNA HOTAIR induces in *trans*, on the basis of epigenetic mechanisms, the silencing of distal genes located in different chromosomes. But which is the relationship between HOTAIR and cancer? Interestingly, Chang and collaborators demonstrated that HOTAIR expression is increased in primary breast tumors and metastasis suggesting that HOTAIR can promote cancer metastasis [81]. What turns to be surprising is that HOTAIR is one lincRNA among 233 regions that are transcribed in the HOX loci, with 170 defined as noncoding RNAs [80, 81]. The interpretation of these observations is that HOTAIR over-expression induce abnormal targeting of the Polycomb Repressive Complex 2 and gain of H3K27me3 repressive histone mark in a large number of genes genome-wide, increasing cancer invasiveness and metastasis. In other words, the abnormal over-expression of HOTAIR generates an aberrant amplification signal that causes the epigenetic silencing of many genes. This is only one example among many others lncRNAs like ANRIL or the lincRNA-p21, which is transcriptionally regulated by p53 [82, 83]. Interestingly, the counterpart to HOTAIR has been discovered and is known as *HOXA transcript at the distal tip* or *HOTTIP*, which binds to the WD repeat domain of the WDR5 protein, a cofactor of the Trithorax group of protein MLL who is a histone methyltransferase that mediates the open histone mark H3K4me3 associated with gene activation [84].

Finally, recent published data is linking lncRNAs and microRNAs through an activity defined as "sponge" where the lncRNAs can reduce or titrate the concentration of particular groups of microRNAs [85]. Thus, these observations support the interdependence of genetic and epigenetic regulators and their potential amplification role in normal and cancer cells in association with noncoding RNAs.

6 Spatial Organization of the Genome, Chromosomal Domains, and Cancer

The progress in the capacity to label molecules with fluorescent reagents complemented with sophisticated and day-by-day improved fluorescent microscopy techniques enabled to get together two historical research fields: the cytological studies and the molecular and biochemical aspects of the genome. Such progress converged in a critical observation suggesting that interphase chromosomes occupy determined areas inside the nuclear space leading to the proposal of the distribution of chromosomes in territories [86]. This organization implies the concentration of the transcription machinery and RNA processing in the inter-chromatic compartment and the conformation of active chromatin domains, through the chromatin loop formation, in the periphery of the chromosome territories [86, 87]. This vision has an attractive implication, because we can then imagine that territories formation is a tight controlled process and, even more, the linear distribution of genes is not as random as we think [88].

From the clinical point of view, and in particular in cancer, pathologists concentrate their effort and expertise in defining changes in cancer cells based in abnormal nuclear morphology. Such examination allows the pathologist to define the grade, stage and even to specify the type of cancer. On our days we need to enlarge our vision not only at the genomic scale but also at the epigenomic scale, taking into consideration local and global changes. Therefore, in cancer it is necessary to understand how genome organization is perturbed and also, make associations with the ploidy of the cancerous cells since such changes in the amount of genome may create an abnormal imbalance not only at the level of the genome organization inside the nucleus, but also in the nuclear architecture [4, 89].

With the advent of genome-wide methodologies to define genomic, epigenomic and nuclear organization components, we start to have examples in which human cancer cells harbor global epigenetic abnormalities and that such alterations may contribute to the initial and/or progression of cancer. One recent example is the study of epigenetic features covering large genomic domains defined as *long-range epigenetic silencing* (LRES) or *activating* (LREA) domains in prostate cancer [90, 91]. LRES/LREA are genomic regions from tens of kilo bases to several megabases containing intergenic sequences and key genes associated to cancer, including tumor suppressor genes. These genomic regions are classified in terms of epigenetic modifications (DNA methylation and histone modifications) including three types of domains: the one who gain in cancer repressive chromatin features, the one that exchange repressive and active marks, and the one who only gain active marks. All these evidence place us in a good position to predict at the level of large genomic areas, epigenetic abnormalities that can be associated to a particular type of cancer.

Another relevant aspect is the definition of the so-called *Lamina Associated Domains* or LADs who correspond to genomic regions ranging from 100 kb to few megabases that are associated to the nuclear envelope through the interaction with the lamin proteins [92, 93]. Genes associated to LAD regions are mainly silenced due to their heterochromatin context, in contrast, the genes located distal to the nuclear envelope are prone to be active transcriptionally. Thus, in cancer, the abnormal associations of genomic regions to the nuclear periphery may induce the epigenetic silencing of cancer-associated genes. In contrast, release of genomic regions from the heterochromatin nuclear periphery could lead to the inappropriate activation of genes that can contribute to cell transformation.

In conclusion, the study of the spatial organization of the genome inside the cell nucleus, the dynamics of local and long-range chromatin domains and the association of specific genomic regions to the nuclear periphery, are relevant to understand the dynamics of the genome and, as a consequence of aberrant regulatory processes that remain to be discovered, influencing the development of cancerous cells.

7 Epigenetic Therapies

At this point it is becoming increasingly clear that cancer initiation and progression are not caused only by genetic defects but also by epigenetic deregulation (Fig. 1) [3, 4]. In view of the numerous epigenetic abnormalities found in cancer and considering that they can be rendered reversible through molecular manipulation it seemed possible to design and development new therapeutic strategies.

In most cases radiotherapy and chemotherapy clinical protocols are insufficient to eradicate a tumor. There is more and more discussion about complementary epigenetic therapies focused on the chromatin structure as the main target. For example, one theoretical prediction is that the induction to distinct degrees of chromatin structure relaxation will allow more efficient action of the radiotherapy and chemotherapy therapies.

Based on the prevailing scenarios it seems advisable to integrate epigenetic and genetic anomalies involved in distinct types of cancers that could allow a molecular classification of each tumor type. In consequence, a systematic classification of tumors based on epigenetic features may probably allow an early detection, diagnosis and tumor treatment.

Finally, and as I mentioned before, the epigenetic way of action in human diseases should let the design and incorporation of new cancer therapies. Efforts have been made to design and test new drugs that modify or regulate epigenetic marks, in particular, the use of DNA methylation and histone deacetylase inhibitors [46, 94]. For example, the two widely used DNA methylation inhibitors, the cytidine analogues: 5-azacytidine or 5-aza-2′-deoxycytidine, had been the most commonly used in clinical protocols. Furthermore, zebularine has been one of the most recently cytidine analogue developed [94]. Interestingly, zebularine has shown to be less toxic, with few side effects but unfortunately it has been tested with limited success in clinical trials. Then great interest has been focus on the development of more specific inhibitors, also avoiding unspecific effects, i.e., reactivation of endogenous normally silenced genes. Something similar occurs with histone deacetylase inhibitors like trichostatin-A and butyric acid. Even more there are clear examples on the literature that the two types of inhibitors are required to reach convenient degrees of reactivation [24].

In conclusion, now days the use of drugs designed to correct or reverse epigenetic anomalies in cancer are limited. Further biochemical and protein engineering investigations are required to reach therapeutic standards and mainly therapeutic specificities.

8 Prospects and Concluding Remarks

Epigenetic regulation has emerged as a relevant field of research with direct consequences in human diseases. Epigenetics is a growing research topic where new and unprecedented processes are emerging. I believe that in cancer epigenetics the main challenge will be the logical and ordered integration of all the distinct epigenetic events including their tight association with genetic defects (Fig. 1) [4]. Just as an example, in a very near future and as mentioned before, distinct types of noncoding RNAs and nuclear dynamics should be incorporated or are already considered as part of the epigenetic mechanisms demonstrating the complexity of such processes (Fig. 1) [4, 42, 70, 89].

Epigenetic studies have demonstrated that large scale expression profile studies will be limiting and will need for their interpretation to incorporate the natural environment of the nucleus and genome organization into chromatin for proper interpretation. Thus, we can predict that once the majority of the epigenetic and genetic information is integrated it will probably allow tumor classification that could contribute to early tumor detection and diagnosis with better expectation for treatment.

Finally, epigenetic therapies represent a real challenge basically because of the urgent need to reach specific effects on particular targets [94]. The multistage epigenetic regulation confronts a difficult challenge for the design and development of specific drugs. There is a need to bring together epigenetics, genetics and more recently epigenomics to reach new insights for novel and highly effective therapeutic protocols.

Acknowledgments

This work was supported by the Dirección General de Asuntos del Personal Académico-Universidad Nacional Autónoma de México (IN209403 and IN203811), Consejo Nacional de Ciencia y Tecnología, México (CONACyT; 42653-Q and 128464).

References

1. Hanahan D, Weinberg RA (2011) Hallmarks of cancer: the next generation. Cell 144:646–674
2. Vogelstein B, Papadopoulos N, Velculescu VE, Zhou S, Diaz LA Jr, Kinzler KW (2013) Cancer genome landscapes. Science 339:1546–1558
3. Rodríguez-Paredes M, Esteller M (2011) Cancer epigenetics reaches mainstream oncology. Nat Med 17:330–339
4. You JS, Jones PA (2012) Cancer genetics and epigenetics: two sides of the same coin? Cancer Cell 22:9–20
5. Van Speybroeck L (2002) From epigenesis to epigenetics: the case of C.H. Waddington. Ann N Y Acad Sci 981:61–81
6. Slack JMW (2002) Conrad Hal Waddington: the last renaissance biologist? Nat Rev Genet 3:889–895
7. Lim JP, Brunet A (2013) Bridging the transgenerational gap with epigenetic memory. Trends Genet 29:176–186
8. Pujadas E, Feinberg AP (2012) Regulated noise in the epigenetic landscape of development and disease. Cell 148:1123–1131
9. Portela A, Esteller M (2010) Epigenetic modifications and human disease. Nat Biotechnol 28:1057–1068
10. ENCODE Project Consortium (2012) An integrated encyclopedia of DNA elements in the human genome. Nature 489:57–74
11. Ernst J, Kheradpour P, Mikkelsen TS, Shoresh N, Ward LD, Epstein CB et al (2011) Mapping and analysis of chromatin state dynamics in nine human cell types. Nature 473:43–39
12. Bird A (2002) DNA methylation patterns and epigenetic memory. Genes Dev 16:6–21
13. Smith ZD, Meissner A (2013) DNA methylation: roles in mammalian development. Nat Rev Genet 14:204–220
14. Okano M, Bell DW, Haber DA, Li E (1999) DNA methyltransferases Dnmt3a and Dnmt3b are essential for the *de novo* methylation and mammalian development. Cell 99:247–257
15. Bestor TH (2000) The DNA methyltransferases of mammals. Hum Mol Genet 9:2395–2402
16. Rhee I, Bachman KE, Park BH, Jair KW, Yen RW, Schuebel KE et al (2002) DNMT1 and

DNMT3b cooperate to silence genes in human cancer cells. Nature 416:552–556

17. Li E (2002) Chromatin modification and epigenetic reprogramming in mammalian development. Nat Rev Genet 3:662–675
18. Cordaux R, Batzer MA (2009) The impact of retrotransposons on human genome evolution. Nat Rev Genet 10:691–703
19. Meehan RR, Lweis JD, Bird AP (1992) Characterization of MeCP2, a vertebrate DNA binding protein with affinity to methylated DNA. Nucleic Acids Res 20:5085–5092
20. Hendrich B, Bird AP (1998) Identification and characterization of a family of mammalian methyl-CpG binding proteins. Mol Cell Biol 18:6538–6547
21. Jones PL, Veenstra GJ, Wade PA, Vermaak D, Kass SU, Landsberger N et al (1998) Methylated DNA and MeCP2 recruit histone deacetylase to repress transcription. Nat Genet 19:187–191
22. Nan X, Ng HH, Johnson CA, Laherty CD, Turner BM, Eisenman RN et al (1998) Transcriptional repression by the methyl-CpG-binding protein MeCP2 involves a histone deacetylase complex. Nature 393:386–389
23. Magdinier F, Billard LM, Wittmann G et al (2000) Regional methylation of the 5′ end CpG island of BRCA1 is associated with reduced gene expression in human somatic cells. FASEB J 14:1585–1594
24. Magdinier F, Wolffe AP (2001) Selective association of the methyl-CpG binding protein MBD2 with the silent p14/p16 locus in human neoplasia. Proc Natl Acad Sci USA 98: 4990–4995
25. Ballestar E, Paz MF, Valle L, Wei S, Fraga MF, Espada J et al (2003) Methyl-CpG binding proteins identify novel sites of epigenetic inactivation in human cancer. EMBO J 22: 6335–6345
26. Baubec T, Ivánek R, Lienert F, Schübeler D (2013) Methylation-dependent and -independent genomic targeting principles of the MBD protein family. Cell 153:480–492
27. Antequera F (2003) Structure, function and evolution of CpG island promoters. Cell Mol Life Sci 60:1647–1658
28. Weber M, Hellmann I, Stadler MB, Ramos L, Pääbo S, Rebhan M et al (2007) Distribution, silencing potential and evolutionary impact of promoter DNA methylation in the human genome. Nat Genet 39:457–466
29. Lienert F, Wirbelawer C, Som I, Dean A, Mohn F, Schübeler D (2011) Identification of genetic elements that autonomously determine DNA methylation states. Nat Genet 43: 1091–1097
30. Issa J-P (2004) CpG island methylator phenotype in cancer. Nat Rev Cancer 4:988–993
31. Recillas-Targa F, de la Rosa-Velázquez IA, Soto-Reyes A (2011) Insulation of tumor supresor genes by the nuclear factor CTCF. Biochem Cell Biol 89:479–488
32. Irizarry RA, Ladd-Acosta C, Wen B, Wu Z, Montano C, Onyango P et al (2009) The human colon cancer methylome shows similar hypo- and hypermethylation at conserved tissue-specific CpG island shores. Nat Genet 41:178–186
33. Lister R, Pelizzola M, Dowen RH, Hawkins RD, Hon G, Toni-Filippini J et al (2009) Human DNA methylomes at base resolution show widespread epigenomic differences. Nature 462:315–322
34. Hemberger W, Dean W, Reik W (2009) Epigenetic dynamics of stem cells and cell lineage commitment: digging Waddingston's canal. Nat Rev Mol Cell Biol 10:526–537
35. Wu H, Zhang Y (2011) Mechanisms and functions of Tet protein-mediated 5-methylcytosine oxidation. Genes Dev 25:2436–2452
36. Tahiliani M, Koh KP, Shen Y, Pastor WA, Bandukwala H, Bridno Y et al (2009) Conversion of 5-methylcytosine to 5-hydroxymethylcytosine in mammalian DNA by MLL partner TET1. Science 324:930–935
37. Lund AH, van Lohuizen M (2004) Epigenetics and cancer. Genes Dev 18:315–2335
38. Zenter GE, Henikoff S (2013) Regulation of nucleosome dynamics by histone modification. Nat Struct Mol Biol 20:259–266
39. Fraga MF, Balletar E, Villar-Garea A, Boix-Chornet M, Espada Y, Schotta G et al (2005) Loss of acetylation at Lys16 and trimethylation at Lys20 of histone H4 is a common hallmark of human cancer. Nat Genet 37:91–400
40. Zink D, Fischer AH, Nickerson JA (2004) Nuclear structure in cancer cells. Nat Rev Cancer 4:677–687
41. Kondo Y, Shen L, Cheng AS, Ahmed S, Boumber Y, Charo C et al (2008) Gene silencing in cancer by histone H3 lysine 27 trimethylation independent of promoter DNA methylation. Nat Genet 40:741–750
42. Guil S, Esteller M (2012) *Cis*-acting noncoding RNAs: friends and foes. Nat Struct Mol Biol 19:1068–1075
43. Recillas-Targa F (2002) DNA methylation, chromatin boundaries and mechanisms of genomic imprinting. Arch Med Res 33:428–438
44. Henckel A, Nakabayashi K, Sanz LA, Feil R, Hata K, Arnaud P (2009) Histone methylation is mechanistically linked to DNA methylation

at imprinting control regions in mammals. Hum Mol Genet 18:3375–3383

45. Ratnakumar K, Berstein E (2013) ATRX: the case of a peculiar chromatin remodeler. Epigenetics 8:3–9
46. Esteller M (2005) DNA methylation and cancer therapy: new developments and expectations. Curr Opin Oncol 17:55–60
47. Esteller M, Levine R, Baylin SB, Ellenson LH, Herman JG (1998) MLH1 promoter hypermethylation is associated with the microsatellite instability phenotype in sporadic endometrial carcinomas. Oncogene 17:2413–2417
48. Esteller M (2005) Dormant hypermethylated tumour suppressor genes: questions and answers. J Pathol 205:172–180
49. DeSmet C, DeBacker O, Faraoni I, Lurquin C, Brasseur F, Boon T (1996) The activation of human gene MAGE-1in tumor cells is correlated with genome-wide demethylation. Proc Natl Acad Sci USA 93:7149–7153
50. Adorjan P, Distler J, Lipscher E, Model F, Muller J, Pelet C et al (2002) Tumour class prediction and discovery by microarray-based DNA methylation analysis. Nucleic Acids Res 30:e21
51. Oshimo Y, Nakayama H, Ito R, Kitadai Y, Yoshida K, Chayama K et al (2003) Promoter methylation of cyclin D2 gene in gastric carcinoma. Int J Oncol 6:1663–1670
52. Sakai T, Toguchida J, Ohtani N, Yandell DW, Rapaport JM, Dryja TP (1991) Allele-specific hypermethylation of the retinoblastoma tumor-suppressor gene. Am J Hum Genet 48: 880–888
53. Herman JG, Latif F, Weng Y, Lerman MI, Zbar B, Liu S et al (1994) Silencing of the VHL tumor-suppressor gene by DNA methylation in renal carcinoma. Proc Natl Acad Sci USA 91:9700–9704
54. Merlo A, Herman JG, Mao L, Lee DJ, Gabrielson E, Burger PC et al (1995) 5′CpG island methylation is associated with transcriptional silencing of the tumour suppressor p16/CDKN2/MTS1 in human cancer. Nat Med 1:686–692
55. De La Rosa-Velázquez IA, Rincón-Arano H, Benítez-Bribiesca L, Recillas-Targa F (2007) Epgienetic regulation of the human retinoblastoma tumor supresor gene promoter by CTCF. Cancer Res 67:2577–2585
56. Soto-Reyes E, Recillas-Targa F (2010) Epigenetic regulation of the human p53 gene promoter by the CTCF transcription factor in transformed cell lines. Oncogene 29:2217–2227
57. Cunningham JM, Christensen ER, Tester DJ, Kim CY, Roche PC, Burgart LJ et al (1998) Hypermethylation of the hMLH1 promoter in colon cancer with microsatellite instability. Cancer Res 58:3455–3460
58. Chen WY, Cooper TK, Zahnow CA, Overholtzer M, Zhao Z, Ladanyi M et al (2004) Epigenetic and genetic loss of *Hic1* function accentuates the role of p53 in tumorigenesis. Cancer Cell 6:387–398
59. Narita M, Nunez S, Heard E, Narita M, Lin AW, Hearn SA et al (2003) Rb-mediated heterochromatin formation and silencing of E2F target genes during cellular senescence. Cell 113:703–716
60. Liu DX, Nath N, Chellappan SP, Greene LA (2005) Regulation of neuron survival and death by p130 and associated chromatin modifiers. Genes Dev 19:719–732
61. Margueron R, Reinberg D (2011) The Polycomb complex PRC2 and its mark in life. Nature 469:343–349
62. Hansen KH, Bracken AP, Pasini D, Dietrich N, Gehani SS, Monrad A et al (2008) A model for transmission of the H3K27me3 epigenetic mark. Nat Cell Biol 10:1291–1300
63. Varambally S, Dhanasekaran SM, Zhou M, Barrette TR, Kumar-Sinha C, Sanda MG et al (2002) The polycomb group protein EZH2 is involved in progression of prostate cancer. Nature 419:624–629
64. Kleer CG, Cao Q, Varambally S, Shen R, Ota I, Tomlins SA et al (2003) EZH2 is a marker of aggressive breast cancer and promotes neoplastic transformation of breast epithelial cells. Proc Natl Acad Sci USA 100:11606–11611
65. Asangani IA, Ateeq B, Cao Q, Dodson L, Pandhi M, Kunju LP et al (2013) Characterization of the EZH2-MMSET histone methyltransferase regulatory axis in cancer. Mol Cell 49:741–750
66. Dávalos-Salas M, Furlan-Magaril M, González-Buendía E, Valdes-Quezada C, Ayala-Ortega E, Recillas-Targa F (2011) Gain of DNA methylation is enhanced in the absence of CTCF at the human retinoblastoma gene promoter. BMC Cancer 11:232
67. DiCroce L, Raker VA, Corsaro M, Fazi F, Fanelli M, Faretta M et al (2002) Methyltransferase recruitment and DNA hypermethylation of target promoters by an oncogenic transcription factor. Science 295:1079–1082
68. DiCroce L (2005) Chromatin modifying activity of leukaemia associated fusion proteins. Hum Mol Genet 14:R77–R84
69. Peterson LF, Zhang DE (2004) The 8; 21 translocation in leukaemogenesis. Oncogene 23:4255–4262
70. Rinn JL, Chang HY (2012) Genome regulation by long noncoding RNAs. Annu Rev Biochem 81:145–166

71. Mercer TR, Mattick JS (2013) Structure and function of long noncoding RNAs in epigenetic regulation. Nat Struct Mol Biol 20:300–307
72. Ryan BM, Robles AD, Harris CC (2010) Genetic variation in microRNA networks: the implications for cancer research. Nat Rev Cancer 10:389–402
73. Varambally S, Cao Q, Mani RS, Shankar S, Wang X, Ateeq B et al (2008) Genomic loss of microRNA.101 leads to overexpression of histone methyltransferase EZH2 in cancer. Science 322:1695–1699
74. Friedman JM, Liang G, Liu CC, Wolff EM, Tsai YC, Ye W et al (2009) The putative tumor suppressor microRNA-101 modulates the cancer epigenome by repressing the polycomb group protein EZH2. Cancer Res 69:2623–2629
75. Fabbii M, Garzon R, Cimmino A, Liu Z, Zanesi N, Callegari E et al (2007) MicroRNA-29 family reverts aberrant methylation in lung cancer by targeting DNA methyltransferases 3A and 3B. Proc Natl Acad Sci USA 104:15805–15810
76. Baer C, Claus R, Plass C (2012) Genome-wide epigenetic regulation of microRNAs in cancer. Cancer Res 73:473–477
77. Lujambio A, Ropero S, Ballestar E, Fraga MF, Cerrato S, Setien F et al (2007) Genetic unmasking of an epigenetically silenced microRNA in human cancer cells. Cancer Res 67:1424–1429
78. Derrien T, Johnson R, Bussotti G, Tanzer A, Djebali S, Tilgner H et al (2012) The GENCODE v7 catalog of human long noncoding RNAs; analysis of their gene structure, evolution and expression. Genome Res 22:1775–1789
79. Escamilla-Del-Arenal M, da Rocha ST, Heard E (2011) Evolutionary diversity and developmental regulation of X-chromosome inactivation. Hum Genet 130:307–327
80. Rinn JL, Kertesz M, Wang JK, Squazzo SL, Xu X, Brugmann SA et al (2007) Functional demarcation of active and silent chromatin domains in human *HOX* loci by noncoding RNAs. Cell 129:1311–1323
81. Gutpa RA, Shah N, Wang KC, Kim J, Horlings HM, Wong DJ et al (2010) Long non-coding RNA HOTAIR reprograms chromatin state to promote cancer metastasis. Nature 464:1071–1076
82. Yap KL, Lis S, Muñoz-Cabello AM, Raguz S, Zeng L, Mujtaba S et al (2010) Molecular interplay of the noncoding RNA ANRIL and methylated histone H3 lysine 4 by polycomb CBX7 in transcriptional silencing of INK4a. Mol Cell 38:662–674
83. Huarte M, Guttman M, Feldser D, Garber M, Koziol MJ, Kenzelmann-Broz D et al (2010) A large intergenic noncoding RNA induced by p53 mediates global gene repression in the p53 response. Cell 142:409–419
84. Wang KC, Yang YW, Liu B, Sanyal A, Corces-Zimmerman R, Chen Y et al (2011) A long noncoding RNA maintains active chromatin to coordinate homeotic gene expression. Nature 472:120–124
85. Wang Y, Xu Z, Jiang J, Xu C, Kang J, Xiao L et al (2013) Endogenous miRNA sponge lincRNA-RoR regulates Oct4, Nanog, and Sox2 in human embryonic stem cell self-renewal. Dev Cell 25:69–80
86. Cremer T, Cremer C (2001) Chromosomes territories, nuclear architecture and gene regulation in mammalian cells. Nat Rev Genet 2:292–301
87. Fraser P, Bickmore W (2007) Nuclear organization of the genome and the potential for gene regulation. Nature 447:413–417
88. Bulger M, Groudine M (1999) Looping versus linking: toward a model for long-distance gene activation. Genes Dev 13:2465–2477
89. Reddy KL, Feinberg AP (2012) Higher order chromatin organization in cancer. Semin Cancer Biol 23:109–115
90. Coolen MW, Stirzaker C, Song JZ, Statham AL, Kassir Z, Moreno CS et al (2010) Consolidation of the cancer genome into domains of repressive chromatin by long-range epigenetic silencing (LRES) reduces transcriptional plasticity. Nat Cell Biol 12:235–246
91. Bert SA, Robinson MD, Strbenac D, Statham AL, Song JZ, Hulf T et al (2013) Regional activation of the cancer genome by long-range epigenetic remodeling. Cancer Cell 2:9–22
92. Peric-Hupkes D, Meuleman W, Pagie L, Bruggeman SW, Solovei I, Brugman W et al (2010) Molecular maps of the reorganization of genome-nuclear lamina interaction during differentiation. Mol Cell 38:603–613
93. Guelen L, Pagie L, Brasset E, Meuleman W, Faza MB, Talhout W et al (2008) Domain organization of human chromosomes revealed by mapping of nuclear lamina interactions. Nature 453:948–951
94. Azad N, Zahnow CA, Rudin CM, Baylin SB (2013) The future of epigenetic therapy in solid tumors-lessons from the past. Nat Rev Clin Oncol 10:256–266

Chapter 5

Experimental Strategies to Manipulate the Cellular Levels of the Multifunctional Factor CTCF

Edgar González-Buendía, Rosario Pérez-Molina, Erandi Ayala-Ortega, Georgina Guerrero, and Félix Recillas-Targa

Abstract

Cellular homeostasis is the result of an intricate and coordinated combinatorial of biochemical and molecular processes. Among them is the control of gene expression in the context of the chromatin structure which is central for cell survival. Interdependent action of transcription factors, cofactors, chromatin remodeling activities, and three-dimensional organization of the genome are responsible to reach exquisite levels of gene expression. Among such transcription factors there is a subset of highly specialized nuclear factors with features resembling master regulators with a large variety of functions. This is turning to be the case of the multifunctional nuclear factor CCCTC-binding protein (CTCF) which is involved in gene regulation, chromatin organization, and three-dimensional conformation of the genome inside the cell nucleus. Technically its study has turned to be challenging, in particular its posttranscriptional interference by small interference RNAs. Here we describe three main strategies to downregulate the overall abundance of CTCF in culture cell lines.

Key words CTCF, Lentivirus, Interference RNA, Inducible vectors, Lipofection, Electroporation, Fluorescent cytometry

1 Introduction

Over the years the progress in the capacity to manipulate the levels of a particular messenger RNA and/or protein synthesis in cultured cells or even in organisms has contributed to the understanding in more detail many cellular processes [1]. From time to time, the RNA interference strategy is inefficient or even deleterious for the cell. One example is the knockdown of the multifunctional 11-zinc-finger protein CTCF [2, 3]. CTCF functions are so varied; it is not surprising that its manipulation can dramatically alter many cellular processes rendering its study difficult.

CTCF was initially discovered as a transcriptional repressor of the *c-myc* gene in chicken embryonic fibroblasts [4]. CTCF versatility has also been documented since it can also act as transcriptional

Martha Robles-Flores (ed.), *Cancer Cell Signaling: Methods and Protocols*, Methods in Molecular Biology, vol. 1165, DOI 10.1007/978-1-4939-0856-1_5,

activator, in particular, based on its capacity to employ distinct combinatorial of the zinc fingers to bind numerous and highly divergent sequences [5–7]. Furthermore, CTCF can interact with itself and with a large variety of cofactors [6, 7]. Another interesting aspect is that the binding of CTCF to a subset of recognition sequences is sensitive to DNA methylation [6, 8, 9].

In addition to the functional consequences of the use of different zinc fingers, CTCF factor can be posttranslationally modified. CTCF can be phosphorylated in the C-terminal domain of the protein and poly-ADP-ribosylated in the N-terminal domain [10, 11]. Interestingly, CTCF poly-ADP-ribosylation is associated with the 180 kDa form that is present in normal breast tissues, in contrast to the 130 kDa protein, which is present in breast tumors [11]. More recently, it has been demonstrated that CTCF can be sumoylated promoting relaxation of the chromatin structure and transcriptional activation [12, 13].

The contribution of CTCF to cell homeostasis and organism development has also been documented. For example, *ctcf*$^{-/-}$ knockout mice show a lethal phenotype in early stages of development (before implantation). In contrast, its over-expression in breast cancer cell lines and tumors causes resistance to apoptosis [3, 14]. Conditional knockout mice have shown that the absence of CTCF in mouse limb affects its development increasing apoptosis of limb cells [15]. CTCF also participates in cell differentiation, as in the case of myogenic differentiation by regulating MyoD myogenic potential [16]. Another related aspect has to do with the interaction between CTCF and the TBP-associated factor TAF3, an interaction that confers differentiation capabilities to stem cells and gives rise to different somatic cell types [17].

Together, CTCF has a central role in the control of many processes that converge in the control of the cell differentiation, organism development, and, furthermore, cell cycle since CTCF has also been involved in cancer cells epigenetically regulating tumor-suppressor genes like *p53*, *Rb*, and *p16*^{INK4a} and the proto-oncogene *c-myc,* among others [18, 19].

Furthermore, CTCF has been linked to the transition between open chromatin and highly compacted genomic regions. Its location in transition zones of open and repressive chromatin led to the idea that CTCF represents a chromatin insulator or boundary element [20]. Based in this view, two functional definitions emerge for the insulator regulatory elements. The first one has to do with the ability of an insulator to block promoter–enhancer communication in a position-dependent manner [21]. The second one defines an insulator as a sequence together with its associated factors (among them CTCF) that can protect a transgene against chromosomal position effects [22–24]. Derived from this second functional definition, CTCF and its insulator properties have been proposed as a barrier element that prevents the spreading of repressive heterochromatin [20]. Therefore, CTCF-dependent insulators contribute to the chromatin organization of the genome.

Nowadays, this view has been reinforced by the experimental evidence showing CTCF as one of the most relevant factors involved in long-range chromatin–fiber interactions. These interactions were initially described in *Drosophila* and now observed in many genomic loci including the *Igf2/H19*, the mouse β-globin locus, and the immunoglobulin heavy-chain locus, among others [25–27]. This alternative activity of CTCF is apparently relevant for the genome organization in the nuclear space through the formation of domains and large chromosomal regions associated with the nuclear lamina (also known as LADs) [28].

Here we present a broad overview of CTCF and its varied functions. For several years now, our research group and many others have been working in the generation of different CTCF knockdowns to manipulate the cellular levels of this protein. This useful experimental strategy has allowed us to understand and reveal new functions for CTCF. Unfortunately, we have been confronted with several technical difficulties in addition to the pleiotropic effects due to a reduction in the cellular levels of the CTCF factor. Thus, in our experience, we observed that the transduction of interference RNAs against CTCF requires several considerations that we discuss here in detail. We address these aspects from the perspective of the generation of stable cell lines expressing small hairpin interference RNAs (shRNAi), their transduction taking advantage of lentiviral particles, and, alternatively, the use of an inducible system. These three different strategies are described considering their advantages and limitations.

1.1 CTCF Knockdown Stable Transfection

Stable transfection can provide long-term expression of the integrated DNA, in contrast to transiently transfected DNA. Stably transfected cells segregate the foreign DNA to their progeny, because the transfected DNA is apparently incorporated randomly into the genome. This kind of transfection is useful for production of recombinant proteins and analysis of short- and long-term effects of exogenous transgene expression [29].

Nowadays, the most common way to deliver DNA into a cell is via the lipofectamine reagent. Lipofectamine is a polycationic synthetic lipid mixed with a fusogenic lipid with an amine group. This lipid mixture forms DNA–lipid complexes due to ionic interactions between the head group of the lipid with a strong positive charge, which neutralizes the negative charge of phosphate groups of DNA. Lipid–DNA mixture results in the formation of structures that fuse and pass across the plasma membrane to deliver DNA into the cell [30].

Electroporation is another strategy to introduce DNA in a cell. It is a mechanical method that uses an electrical pulse to create temporary pores in cell membranes through which substances like nucleic acids can be introduced into cells. Host cells and selected molecules must be suspended in a conductive solution, where an electrical circuit is closed around the mixture, creating an electrical pulse during a short period of time. This disturbs the phospholipid bilayer of the membrane and causes the formation of temporary pores through which charged molecules, such as DNA, are driven across the membrane. This method is mainly recommended for tissue culture cells [30]. It is worth mentioning that electroporation can cause significant cell death due to the electrical pulse.

Whichever the method, the next key step in a stable transfection is to select cells in which the transgene is stably integrated. Usually, transfected DNA includes an antibiotic resistance gene as part of the construct in order to perform drug selection of the cells after a short time of recovery. Cells that were not transfected will die, and those that express the antibiotic resistance gene at sufficient levels will survive. G418 (Geneticin), hygromycin B, puromycin, blasticidin, and Zeocin compounds are the most commonly used antibiotics to select stably transfected cells.

1.2 Manipulating the Levels of the Nuclear Factor CTCF

Different study groups have performed stable transfection of shRNAi against CTCF in several cell lines. The effect of these transfections appears to be quite variable depending on the system used as well as the cell type. For example, we have found that the stable transfection of shRNAi against CTCF is for example very unstable in HeLa cells. In order to investigate the role of CTCF in the maintenance of expression of latency genes in Epstein–Barr virus-infected B-cell lines (Kem I and Mutu I cell lines), Sample and collaborators carried out a knockdown of CTCF in Kem I and Mutu I cells using SureSilencing shRNA expression plasmids (SABiosciences) that also encodes for hygromycin or puromycin resistance [31]. The SureSilencing shRNA plasmids are designed to knock down individual genes under stable transfection conditions after appropriate selection procedures. Each vector contains the shRNA under the control of the U1 promoter and a neomycin or a puromycin resistance gene (http://www.sabiosciences.com/RNAi). They transfected 3×10^6 cells with a plasmid via Amaxa Nucleofection, which is an electroporation commercial method. Cells were transfected with four different shRNA plasmids, using 5 μg of each, and control cells received 20 μg of control shRNA plasmid. Two days after Nucleofection, cells were selected with puromycin (Kem I 200 ng/ml and Mutu I 2 μg/ml). In both cell lines the knockdown was confirmed by immunoblotting. According to the manufacturer of Amaxa Nucleofection, this method has little effect on the viability of the cells, although there are no comments regarding it.

MCF-7 breast cancer cell line has also been transfected with siRNA against CTCF. Since CTCF is distributed differentially in a cell type-specific manner along the *kcnq5* locus, this research group wanted to know if the distribution of CTCF determined the interactions occurring in the gene locus [32]. The RNA interference experiments were performed according to the siRNA strategy designed by Thermo Scientific Dharmacon (ON-TARGET plus siRNA Reagents). This siRNA technology is different from others because it has a dual-strand modification that allows the reduction of nonspecific targets. First, the sense strand is modified to prevent uptake by the RISC complex, which favors the antisense strand loading into the complex [33, 34]. Next, the antisense strand is modified to destabilize off-target activity and enhance target specificity. This is mainly achieved by excluding siRNA with known microRNA seed region motifs and preferentially choosing lower frequency seed regions [35, 36]. They used the ON-TARGET plus non-targeting siRNA as a control. The siRNA transfection of MCF-7 cells was performed using DharmaFECT 1 in 6-well plates. The DharmaFECT 1 transfection protocol is based on the lipofectamine reagent (Dharmacon Inc. 2006).

This system was also used in primary cells that were transfected with siRNA against CTCF. Such is the case of erythroid cells extracted from spleen of transgenic mice that were genetically modified with an extra CTCF site between their endogenous CTCF insulator (HS5) and the α-globin locus [37]. In order to probe the CTCF dependence of the insulator loop, they isolated erythroid cells from mice and cultured them. These cells were also transfected with the mouse CTCF ON-TARGET plus siRNA system with the Amaxa Biosystems Nucleofector, as described previously. In this case, the cells were harvested 72 h after the transfection.

The knockdown of CTCF by stable transfection of a siRNA has also been carried out in embryonic stem cells (ESCs). It is well known that CTCF has a cell type-specific distribution, including human ESCs. Single cells were collected and transfected with a predesigned Silencer Select siRNA (Ambion) against CTCF [38]. The Silencer Select siRNA system presumably incorporated improvements in siRNA design, such as prediction algorithms that allow the siRNA to recognize its target in a more specific manner and reduce off-targets, increasing their efficiency up to a 100-fold. In this system siRNAs also have a locked nucleic acid (LNA) chemical modification that also influences the off-target reduction (www.invitrogen.com). For the transfection they employed RNAiMAX (Invitrogen)-based lipofectamine. Importantly, a second round of transfection was performed 24 h later. They proved that proliferation of hESCs was affected by the ablation of CTCF since they incorporated 60 % less BrdU, and it also decreases the expression of pluripotency and self-renewal genes. This data suggests that CTCF may have different roles in the regulation of stem cells, which is no wonder given its varied distribution along the genome and the amount of genes and nuclear organization it may be regulating.

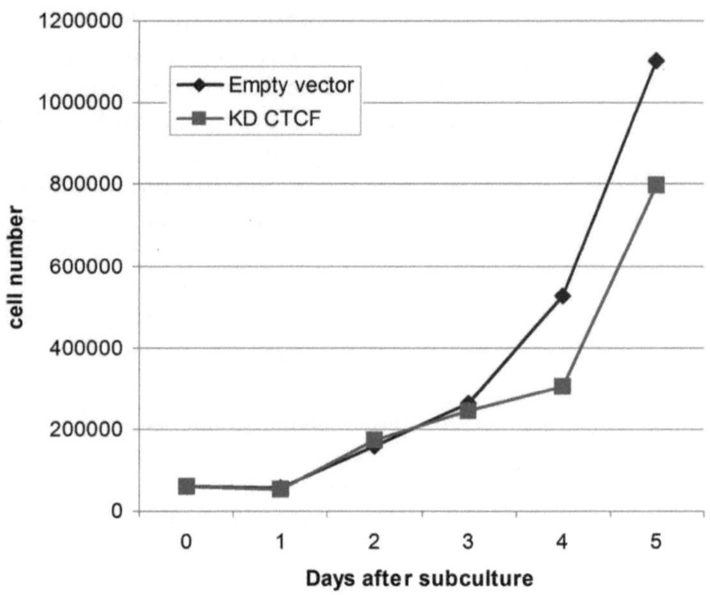

Fig. 1 Growth curves for a control HeLa cell line and a CTCF knockdown HeLa cell line grown in culture. There is a decrease in the population doubling time in the CTCF knockdown cell line compared with the control cell line

In conclusion, there are numerous ways to carry out a stable transfection to perform a CTCF knockdown. But, it is important to mention that an effective knockdown generated by a stable transfection often depends on the transfection capacity of a cell type and its own phenotype. For example we have reproducibly observed differences in the timing of cell division in stable CTCF knockdown context versus a control (Fig. 1). It is also important to design a highly specific siRNA with a strong ability to decrease its target gene without affecting other genes. The site of integration of the transgene has also shown to be quite influential in the knockdown process. Next, we present the detailed protocol for the generation of a stable knockdown against CTCF.

2 Materials

2.1 Stable Transfection of Interference RNA Against CTCF

2.1.1 Cell Culture

1. HeLa cell culture is performed in a fresh Dulbecco MEM medium (with 10 % fetal bovine serum (FBS) and 1 % penicillin/streptomycin).
2. Cells are incubated at 37 °C and 5 % CO_2.

2.1.2 Plasmid Preparation

1. Linearize the plasmid with the convenient enzyme (*see* **Note 1**).

2.2 Pseudo-Lentiviral Particle Production

1. Transfer, envelope, and packaging plasmids must have high quality for transfection. Use commercial kits to purify DNA and avoid phenol–chloroform extractions.

2.2.1 Plasmids for Lentiviral Particle Production (See **Note 2**)

2.2.2 Cell Culture

1. HEK293T (high titer of infectious virus) or HEK293 (parental cell line) are human embryonic kidney cell lines. Both are commonly used for virus production, and their low passing number in culture is critical to obtain high viral titers.
2. HEK293T are cultured in DMEM: Dulbecco's Modified Eagle Medium (GIBCO), supplemented with 10 % FBS, pyruvate, glutamine, and penicillin/streptomycin 1× antibiotics at 37 °C, 5 % CO_2, in a 150 mm dish.

2.2.3 Solutions and Materials

0.1 % Sodium hypochlorite solution.

1.25 M calcium chloride solution in H_2O, filter sterilized and aliquoted.

0.25 % Trypsin–EDTA.

15 ml centrifuge tube.

HeBS 2×.

Ultra-clear centrifuge tubes.

Swinging bucket rotor (Beckman SW28 or SW32).

Polybrene (hexadimethrine bromide).

Puromycin dihydrochloride or the corresponding antibiotic.

Epifluorescence microscope.

2.3 Inducible System

In many fields of basic research, temporal expression of a gene can be needed depending on the experimental setting or, in worse cases, large exposures or high amounts of a transcript can be toxic to the cell, so it would be necessary to turn on and off their expression [41]. The tetracycline-dependent transcriptional regulatory system is one of the best studied inducible systems with proven efficacy in vitro and in vivo [42]. This system is based on the *E. coli* Tn10 tetracycline resistance operator, and its components were improved by mutagenesis and codon optimization to obtain variants with reduced basal activity and greater sensitivity to doxycycline [43].

Currently, lentiviral vectors were developed to incorporate all elements of the doxycycline-inducible system in one or two vectors, and choosing between them will depend on individual applications. One method that is frequently used in this inducible lentiviral vector is the silencing of a gene of interest by interference RNA. The latest generation of this kind of vectors uses microRNA sequences that have been adapted to incorporate siRNA sequences [44]. One example of this kind of vectors is the pTRIPZ lentiviral inducible vector that combines the design advantages of microRNA-adapted shRNA (shRNAmir) with a lentiviral inducible vector to create a powerful RNAi trigger capable of producing RNAi in most cell types (Open Biosystems). The vector is engineered to be Tet-On and produces tightly regulated induction of shRNAmir expression in the presence of doxycycline. With the aim to generate a knockdown of CTCF under controlled conditions, it seems to be convenient to use an inducible system to turn on/off the interference RNA expression whenever the inductor is present or not (Fig. 1). Furthermore, when CTCF is knocked down in a stable context it is harder to sustain shRNAi expression for several days and even more difficult to recover the transgenic line after freezing the cells. In conclusion it seems that this way to perform the knockdown of an essential protein is more effective and offers many advantages over a constitutive knockdown.

2.3.1 Solutions

1 mg/ml Doxycycline solution in H_2O: Sterilize the solution by filtration and store aliquoted at −20 °C.

3 Methods

3.1 Stable Transfection of an Interference RNA Against CTCF

Day 1

1. For each transfection 3×10^5 cells are placed in 3 ml of complete DMEM; we recommend to perform it in triplicates (*see* **Note 3**).

Day 2

1. Remove the medium of the cells, and add 800 μl of DMEM lacking FBS and antibiotic (DMEM (−)).
2. Mix 2 μl of lipofectamine and 98 μl of DMEM (−) and incubate for 15 min (*see* **Note 4**).
3. Next add the desired amount of DNA in a volume of 100 μl of DMEM (−) to the lipofectamine mix and incubate for 30 min (*see* **Note 5**).
4. Add the final mix to the cells to reach a final volume of 1 ml and incubate at 37 °C and 5 % CO_2 for at least 6 h.
5. Add 3 ml of complete medium.

Day 4

1. Change the cells to a selective medium containing antibiotic at the previously established concentration (*see* **Note 6**).

Day 5

1. Harvest the cells, and perform a total protein extraction to verify the abundance of CTCF by Western blot (*see* **Note 7**).

3.2 Lentiviral Particle Production

Extreme caution is recommended when working with lentivirus.
(http://oba.od.nih.gov/oba/rac/Guidance/LentiVirus_Containment/pdf/Lenti_Containment_Guidance.pdf.)

3.2.1 Cell Culture

Day 1

1 Plate 4×10^6 HEK293T cells in two 150 mm culture dishes each, at least 12 h before the viral plasmid co-transfection.

2 The confluence of cells must be 60–70 %; this is a critical aspect for the efficient production of lentivirus particles.

3.2.2 Co-transfection of HEK293T Cells

Day 2

1. Mix 15 μg transfer DNA, 10 μg gag/pol-expressing plasmid, 10 μg rev/tat-expressing plasmid, and 5 μg VSV-G envelope plasmid in a 15 ml centrifuge tube, and make up to 450 μl with sterile water.
2. Add dropwise constantly 50 μl of 2.5 M $CaCl_2$.
3. Add 500 μl HeBS 2× in a dropwise fashion.
4. Incubate at room temperature for 20 min.
5. Add dropwise the transfection mixture across the plate of HEK293T.
6. Incubate the cells at 37 °C, 5 % CO_2, for 8 h.
7. Aspirate medium, wash with PBS 1×, and add 16 ml of fresh medium.

3.2.3 Collect Viral Supernatants

Day 4

1. Centrifuge supernatant at $778 \times g$ for 5 min to clear debris (*see* **Note 8**).
2. Supernatant can be stored at 4 °C for a couple of days prior to concentration.

Day 5

1. Collect and centrifuge supernatant at $778 \times g$ for 5 min to remove debris.
2. The total volume of collected supernatant along this time must be cleared with a 0.45 μm filter to remove viral aggregates and debris (*see* **Note 9**).
3. At this point the supernatant can be used to infect target cells or concentrate the viral particles (*see* **Note 10**).
4. Preferably aliquot the supernatant before storage to avoid thawing several times.
5. Store at −80 °C (*see* **Note 10**).

3.2.4 Lentivirus Concentration

Day 6

1. Pre-chill rotor and swinging buckets.
2. Add 32 ml of the total amount of lentiviral supernatant in each ultra-clear centrifuge tube.
3. Centrifuge at 82,705 × *g* for 2–2.5 h at 4 °C (*see* **Note 11**).
4. Discard the supernatant by decantation in a container with 10 % bleach.
5. Resuspend the lentiviral pellet (it may be translucent or invisible) with 60 μl of chilled PBS 1× pH 7.4 and let stand for 8–12 h at 4 °C.
6. Make 10 μl aliquots in clearly labeled microcentrifuge tubes.
7. Froze in liquid nitrogen, and store at −80 °C.
8. Calculate the titer of the lentivirus by transducing target cells with serially diluted viral preparations.

3.2.5 Titration of Lentivirus Expressing Fluorescent Proteins by Flow Cytometry (FACS)

Days 7, 8

1. Plate HEK293T cell line at a density of 1×10^5 cells per well in 6-well plate 12–24 h before the transduction.
2. Remove the medium from the wells and wash twice with PBS 1× pH 7.4.
3. Add 0.5 ml of serum- and antibiotic-free fresh medium with 8 μg/ml of polybrene.
4. Generate a tenfold serial dilution of the lentiviral preparation: Mock, undiluted, 10, 100, 1,000, and 10,000 (*see* **Note 12**).
5. Grow the cells for 48 h.
6. Collect the cells by trypsinization.
7. Determine the percentage of fluorescent cells by FACS (*see* Figs. 2 and 3).
8. Calculate lentiviral titer according of the following formula:

   TU/μl = (%GFP-positive cells × 100,000)/(100× volume added to each μl per well × DF)

   Where TU = transduced units (Fig. 3).

   DF = Dilution factor.

Dilution	Dilution factor
Undiluted	1
1/10	10^{-1}
1/100	10^{-2}
1/1,000	10^{-3}
1/10,000	10^{-4}

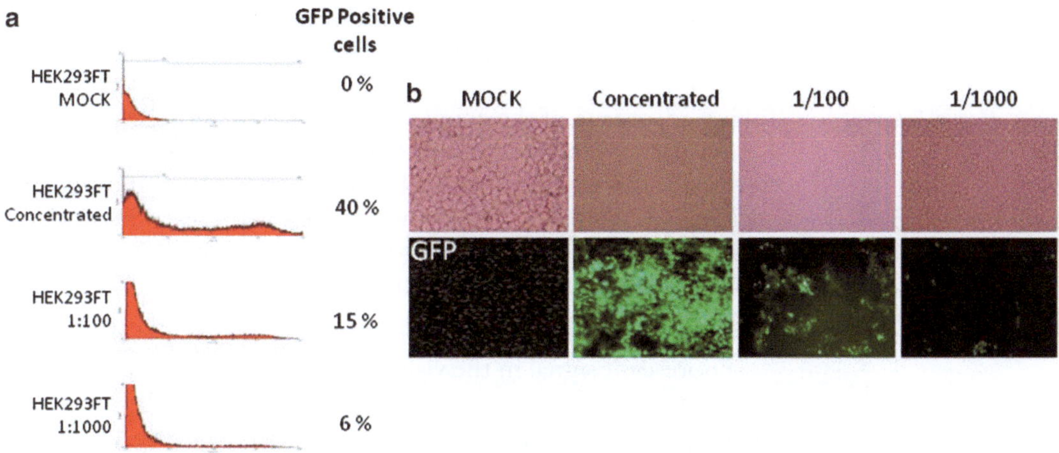

Fig. 2 HEK293-infected cell line with *GFP*-expressing lentivirus. (**a**) Flow cytometry profiles of the different dilutions used. The percentages of GFP-positive cells of each dilution are shown. (**b**) Epifluorescence microscope images of the expression of *GFP*, which correspond to each condition described in (**a**)

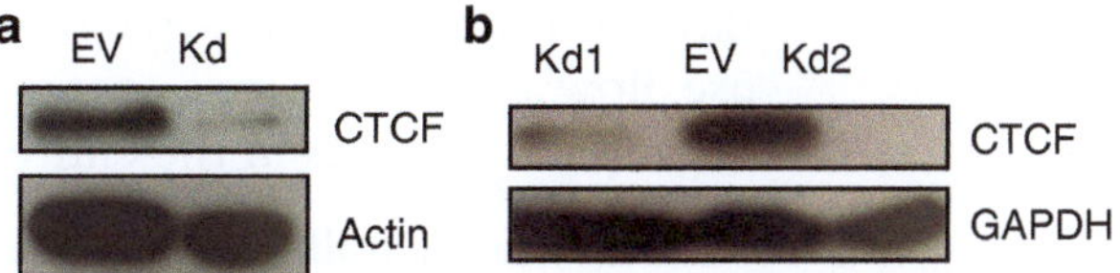

Fig. 3 Western blot of knockdown of CTCF. (**a**) Infection with lentivirus in U87MG glioblastoma cell line. Actin antibody was used as a loading control. (**b**) Western blot of knockdown of CTCF by infection with lentivirus in U87MG glioblastoma cells. The first line corresponds to one round of infection, the second line is the control of empty vector, and the third line corresponds to U87MG cells with two rounds of infection with lentivirus

9. To obtain an accurate titer, average values from at least two lentiviral dilutions are required. TU/μl is the theoretical quantity of virus particles capable of infecting the same number of cells. For example, to integrate one copy of lentivirus vector in the genome of 100,000 cells it is necessary to use the volume (μl) corresponding to 100,000 TU. Multiplicity of infection (MOI) is the calculated number of lentiviral copies per host genome; thus, MOI depends on TU/μl and the total cell number used during infection. MOI values between 10 and 100 MOIs are the most frequently used.

3.2.6 Selection and Amplification of the Transduced Cell Line

Day 9

After transduction, cells must be drug selected. To accomplish this, many lentiviral systems contain the cDNA of fluorescent proteins and/or the resistance cassette for any antibiotic (i.e., puromycin, ampicillin, G418). There are many ways to perform the selection of positive clones; we will briefly describe three of them.

3.2.7 Manual Selection

Taking advantage of the coupled expression of fluorescent proteins it is possible to enrich the infected population by picking the spot with the highest fluorescent cells identified by an epifluorescence microscope. A single cell with high fluorescence is an indicative of a multiple infection event. This kind of selection allows the isolation of colonies with the best efficiency of infection.

1. Using an epifluorescence microscope, search for the highest fluorescent signal in the visual field and spot it in the plate with a marker.
2. Remove the medium, and wash the cells with PBS.
3. Using sterile tweezers place a cloning disc soaked in trypsin over the previously marked spot, and wait for around 5 min.
4. Remove the cloning disc, and rinse it in a new plate with fresh medium in order to expand the selected clone (*see* **Note 13**).

3.2.8 Sorting by Flow Cytometry (FACS)

Based on the same principle described previously, it is possible to use flow cytometry cell sorting for a quantitative selection of fluorescent colonies in the entire infected culture (Fig. 2a).

1. Harvest ~1 million infected cells in PBS.
2. Determine the profile for the positive fluorescence cells in the FACS.
3. Sort the cells, selecting those with the highest fluorescence emission.
4. Expand enriched cell culture.
5. Days later check again the selected culture in the FACS to corroborate that the levels of fluorescence are maintained (Fig. 2b).

3.2.9 Selection by Antibiotic Resistance (See **Note 14**)

1. After infection cells are recovered and cultured in the presence of 1 μg/ml of puromycin or the appropriate antibiotic (*see* **Note 15**).
2. In parallel, add the same concentration of antibiotic to a non-infected culture, as a control to define the time of death.
3. The infected cells will be selected when the control cell culture (without resistance gene) is death.

3.3 Inducible System

The methodology used for transduction of this lentiviral system is the same as described above. Depending on the resistance cassette of the vector, select the transduced culture with the proper antibiotic in the previously established concentration. After 3–5 days of drug selection, proceed with the following steps for induction.

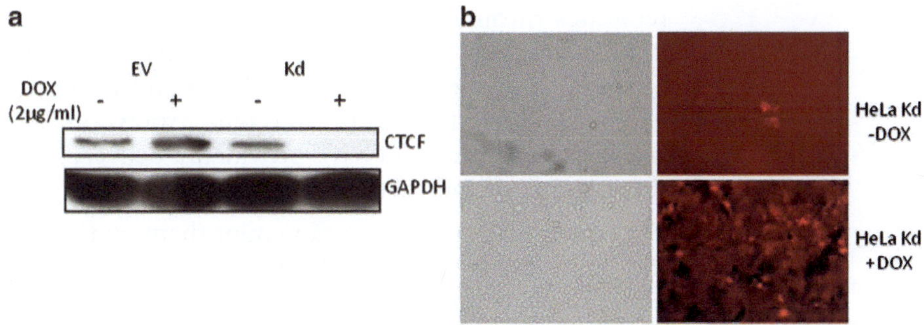

Fig. 4 Inducible system for the CTCF knockdown. (**a**) Western blot for CTCF of the different HeLa cell lines with or without induction of doxycycline with their respective loading control using a GAPDH antibody. (**b**) Induction of shRNAmir (tracked by TurboRFP) with doxycycline is tightly regulated. HeLa cells were transduced with the different plasmids (empty vector *upper panels* and the shRNAi against CTCF at the *bottom*) selected with puromycin (2 μl/ml) for 96 h after which 2 μg/ml of doxycycline was added to the cells and TurboRFP expression was assessed at 72 h

Day 1

1. Remove the medium with the selection antibiotic, and add fresh medium with the desired concentration of induction reagent, doxycycline (0.1–2 μg/ml) (*see* **Note 16**).
2. Repeat the process every 24 h for the next 48–72 h (*see* **Note 17**).

Days 3, 4

1. If the vector employed has a reporter gene (TurboRFP, GFP, etc.), it is possible to evaluate in a fluorescent microscope the expression of the transgene after induction, comparing with a non-induced infected cell context (Fig. 3).
2. Evaluate by the desired method (Western blot, RT-PCR) the knockdown after induction, again, comparing with non-induced infected cells (Fig. 4).

4 Notes

1. The plasmid we use for CTCF knockdown is a siRNA against CTCF cloned in a pSilencer vector (Ambion).
2. Viral vectors emerge as a method for modifying genetically target cells in vitro and in vivo. There are different types of virus used to transfer genetic information including adenoviruses, retroviruses, adeno-associated viruses, herpes simplex virus, and flaviviruses [39]. In particular, lentiviruses are members of the *Retroviridae* family that infect primarily vertebrates and are associated with chronic immune disease. Most of the lentiviral vectors used are derived from the human HIV-1. This family

replicates through the conversion of viral RNA genome into double-stranded DNA. Lentiviruses have the capability to integrate into the genome of dividing and nondividing cells and show stable long-term transgene expression. During the last decade, lentiviruses have provided a delivery platform for several gene therapy clinical approaches [39, 40].

3. For cells in suspension you can count them on the same day of the transfection, and culture them in their respective non-complemented medium.
4. To establish the amount of lipofectamine required, we begin by using a proportion of 1:2 (1 μg of DNA:2 μg of lipofectamine). Otherwise, different concentrations can be tested.
5. We recommend using 1 μg of plasmid for each transfection.
6. Antibiotic concentration for selection depends on the cell type; we suggest using the highest possible avoiding massive cell death.
7. In the knockdown cell line the maximum decrease reached in this kind of assays is about 30–40 %. In our experience it seems that the expression of the plasmid is very unstable since in a couple of days the protein levels of CTCF become similar to those observed in the control line transfected with an empty vector, although the cells continue to survive in the selective medium. The same effect was observed when we froze the knockdown CTCF cell line and later tried to use it for different purposes.
8. Since the highest production of viral particles occurs between 48 and 72 h post-transfection, 16 ml should be sufficient to maintain the cell culture in optimal conditions. Sometimes the cell culture reaches a high confluence earlier; in that case collect virus supernatant and replace with fresh medium.
9. The supernatant contains viral particles, and you must be careful with handling.
10. Virus can be harvested up to 4 days post-transfection.
11. Transport the tubes carefully to avoid disturbing the viral pellet.
12. Avoid pipetting volumes <10 μl to minimize dilution errors.
13. Repeat the procedure with as many different individual colonies as you want to test.
14. This method is the most practical and direct, although the selected cell culture possesses a more variable phenotype due to the different efficiencies of infection for each cell type.
15. It is possible to carry out a concentration curve to determine the ideal selection condition, in which the cell viability is not greatly affected. Depending on the efficiency of infection and the cell type, the concentration of antibiotic can be increased. We suggest testing a range going from 1 to 10 μg/ml.

16. The amounts of doxycycline necessary vary due to the infection efficiency of the particular cell line as well as the amounts of DNA used.
17. In the case of CTCF knockdown, by 48 h after induction it is possible to observe an effect, although the highest levels of expression of the transgene have been detected at 72 h post-induction.

Acknowledgements

We acknowledge Ricardo Saldaña-Meyer for comments and critical reading of the manuscript. This work was supported by the Dirección General de Asuntos del Personal Académico-Universidad Nacional Autónoma de México (IN209403 and IN203811), Consejo Nacional de Ciencia y Tecnología, México (CONACyT; 42653-Q and 128464), and a Ph.D. fellowship from CONACyT and Dirección General de estudios de Posgrado-Universidad Nacional Autónoma de México (DGEP) (EG-B: 207989, RP-M: 173754, and EA-O: 256339). Additional support was provided by the Ph.D. Graduate Program, "Doctorado en Ciencias Biomédicas and Ciencias Bioquímicas." We thank Fernando Suaste-Olmos for technical assistance.

References

1. Boudreau RL, Davidson BL (2012) Generation of hairpin-based RNAi vectors for biological and therapeutic applications. Methods Enzymol 507:275–296
2. Phillips JE, Corces VG (2009) CTCF: master weaver of the genome. Cell 137:1194–1211
3. Herold M, Bartkuhn M, Renkawitz R (2012) CTCF: insights into insulator function during development. Development 139:1045–1057
4. Lobanenkov VV, Nicolas RH, Adler VV, Paterson H, Klenova EM, Polotskaja AV et al (1990) A novel sequence-specific DNA binding protein which interacts with three regulatory spaced direct repeats of the CCCTC-motif in the 5′ flanking sequence of the chicken c-myc gene. Oncogene 5:1743–1753
5. Philippova GN, Qi CF, Ulmer JE, Moore JM, Wards MD, Hu YJ et al (2002) Tumor-associated zinc finger mutations in the CTCF transcription factor selectively alter its DNA-binding specificity. Cancer Res 62:48–52
6. Ohlsson R, Lobanenkov V, Klenova E (2010) Does CTCF mediate between nuclear organization and gene expression? BioEssays 32:37–50
7. Valadez-Graham V, Razin SV, Recillas-Targa F (2004) CTCF-dependent enhancer blocker at the upstream region of the chicken α-globin gene domain. Nucleic Acids Res 32:1354–1362
8. Engel N, West AG, Felsenfeld G, Bartolomei MS (2004) Antagonism between DNA hypermethylation and enhancer-blocking activity at the H19 DMD is uncovered by CpG mutations. Nat Genet 36:883–888
9. De La Rosa-Velázquez IS, Rincón-Arano H, Benítez-Bribiesca L, Recillas-Targa F (2007) Epigenetic regulation of the human retinoblastoma tumor suppressor gene promoter by CTCF. Cancer Res 67:2577–2585
10. El-Kady A, Klenova E (2005) Regulation of the transcription factor, CTCF, by phosphorylation with protein kinase CK2. FEBS Lett 579:1424–1434
11. Docquier F, Jita GX, Farrar D, Jat P, O'Hare M, Chernukhin I et al (2009) Decreased poly(ADP-ribosyl)ation of CTCF, a transcription factor, is associated with breast cancer phenotype and cell proliferation. Clin Cancer Res 15:5762–5771

12. Kitchen NS, Schoenherr CJ (2010) Sumoylation modulates a domain in CTCF that activates transcription and decondenses chromatin. J Cell Biochem 11:665–675
13. MacPherson MJ, Beatty LG, Zhoi W, Du M, Sdowski PD (2009) The CTCF insulator protein posttranslationally modified by SUMO. Mol Cell Biol 29:714–725
14. Docquier F, Farrar D, D'Arcy V, Chernukhin I, Robinsin AF, Loukinov D et al (2005) Heightened expression of CTCF in breast cancer cells is associated with resistance to apoptosis. Cancer Res 65:5112–5122
15. Soshnikova N, Montavon T, Leleu M, Galjart N, Duboule D (2010) Functional analysis of CTCF during mammalian limb development. Dev Cell 19:819–830
16. Delgado-Olguín P, Brand-Arzamendi K, Scott IC, Jungblut B, Stainier DY, Bruneau BG et al (2011) CTCF promotes muscle differentiation by modulating the activity of myogenic regulatory factors. J Biol Chem 286:12483–12484
17. Liu Z, Scannell DR, Eisen MB, Tjian R (2011) Control of embryonic stem cell lineage commitment by core promoter factor, TAF3. Cell 146:720–731
18. Fiorentino FP, Giordano A (2011) The tumor suppressor role of CTCF. J Cell Physiol 227: 479–492
19. Recillas-Targa F, De La Rosa-Velázquez IA, Soto-Reyes E (2011) Insulation of tumor suppressor genes by the nuclear factor CTCF. Biochem Cell Biol 89:479–488
20. Barkess G, West AG (2012) Chromatin insulator elements: establishing barriers to set heterochromatin boundaries. Epigenomics 4:67–80
21. Burgess-Beusse B, Farrell C, Gaszner M, Litt M, Mutskov V, Recillas-Targa F et al (2002) The insulation of genes from external enhancers and silencing chromatin. Proc Natl Acad Sci USA 4:16433–16437
22. Pikaart M, Recillas-Targa F, Felsenfeld G (1998) Loss of transcriptional activity of the transgene is accompanied by DNA methylation and histone deacetylation and is prevented by insulators. Genes Dev 12:2852–2862
23. Recillas-Targa F, Valadez-Graham V, Farrell CM (2004) Prospects and implications of using chromatin insulators in gene therapy and transgenesis. BioEssays 26:796–807
24. Furlan-Magaril M, Rebollar E, Guerrero G, Fernández A, Moltó E, González-Buendía E et al (2011) An insulator embedded in the Chicken α-globin locus regulates chromatin domain configuration and differential gene expression. Nucleic Acids Res 39:89–103
25. Splinter E, Heath H, Kooren J, Palstra RJ, Klaus P, Grosveld F et al (2006) CTCF mediates long-range chromatin loop and local histone modification in the β-globin locus. Genes Dev 20:2349–2354
26. Kurukuti S, Tiwari VK, Tavoosidana G, Pugacheva E, Murrell A, Zhao Z et al (2006) CTCF binding at the H19 imprinting control region mediates maternally inherited higher-order chromatin conformation to restrict enhancer access to Igf2. Proc Natl Acad Sci USA 103:10684–10689
27. van Bortle K, Corces VG (2013) The role of chromatin insulators in nuclear architecture and genome function. Curr Opin Genet Dev 23(2):212–218
28. Guelen L, Pagie L, Brasset E, Meuleman W, Faza MB, Talhout W et al (2008) Domain organization of human chromosomes revealed by mapping the nuclear lamina interactions. Nature 453:948–951
29. Bonetta L (2005) The inside scoop-evaluating gene delivery methods. Nat Methods 2: 875–883
30. Sambrook J, Russell DW. Molecular cloning, a laboratory manual, 2(4):15.1–15.5
31. Hughes D, Marendy E, Dickerson C, Yetming K, Sample C, Sample J (2012) Contribution of CTCF and DNA methyltransferases DNMT1 and DNMT3B to Epstein-Barr Virus restricted latency. J Virol 86:1034–1045
32. Ren L, Wang Y, Shi M, Wang X, Yang Z, Zhao Z (2012) CTCF mediates the cell-type specific spatial organization of the *Kcnq5* locus and the local gene regulation. PLoS One 7:e31416
33. Jackson AL, Burchard J, Leake D, Reynolds A, Schelter J, Guo J et al (2006) Position-specific chemical modification increases specificity of siRNA-mediated gene silencing. RNA 12: 1197–1205
34. Chen PY, Weinmann L, Gaidatzis D, Pei Y, Zavolan M, Tuschl T et al (2008) Strand-specific 5′O-methylation of siRNA duplexes controls guide strand selection and target specificity. RNA 2:263–274
35. Birmingham A, Anderson EM, Reynolds A, Ilsley-Tyree D, Leake D, Fedorov Y et al (2006) 3′-UTR seed matches, but not overall identity, are associated with RNAi off-targets. Nat Methods 3:199–204
36. Anderson EM, Birmingham A, Baskerville S, Reynolds A, Maksimova E, Leake D et al (2008) Experimental validation of the importance of seed frequency to siRNA specificity. RNA 14:853–861
37. Hou C, Zhao H, Tanimoto K, Dean A (2008) CTCF-dependent enhancer-blocking by alter-

native chromatin loop formation. Proc Natl Acad Sci USA 105:20398–20403

38. Balakrishnan S, Witcher M, Berggren T, Emerson B (2012) Functional and molecular characterization of the role of CTCF in human embryonic stem cell biology. PloS One 7:e42424
39. Klimatcheva E, Ronsenblatt JD, Planelles V (1999) Lentiviral vectors and gene therapy. Front Biosci 4:D481–96
40. Biasco L, Baricordi C, Aiuti A (2012) Retroviral integrations in gene therapy trials. Mol Ther 20:709–716
41. Markusic D, Seppen J (2010) Doxycycline regulated lentiviral vectors. Methods Mol Biol 614:69–76
42. Gossen M, Bujard H (1992) Tight control of gene expression in mammalian cells by tetracycline-responsive promoters. Proc Natl Acad Sci USA 89:5547–5551
43. Urlinger S, Baron U, Thellmann M, Hasan MT, Bujard H, Hillen W (2000) Exploring the sequence space for tetracycline-dependent transcriptional activators: novel mutations yield expanded range and sensitivity. Proc Natl Acad Sci USA 97:7963–7968
44. Zeng Y, Wagner EJ, Cullen BR (2002) Both natural and designed micro RNAs can inhibit the expression of cognate mRNAs when expressed in human cells. Mol Cell 9:1327–1333

Further Reading

DharmaFECT General Transfection Protocol, Dharmacon Inc. 2006, 00033–05-F-03-U

Ambion by Life Technologies, Silencer Select Pre-designed, Validated and Custom Designed siRNA, Custom Select siRNA, Ambion In Vivo Pre-designed, Custom Designed and Custom siRNA, 2011, Insert PN4457171 Rev B

http://www.invitrogen.com/site/us/en/home/Products-and-Services/Applications/rnai/Synthetic-RNAi-Analysis/Ambion-Silencer-Select-siRNAs/silencer-select-sirna.html.html

Chapter 6

Use of Serum-Circulating miRNA Profiling for the Identification of Breast Cancer Biomarkers

Fermín Mar-Aguilar, Cristina Rodríguez-Padilla, and Diana Reséndez-Pérez

Abstract

MicroRNAs (miRNAs) are important regulatory molecules involved in disease pathogenesis. miRNAs are very stable in bodily fluids and can be detected in serum, plasma, saliva, and urine, among other fluids. Several studies have demonstrated the usefulness of serum miRNAs as potential biomarkers for detecting and monitoring cancer progression. Here, we describe in detail the experiment protocol we used to profile miRNA expression in the serum of breast cancer patients, including RNA extraction from serum, RT-qPCR quantification, and analysis of the deregulated miRNAs. Detection of circulating miRNAs may be a useful, noninvasive diagnostic tool for breast cancer.

Key words miRNAs, Circulating microRNA, Serum, RT-qPCR, Blood, Biomarker, Breast cancer

1 Introduction

A novel class of small regulatory RNAs [1] has been the focus of intensive investigations. These so-called microRNAs (miRNAs) are single-stranded RNA molecules, approximately 23 nucleotides long, that play important regulatory roles in animals and plants by pairing to the messenger RNAs (mRNAs) of protein-coding genes. These molecules direct posttranscriptional repression by promoting mRNA degradation or blocking translation by binding to complementary sequences in the 3′ untranslated regions (3′ UTR) of mRNAs [2, 3].

Several studies have demonstrated the existence of abundant extracellular miRNAs that circulate in serum, plasma, saliva, urine, and other bodily fluids [4–6]. These circulating miRNAs are highly stable and protected from endogenous RNase activity because they are either bound to proteins or packaged into microparticles, which include exosome-like particles, microvesicles, apoptotic bodies, and apoptotic microparticles [7–9]. The functions of exosomal miRNAs

Martha Robles-Flores (ed.), *Cancer Cell Signaling: Methods and Protocols*, Methods in Molecular Biology, vol. 1165,
DOI 10.1007/978-1-4939-0856-1_6,

are poorly understood, but some reports have shown their essential roles in cancer development. miRNAs have been found in other bodily fluids besides serum including urine and tears as well as synovial, ascetic, amniotic, and cerebrospinal fluids using established techniques such as quantitative reverse transcription polymerase chain reaction (RT-qPCR).

Breast cancer causes 400,000 deaths annually, accounting for 23 % of all cancers in women worldwide [10]. Early detection of breast cancer is vital to reduce the mortality of this disease [11, 12]. However, even with the most widely used methods for breast cancer detection such as mammography, ultrasonography, and magnetic resonance imaging, concerns remain for the high rates of misdiagnosis, missed diagnoses, and overdiagnosis of this malignancy [12]. Several reported studies have investigated serum/plasma miRNA levels in breast cancer patients and have established the potential use of these molecules as biomarkers for this disease [13–16].

One of the biggest challenges in medicine is to detect cancer at an early stage and, ideally, find the means to predict who will develop cancer. Current diagnostic techniques, such as histological evaluation of biopsies, remain the gold standard. Although tumor markers (e.g., HER-2/neu, estrogen receptor, and progesterone receptor) greatly improve diagnosis, the invasive, unpleasant, and inconvenient nature of current diagnostic procedures such as biopsies limits their clinical application. Therefore, it is highly desirable to develop tumor markers that can be collected by noninvasive procedures [17].

Available evidence suggests that serum miRNAs may be used as breast cancer biomarkers in noninvasive diagnostic procedures. This chapter describes the method to measure serum miRNA expression levels, from miRNA extraction to RT-qPCR and analysis of the deregulated miRNAs in breast cancer samples.

2 Materials

2.1 Serum Isolation

1. 16 × 100 mm × 8.5 mL plastic serum tube with clot activator and gel for serum separation.
2. 1 mL micropipette.
3. 2 mL cryovials.

2.2 miRNA Extraction from Serum

1. Reactives for total RNA extraction including miRNAs (miRNeasy kit) (*see* **Note 1**).
2. Molecular biology-grade chloroform (without added isoamyl alcohol).
3. Molecular biology-grade ethanol (absolute).
4. RNase-free 1.5 mL microcentrifuge tubes.

5. DNAse- and RNAse-free water.
6. RNase-free pipette tips.
7. Refrigerated centrifuge.
8. Disposable gloves.

2.3 miRNA Quantification by RT-qPCR

1. Specific primers for amplification by real-time PCR of targets (miRNAs, TaqMan MicroRNA assays). Store at −20 °C (*see* **Note 2**).
2. Reactives for reverse transcription (TaqMan MicroRNA Reverse Transcription Kit). Store at −20 °C (*see* **Note 3**).
3. Reactives for real-time PCR (TaqMan Gene Expression Master Mix) (*see* **Note 4**), ultrapure (UP), uracil-DNA glycosylase (UDG) (*see* **Note 5**), deoxyribonucleotide triphosphates (dNTPs) with deoxyuridine triphosphate (dUTP), and ROX passive reference (*see* **Note 6**). Store at 4 °C.
4. Nuclease-free 0.2 mL, 8-strip PCR tubes and optical caps (*see* **Note 7**).
5. Disposable powder-free gloves (*see* **Note 8**).
6. Real-time PCR thermal cycler.
7. End point thermal cycler.

3 Methods

3.1 Serum Isolation

1. Blood specimens must be drawn using a gel-separation vacutainer blood collection tube.
2. Centrifuge the blood samples in the gel serum separator tubes at 1,000 × *g* and a temperature of 22–25 °C for 15 min in a swinging bucket rotor; the gel layer will become a barrier and separate out the serum.
3. Use a micropipette to aliquot the serum into 500 μL volumes in 1.5 mL microcentrifuge tubes. Avoid disturbing the gel layer with the pipette.
4. Serum can be used immediately or can be frozen at −80 °C for future use.

3.2 RNA Extraction

RNA extraction is performed following the manufacturer's instructions in the miRNeasy kit as follows:

1. Prepare serum or thaw frozen samples on ice.
2. Add 700 μL of QIAzol Lysis Reagent for every 200–400 μL of serum. Mix by vortex or by pipetting up and down (*see* **Note 9**).
3. Let the tube stand at room temperature (20–25 °C) for 5 min.
4. Add 140 μL of chloroform to the tube, and cap it securely. Mix by vortexing for 15 s. Let the tube stand at room temperature for 2–3 min.

5. Centrifuge for 15 min at 12,000 × *g* and 4 °C.
6. Transfer the upper aqueous phase to a new collection tube, add 1.5 volumes of absolute ethanol at room temperature, and mix thoroughly by pipetting up and down several times.
7. Pipette 700 μL (*see* **Note 10**) of the sample, including any precipitate that may have formed, into an RNeasy Mini spin column in a 2 mL collection tube (included in the kit). Close the lid, and centrifuge at 8,000 × *g* for 15 s at room temperature (15–25 °C).
8. Add 700 μL buffer RWT to the RNeasy Mini spin column. Close the lid, and centrifuge for 15 s at 8,000 × *g* to wash the column.
9. Pipette 500 μL of buffer RPE onto the RNeasy Mini spin column. Close the lid, and centrifuge for 15 s at 8,000 × *g* to wash the column.
10. Add another 500 μL of buffer RPE to the RNeasy Mini spin column. Close the lid, and centrifuge for 2 min at 8,000 × *g* in order to dry the RNeasy Mini spin column membrane.
11. Transfer the RNeasy Mini spin column to a new 1.5 mL collection tube (supplied). Pipette 30–50 μL RNase-free water directly onto the RNeasy Mini spin column membrane. Close the lid, and centrifuge for 1 min at 8,000 × *g* to elute the RNA (*see* **Note 11**).

3.3 miRNA Quantification by RT-qPCR

Reverse transcription (RT) reactions were performed using the TaqMan miRNA Reverse Transcription Kit. Each 15 μL reaction should contain 100 mM dNTPs (with dTTP) (0.15 μL), multiscribe reverse transcriptase 50 U/μL (1 μL), 10× RT buffer (1.5 μL), RNase inhibitor 20 U/μL (0.188 μL), 5× TaqMan MicroRNA RT Primer (3 μL), and nuclease-free water (to complete 13 μL). Place a tube on ice, and add all the components into a master mix large enough for all of the serum–miRNA samples. Mix gently, and centrifuge to bring the solution to the bottom of the tube.

1. Pipette 13 μL of the master mix into 0.2 mL RNase-free tubes.
2. Add 2 μL of serum miRNA (*see* **Note 12**) to each tube (for a final volume of 15 μL), and spin briefly.
3. Incubate the RT reactions in an end point thermal cycler as follows: 16 °C for 30 min, 42 °C for 30 min, 85 °C for 5 min, and hold at 4 °C.
4. Remove tubes from thermal cycler and continue with qPCR or store at −20 °C.

3.4 Quantitative PCR

Quantitative PCR reactions are performed using the TaqMan Gene Expression Master Mix and TaqMan miRNA assays. Each

10 μL reaction should contain TaqMan Universal PCR Master Mix (5 μL), 20× TaqMan Gene Expression Assay (miRNA-Specific PCR primer) (0.5 μL), and nuclease-free water (to complete 8 μL). Place a tube on ice, and add all the components into a master mix large enough for all of the samples. Mix gently, and collect contents by brief centrifugation.

1. Pipette 8 μL of reaction mix into a 0.2 mL tube.
2. Add 2 μL of cDNA to each tube (for a final volume of 10 μL).
3. Collect contents by brief centrifugation.
4. Incubate the PCR reactions in real-time PCR machine using the following thermal cycling parameters: 50 °C for 2 min (*see* **Note 13**), 95 °C for 10 min, 40 cycles of 95 °C for 15 s, and 60 °C for 1 min (*see* **Note 14**).
5. The relative expression of the mature miRNA may be calculated using the comparative cycle threshold ($2^{-\Delta\Delta CT}$) method (*see* **Note 15**).
6. Compare RT-qPCR data between breast cancer and healthy samples to identify miRNAs that are altered in breast cancer patients (*see* **Note 16**). We investigated whether RT-qPCR miRNA profiling on serum could discriminate between breast cancer patients ($n=61$) and healthy controls ($n=10$). The relative expression values of seven circulating miRNAs are related with breast cancer. All miRNAs were significantly higher in breast cancer serum than in healthy controls ($p<0.001$) as shown in Fig. 1.
7. Construct AROC curves to evaluate the sensitivity and specificity of the assay (*see* **Note 17**). Additionally, you can calculate the efficiency of a diagnostic test (*see* **Note 18**). Figure 2 shows AROC curves constructed for circulating miRNAs tested in this assay, and Table 1 shows the results of sensitivity, specificity, and efficiency of our assay.

4 Notes

1. There are many options of protocols for total RNA extraction, but a protocol that includes miRNAs must be chosen. We choose miRNeasy kit from Qiagen because it is a very convenient tool for isolation of good-quality RNA in a relatively short time.
2. TaqMan MicroRNA Assays employ a novel target-specific stem–loop reverse transcription primer and confer the advantage of specific detection of the mature, biologically active miRNA. Taqman MicroRNA Assay is aligned with release 19 of the miRBase database and provides comprehensive coverage for each of the 193 listed species.

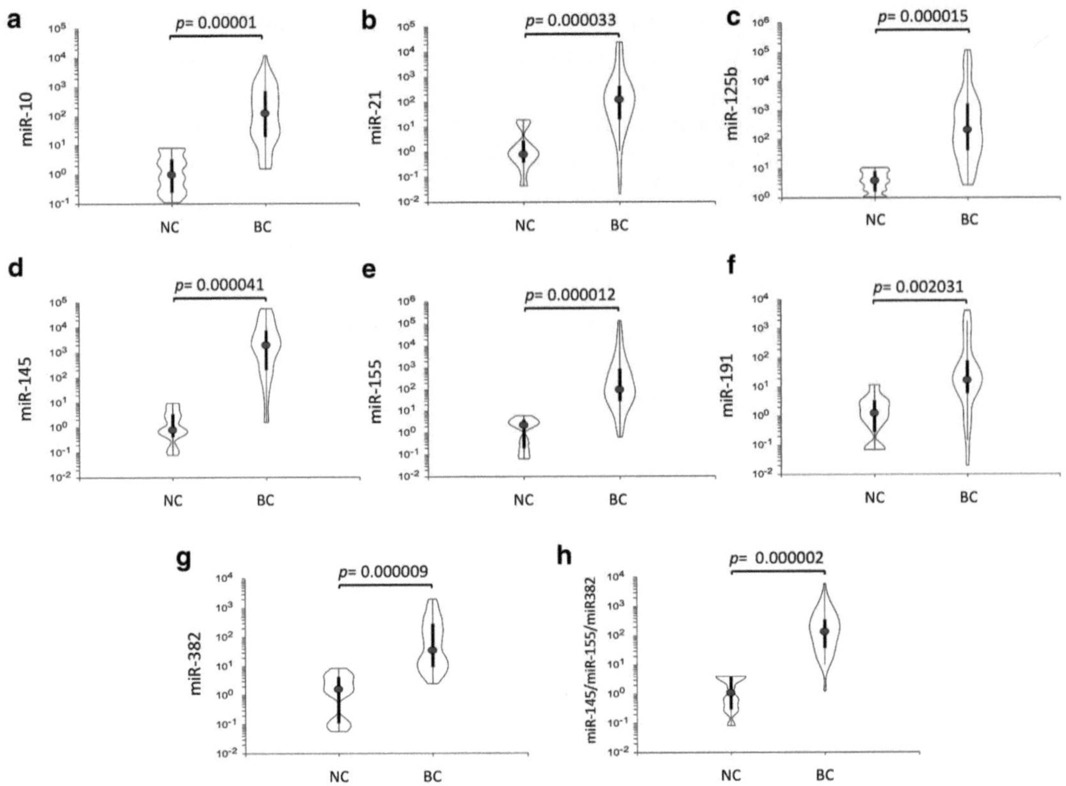

Fig. 1 Relative fold change in the expression levels of (**a**) miR-10b, (**b**) miR-21, (**c**) miR-125b, (**d**) miR-145, (**e**) miR-155, (**f**) miR-191, (**g**) miR-382, and (**h**) miR-145/miR-155/miR-382 in breast cancer (BC) and normal breast tissue (NC). The *p*-values for all miRNAs were significant. The *dot* at the center of violin plot represents the mean fold change for each miRNA. Reprinted with permission from IOS Press © 2013 [16]

3. The TaqMan MicroRNA Reverse Transcription Kit is specifically designed to generate miRNA-specific cDNA for use with the TaqMan MicroRNA Assays. These assays allow the user to detect and accurately quantify specific mature miRNAs.
4. TaqMan Gene Expression Master Mix gets precise and reliable real-time qPCR quantification and is an optimized 2× mix that contains all of the components, excluding the template and primers. TaqMan Gene Expression Master Mix has been validated with TaqMan Gene Expression Assays providing precise quantification for a variety of gene expression-based real-time qPCR applications.
5. UDG prevents re-amplification of carryover PCR products in an assay if all previous PCRs for that assay were performed using a dUTP-containing master mix.
6. The ROX passive reference provides an internal reference to which the reporter-dye signal can be normalized during data analysis. Normalization is necessary to correct for fluorescent fluctuations due to changes in concentration or volume.

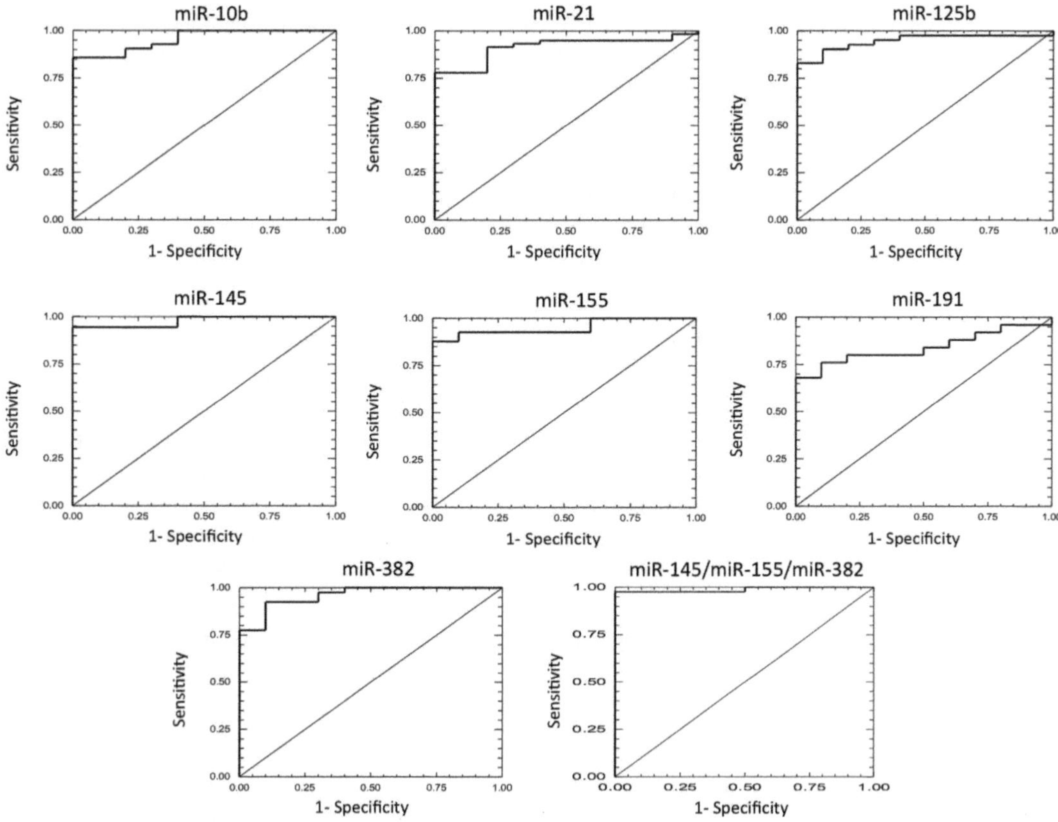

Fig. 2 Area under curve of receiver operating characteristic (ROC) for miR-10b, miR-21, miR-125b, miR-145, miR-155, miR-191, miR-382, and miR-145/miR-155/miR-382. All miRNAs show lower sensitivity and specificity values than the combination of miR-145/miR-155/miR-382. Reprinted with permission from IOS Press © 2013 [16]

Table 1
Sensitivity, specificity, and efficiency values of circulating miRNAs in breast cancer

miRNA	Optimal cutoff	Sensitivity	Specificity	Efficiency
miR-10	59.22	83.3	100	85.7
miR-21	6.48	94.4	80	89.3
miR-125b	8.46	88.9	80	89.3
miR-145	15.93	94.4	100	96.4
miR-155	7.92	94.4	100	96.4
miR-191	11.59	72.2	90	78.6
miR-382	4.85	94.4	90	92.6
miR-145/155/382	10.07	97.60	100	98

7. Most real-time thermal cyclers acquire fluorescence data from the top, but some equipment acquire data from the bottom (e.g., Rotor Gene from Qiagen). If your equipment acquires data from the bottom, you must use an optical tube.

8. It is always recommended to use powder-free gloves to avoid contaminating tube surfaces with background fluorescent materials such as powder, dust, and dead skin cells.

9. Qiagen offers a specific kit for purification of miRNAs from serum or plasma (miRNeasy Serum/Plasma Kit for purification of total RNA). According to the manufacturer's instructions, five volumes of QIAzol per volume of serum must be added to perform lysis of the sample. Since we had previously used the miRNeasy kit for extraction of total RNA from cell culture and tissue, we adapted this protocol for RNA extraction from serum. For this reason we used only 700 μL of QIAzol per 200–400 μL of serum. We tested both protocols, and we did not see any difference for RNA purification. In conclusion, it is possible to do RNA extraction with a small volume of QIAzol without compromising the amount of total RNA recovered.

10. If it is necessary, fill the column and centrifuge several times until the whole mix (aqueous phase + ethanol) has passed the column.

11. The concentration of total RNA extracted from serum is very small, regularly between 3 and 15 ng/μL. The quantification of RNA extracted can be done by spectrophotometry (e.g., NanoDrop); however, it is necessary to consider that 260/280 and 260/230 absorbance ratios will be low (0.3–0.6). These values may occur because the detection of low levels of RNA is not efficient and the 260/280 and 260/230 absorbance ratios do not reflect the true protein content of the sample. In conclusion, it is possible to use spectrophotometry to quantify total RNA extracted from serum samples, but high absorbance ratios should not be expected. Finally, it is advisable to analyze RNA samples with a bioanalyzer because it is the best option to quantify small amounts of nucleic acids. Bioanalyzer is a microfluidics-based platform that allows accurate small RNA quantitation and sizing.

12. In order to add the same amount of RNA to each reaction, we diluted the sample to a concentration of 10 ng/μL (when required) or we added more than 2 μL of RNA when the concentration was lower than 20 ng/μL (if this is the case, you will need to adjust the amount of water in your reaction).

13. Incubation at 50 °C for 2 min is necessary to cleave any dU-containing PCR carryover products. Ten-minute incubation at 95 °C is necessary to substantially reduce UDG activity and to

denature the native DNA in the experimental sample. Because UDG is not completely deactivated during the 95 °C incubation, it is important to keep the annealing temperatures greater than 55 °C and to refrigerate PCR products at 2–8 °C in order to prevent degradation of the amplified product.

14. The acquisition of fluorescent signals depends on the chemistry used in real-time PCR. With Taqman, you must acquire the signal during the extension step (60 °C step).

15. Gene expression can be measured by the quantification of cDNA relative to a calibrator sample, which serves as a physiological reference. The calibrator sample we used is the cDNA from healthy individuals. All quantifications are also normalized to an endogenous control to account for variability in the initial concentration and quality of the total RNA and in the conversion efficiency of the reverse transcription reaction. Some commonly used endogenous genes are Rnu44, Rnu48, and Rnu66 (noncoding nucleolar RNAs), but these are useful only for tissue samples because they are very difficult to detect in serum/plasma. Thus, we used 18S RNA to normalize the samples as previously described by Zhu et al. [18].

16. Relative miRNA quantification shows whether the genes are over-expressed or under-expressed using a healthy sample as reference. If one miRNA is constantly under- or over-expressed in many samples, this miRNA may be used as a biomarker of the disease. However, further validation is needed mainly to obtain a predictive value about the biomarker (AROC curves).

17. Receiver operating characteristic (ROC) curve analysis is an important statistical tool to investigate the accuracy of a quantitative test throughout the whole range of its values and to help to identify the optimal cutoff value. ROC curves are also used to test the prognostic value of biomarkers and to compare their predictive value [19]. The construction of ROC curves is very important to know the sensitivity and specificity of the assay; for example, we have determined the expression profile of seven miRNAs and we were able to distinguish between breast cancer and healthy individuals. Table 1 shows the values of sensitivity and specificity for each miRNA analyzed in breast cancer serum samples. After constructing the area under ROC curve (AROC) we obtained very high values for sensitivity (89.6 %) and for specificity (88.5 %). Finally, when we analyzed a set of three miRNAs, the sensitivity and specificity of the test increased to 97.6 % and 100 %, respectively [16].

18. The efficiency of a diagnostic test is defined by its ability to correctly identify those people who have the condition as well as those who do not have it. We found that this diagnostic method has a very high efficiency: 98 % for the combination of

miR-145, miR-155, and miR-382 (Table 1). These results indicate that the combination signature of those three miRNAs has a strong potential diagnostic value for breast cancer detection.

Acknowledgements

We would like to thank Perla K. Espino-Silva, Sandra K. Santuario-Facio, and Pablo Ruiz-Flores for the samples of breast cancer patients; Ismael Malagón-Santiago for statistical analysis; Alejandra Arreola-Triana for editorial review of this manuscript; and the Teachers' Improvement Program (PROMEP) grants 103.5/07/2523 and 103.5/08/4285 for their financial support.

References

1. Lee RC, Feinbaum RL, Ambros V (1993) The C. elegans heterochronic gene lin-4 encodes small RNAs with antisense complementarity to lin-14. Cell 75:843–854
2. Bartel DP (2009) MicroRNAs: target recognition and regulatory functions. Cell 136: 215–233
3. Du T, Zamore PD (2005) microPrimer: the biogenesis and function of microRNA. Development 132:4645–4652
4. Cortez MA, Calin GA (2009) MicroRNA identification in plasma and serum: a new tool to diagnose and monitor diseases. Expert Opin Biol Ther 9:703–711
5. Mitchell PS, Parkin RK, Kroh EM et al (2008) Circulating microRNAs as stable blood-based markers for cancer detection. Proc Natl Acad Sci U S A 105:10513–10518
6. Park NJ, Zhou H, Elashoff D et al (2009) Salivary microRNA: discovery, characterization, and clinical utility for oral cancer detection. Clin Cancer Res 15: 5473–5477
7. Caby MP, Lankar D, Vincendeau-Scherrer C et al (2005) Exosomal-like vesicles are present in human blood plasma. Int Immunol 17: 879–887
8. Orozco AF, Lewis DE (2010) Flow cytometric analysis of circulating microparticles in plasma. Cytometry A 77:502–514
9. Record M, Subra C, Silvente-Poirot S et al (2011) Exosomes as intercellular signalosomes and pharmacological effectors. Biochem Pharmacol 81:1171–1182
10. Parkin DM, Bray F, Ferlay J et al (2005) Global cancer statistics, 2002. CA Cancer J Clin 55:74–108
11. Rim A, Chellman-Jeffers M (2008) Trends in breast cancer screening and diagnosis. Cleve Clin J Med 75(Suppl 1):S2–S9
12. Nattinger AB (2010) In the clinic. Breast cancer screening and prevention. Ann Intern Med 152:ITC41
13. Lodes MJ, Caraballo M, Suciu D et al (2009) Detection of cancer with serum miRNAs on an oligonucleotide microarray. PLoS One 4:e6229
14. Zhao H, Shen J, Medico L et al (2010) A pilot study of circulating miRNAs as potential biomarkers of early stage breast cancer. PLoS One 5:e13735
15. Asaga S, Kuo C, Nguyen T et al (2011) Direct serum assay for microRNA-21 concentrations in early and advanced breast cancer. Clin Chem 57:84–91
16. Mar-Aguilar F, Mendoza-Ramirez JA, Malagon-Santiago I et al (2013) Serum circulating microRNA profiling for identification of potential breast cancer biomarkers. Dis Markers 34:163–169
17. Wittmann J, Jack HM (2010) Serum microRNAs as powerful cancer biomarkers. Biochim Biophys Acta 1806:200–207
18. Zhu W, Qin W, Atasoy U et al (2009) Circulating microRNAs in breast cancer and healthy subjects. BMC Res Notes 2:89
19. Tripepi G, Jager KJ, Dekker FW et al (2009) Diagnostic methods 2: receiver operating characteristic (ROC) curves. Kidney Int 76:252–256

Chapter 7

Prostate Cancer Detection Using a Noninvasive Method for Quantifying miRNAs

Mauricio Rodríguez-Dorantes, A. Ivan Salido-Guadarrama, and Pilar García-Tobilla

Abstract

Cancer is one of the diseases with more incidence in industrialized countries. Early detection is important for patient survival in terms of treatment and clinical decisions. Several methods have been described to improve detection, diagnostic, and treatment in cancer, but the diagnostic methods are very aggressive for patients. Biopsies have been the gold standard to determine the clinical pathological characteristics of cancer tissues for many years. The biopsies procedure is very uncomfortable for the patients, and in many cases the results are negative to cancer. Therefore, patients are submitted to a second round of biopsies, because their clinical conditions implicate the presence of neoplasia. During the last years, noninvasive methods have shown relevance because they could be good indicators for cancer detection at molecular level. Molecular signatures have been included to characterize different stages of the disease, and thereby the objective is to prevent unnecessary biopsies. Several body fluids as urine, serum, blood, semen, saliva, etc. have been used. In fluids, miRNA detection provides a promising tool to obtain molecular signatures for different types of cancer. Urine represents a very good option to find molecules representative of tumor status in prostate cancer. The presence or the absence of miRNA involved in the development of the disease has been demonstrated. In this chapter we describe a method to quantify mature miRNA signatures using a noninvasive test using a body fluid such as urine.

Key words Body fluids, miRNAs, Microarrays, Molecular signature, Biopsies

1 Introduction

Cancer is a leading cause of death worldwide and accounted for 7.6 million deaths in 2008. Important risk factors for cancer include tobacco use, unhealthy diet, physical inactivity, and the use of alcohol [1]. Cancer embraces a diversity of diseases that involves proliferation of aberrant cells, distinct from the ordinary cell types [2]. Cancer cells exhibit genetic changes and harbor characteristics defined by Hanahan and Weinberg as "the hallmarks of cancer", including, apoptosis evasion, self-sufficiency in growth signals, insensitivity to anti-growth signals, sustained angiogenesis, limitless

Martha Robles-Flores (ed.), *Cancer Cell Signaling: Methods and Protocols*, Methods in Molecular Biology, vol. 1165,
DOI 10.1007/978-1-4939-0856-1_7, © Springer Science+Business Media New York 2014

replicative potential, tissue invasion or metastasis, deregulated cellular energetics, avoidance of immune destruction, tumor-promoting inflammation, and genome instability or mutations [3, 4].

Particularly, prostate cancer (PCa) is the second most common type of cancer among men worldwide, comprising 13.6 % overall diagnosed tumors [5]. The clinical course of metastatic disease is characteristically progressive and fatal with a median overall survival between 24 and 36 months [6]. PCa detection involves serological measurement of prostate-specific antigen (PSA) and digital rectal examination (DRE) to detect indurations, asymmetry, and nodularity. Currently the gold standard method for PCa diagnostic is the *trans*-rectal ultrasound-directed biopsy followed by histopathological evaluation [6]. Biopsies are painful procedures, and some cancer tumors are not detected by these technique.

The search of new diagnostic methods in cancer detection has been increased during the last years. Novel molecular markers to diagnose cancer tumors at early stages have been developed to have a better tumor stratification [7].

Cancer tumors are capable of shedding nucleic acids (DNA or RNA) into the bloodstream and other body fluids [8]. Alterations in circulating cell-free DNA make it a promising candidate as a noninvasive biomarker of cancer [9, 10]. Numerous studies of both serum- and urine-based PCa to find biomarker candidates have been performed [11–13]. Extracellular miRNAs in serum, plasma, saliva, and urine have recently been shown to be associated with various pathological conditions including cancer [14]. MiRNAs are a class of small noncoding RNA molecules that regulate gene expression by binding to 3′untranslated region (UTR) of target mRNA, inducing translational repression and destabilization of RNA transcripts [15, 16]. Increasing evidence shows that miRNAs are involved in the regulation of a range of physiological and cellular processes conserved among different species including cellular differentiation, proliferation, and cell cycle control [17–20]. Calin and colleagues first demonstrated that a large number of miRNAs are located in cancer-associated genomic regions prone to deletions, amplifications, or recombinations [21, 22]. More recently, it has been revealed that altered miRNA profiling can classify cancers and predict outcome with high accuracy [23–30] and many studies have identified several miRNAs significantly altered in PCa. Because of their relative small size and the fact that they are frequently secreted in exosomes [31], miRNAs may be very useful biomarkers, readily detectable in body fluids. The altered expression of miRNAs in serum from men with PCa in comparison to healthy controls has recently been demonstrated [32]. Here we describe a protocol to perform a high-throughput expression analysis of miRNAs using urine samples in order to identify changes in miRNA levels.

2 Materials

2.1 RNA Extraction

1. TRIzol® Reagent.
2. EtOH solutions at 75 and 85 % and EtOH absolute (100 %).
3. Microcentrifuge.
4. Glycogen.
5. Spectrophotometer.

2.2 cDNA Preparation Specific for miRNAs

1. TaqMan® MicroRNA reverse transcription kit (Applied Biosystems).
2. Megaplex™ RT primers, human pool A v2.0 (Applied Biosystems).
3. Thermocycler.

2.3 cDNA Preamplification

1. Preamp Master Mix (Applied Biosystems).
2. Megaplex preamp primer mix (Applied Biosystems).
3. Thermocycler.

2.4 High-Throughput Quantitative PCR for miRNAs

1. TaqMan® Array Human MicroRNA A Cards v2.0 (Applied Biosystems).
2. TaqMan® Master Mix, no UNG (Applied Biosystems).
3. Real-time PCR system.

3 Methods

3.1 RNA Extraction

1. Whole-urine sample (~30 ml) (*see* **Note 1**) is centrifuged at 3,000 × *g* (*see* **Note 2**) for a time lapse of 10 min.
2. Discard supernatant, add 10 ml of phosphate buffer solution at pH 7.2, and shake vigorously in a vortex device until the sediment pellet is completely resuspended. Then centrifuge at 3,000 × *g* for 10 min (*see* **Note 3**), and discard the supernatant. Repeat this step.
3. Add 800 μl of TRIzol® Reagent (Invitrogen) (*see* **Notes 4** and **5**), and add 200 μg glycogen directly to the TRIzol (final glycogen concentration in TRIzol is 250 μg/ml) for each ml of urine sediment and disaggregate by pipetting.
4. Add 200 μl of chloroform (*see* **Note 5**) for each ml of TRIzol® Reagent. Shake in vortex, and centrifuge at 14,000 × *g* for 10 min at 4 °C.
5. Collect the aqueous phase at the upper part of the tube. RNA including the miRNA fraction is concentrated in this phase. This must be done by pipetting to avoid removal and dragging of the underlying residual organic phase.

6. Deposit aqueous phase into a new 1.5 ml microtube. Then add 500 μl of isopropanol for each ml of TRIzol, and place the tube at 4 °C overnight to allow RNA precipitation.
7. Centrifuge at 14,000 ×*g* for 10 min at 4 °C. Decant supernatant, and wash pellet with 500 μl of EtOH 75 %. To do this, shake the tube gently until the pellet is detached and centrifuge again at 14,000 ×*g* for 10 min at 4 °C. Repeat the wash step using EtOH at a concentration of 85 % and finally at 100 %.
8. After decanting the supernatant, allow the residual EtOH to evaporate, for 1 or 2 min, and then redissolve the pellet in 15–20 μl of RNAse-free water (*see* **Note 6**).
9. Quantify RNA solutions spectrophotometrically, and assess purity according to 260/280 and 260/230 ratios (*see* **Note 6**).

3.2 cDNA Preparation Specific for miRNAs

1. Calculate the RNA volume to obtain ~10–100 ng. The aliquot volume should not exceed 3 μl (*see* **Note 7**).
2. The reverse transcription (RT) reaction to obtain cDNA from mature miRNAs can be performed by using the TaqMan® MicroRNA reverse transcription kit and the Megaplex™ RT primers, human pool A/B v2.0 (Applied Biosystems) (*see* **Note 8**). Each 7.5 μl of reaction contains 3 μl of RNA (*see* **Notes 7** and **9**), 0.8 μl 10× RT primer pool A/B, 0.2 μl 25 mmol/l deoxynucleotide triphosphates, 1.5 μl 50 U/μl MultiScribe Reverse Transcriptase, 0.8 μl 10× RT buffer, 0.8 μl 25 mmol/l MgCl2, 0.1 μl 20 U/μl AB RNase inhibitor, and 0.2 μl RNAse-free water. Mix gently, and then centrifuge to bring all of the liquid to the bottom of the tube.
3. Incubate the tubes on ice for 5 min, and then run RT reaction in thermocycler under the following conditions: (16 °C for 2 min, 42 °C for 1 min, and 50 °C for 1 s) for 40 cycles and 85 °C for 5 min. If the RT product is intended to be used within a week it must be stored at 4 °C; otherwise, it must be stored at −20 °C, preferably, for no longer than 3 months.

3.3 cDNA Preamplification (See Note 10)

1. Preamplification reaction for each cDNA contains 2.5 μl of RT product, 12.5 μl 2× Preamp Master Mix, 2.5 μl Megaplex 10× preamp primer mix (250 mmol/l each), and 7.5 μl RNAse-free water. Mix gently, and then centrifuge to bring all of the liquid to the bottom of the tube.
2. Incubate the tubes on ice for 5 min, and then run preamplification reaction in thermocycler under the following conditions: 95 °C for 10 min, 55 °C for 2 min, 72 °C for 2 min, and (95 °C for 15 s, 60 °C for 4 min) for 14 cycles.
3. Dilute the preamplification product four-fold by adding 75 μl of RNAse-free water. If the product is intended to be used within a week it must be stored at 4 °C; otherwise, it must be stored at −20 °C, preferably, for no longer than 3 months.

3.4 High-Throughput Quantitative PCR for miRNAs

1. For the qPCR reaction, 9 μl diluted preamp product (*see* **Note 11**), 450 μl 2× TaqMan Master Mix, no UNG (Applied Biosystems), and 441 μl RNAse-free water are mixed together.
2. Dispense100 μl of the previous mix in each lane (*see* **Note 12**). Centrifuge at 1,000 × *g* for 1 min, and seal the card.
3. qPCR reactions can be done using a real-time PCR thermocycler under the following conditions: 50 °C for 2 min, 95 °C for 15 s, and (95 °C for 1 min, 60 °C for 2 min) for 40 cycles.
4. Results are stored into a SDS file. Data can be processed and analyzed (*see* **Note 13**) using commercially available suites or utility packages from open-source software.

4 Notes

1. Urine samples with volumes less than 20 ml often yield an insufficient amount of RNA. In order to guarantee a proper amount of RNA, it is recommended to use samples with a minimum of 30 ml containing visible sediment.
2. Utilizing urine as a source of DNA or RNA implies that different portions (liquid or solid) from the urine can be used. In this work we implemented a strategy to obtain total RNA including microRNAs from urine sediment.
3. The purpose of this step is to wash the sediment to optimize the yield of RNA in a subsequent step. This can be accomplished by removing as much as possible the salt compounds present in urine, such as urates.
4. RNA can be extracted by using a number of commercially available reagents for RNA extraction. The procedures described here consider the TRIzol® Reagent manufacturer's recommendations.
5. Reagents employed in this procedure, including TRIzol® Reagent and chloroform, are readily volatile and can be toxic if inhaled. In order to reduce exposure to these reagents, they must be manipulated under a laminar flow chamber.
6. RNA yield is highly variable between samples, so it is recommended that the starting volume of water to redissolve the RNA pellet be the minimum to optimize the RNA concentration. However, as the purity of RNA obtained from urine samples may be low, the amount of water can be increased as desired to improve the confidence of RNA concentration measurement, considering a balance between concentration and purity.
7. The experimenter should use approximately the same amount of RNA for each RT reaction. To this purpose, adjust the RNA amount based on the samples with lower concentration. Nonetheless, it is advisable to discard samples with a very low

purity (260/280 and 260/230 ratios < 1) and very low concentration ([RNA] <10 ng/ μl).

8. There are different commercially available methods to simultaneously measure the abundance of a number of transcripts. The Megaplex™ platform developed by Applied Biosystems allows the amplification and quantification of 667 mature miRNAs divided into two sets (A and B).
9. Sometimes, when the amounts of RNA volume or reagents are limited and reduction of inter-individual biological variability is desirable, RNA aliquots belonging to the same sample group can be pooled in preparation for the RT reaction and subsequent amplification and quantitation by qPCR. RNA aliquots from different samples should be pooled in approximately equimolar amounts.
10. One of the main concerns about using fluids, like urine, as a source of RNA is the low amount obtained from this kind of samples. To address this issue, one of the approaches that have been proposed is the use of a preamplification step, in which RT product is used as input. This step increases the ability for signal detection of small quantities of RNA during PCR reaction.
11. If no preamplification step is performed, 6 μl of the RT product is used as input for the qPCR reaction.
12. The card is divided into eight lanes. Each lane delivers the input fluid to a series of wells, each containing the probe and primers for a specific mature miRNA. For each assay, a total of 800 μl of loading mixture is employed. When dispensing, be careful not to introduce air bubbles or shake too strongly the card. Air bubbles can interfere or cause inaccurate signal detection during qPCR procedure.
13. The algorithm used for the selection of an optimal reference gene calculates the difference in *Ct* values (ΔCt) of each control candidate with respect to other reference genes and determines the standard deviation of all values for that specific control ΔCt. It is advisable to consider only those miRNAs with a *Ct* value less than 35.

References

1. WHO (2013) Cancer mortality and morbidity. http://www.who.int/gho/ncd/mortality_morbidity/cancer_text/en/index.html
2. Kufe DW, Pollock RE, Weichselbaum RR, Bast RC, Gansler TS, Holland JF, Frei E (eds) (2003) Cancer medicine, 6th edn. Hamilton, MA: BC Decker
3. Hanahan D, Weinberg RA (2000) The Hallmarks of cancer. Cell 100:57–70
4. Hanahan D, Weinberg RA (2011) Hallmarks of cancer: the next generation. Cell 144:646–674
5. GLOBOCAN (2008) Prostate cancer incidence, mortality and prevalence worldwide in 2008, GLOBOCAN Project. http://globocan.iarc.fr/.
6. Narain V, Cher ML, Wood DP Jr (2002) Prostate cancer diagnosis, staging and survival. Cancer Metastasis Rev 21:17–27

7. Hansen LL (2006) Molecular diagnosis of breast cancer, in prevention and treatment of age-related diseases. Springer, The Netherlands
8. Kohler C, Barekati Z, Radpour R et al (2011) Cell-free DNA in the circulation as a potential cancer biomarker. Anticancer Res 31: 2623–2628
9. Pathak AK, Bhutani M, Kumar S et al (2006) Circulating cell-free DNA in plasma/serum of lung cancer patients as a potential screening and prognostic tool. Clin Chem 52:1833–1842
10. Wroclawski ML, Serpa-Neto A, Fonseca NL et al (2013) Cell-free plasma DNA as biochemical biomarker for the diagnosis and follow-up of prostate cancer patients. Tumour Biol 34:2921–2927
11. Allegra A, Alonci A, Campo S et al (2012) Circulating microRNAs: New biomarkers in diagnosis, prognosis and treatment of cancer. Int J Oncol 41:1897–1912
12. Filella X, Foj L, Milá M et al (2013) PCA3 in the detection and management of early prostate cancer. Tumour Biol 34:1337–1347
13. Roobol MJ, Haese A, Bjartell A (2011) Tumour markers in prostate cancer III: biomarkers in urine. Acta Oncol 50(Suppl 1):85–89
14. Weber JA, Baxter DH, Zhang S et al (2010) The microRNA spectrum in 12 body fluids. Clin Chem 56:1733–1741
15. Filipowicz W, Bhattacharyya SN, Sonenberg N (2008) Mechanisms of post-transcriptional regulation by microRNAs: are the answers in sight? Nat Rev Genet 9:102–114
16. Rana TM (2007) Illuminating the silence: understanding the structure and function of small RNAs. Nat Rev Mol Cell Biol 8:23–36
17. Brennecke J, Hipfner DR, Stark A et al (2003) Bantam encodes a developmentally regulated microRNA that controls cell proliferation and regulates the proapoptotic gene hid in Drosophila. Cell 113:25–36
18. Bushati N, Cohen SM (2007) microRNA functions. Annu Rev Cell Dev Biol 23:175–205
19. Lee RC, Feinbaum RL, Ambros V (1993) The C. elegans heterochronic gene lin-4 encodes small RNAs with antisense complementarity to lin-14. Cell 75:843–854
20. Rajewsky N, Socci ND (2004) Computational identification of microRNA targets. Dev Biol 267:529–535
21. Calin GA, Dumitru CD, Shimizu M et al (2002) Frequent deletions and down-regulation of micro-RNA genes miR15 and miR16 at 13q14 in chronic lymphocytic leukemia. Proc Natl Acad Sci U S A 99:15524–15529
22. Calin GA, Sevignani C, Dumitru CD et al (2004) Human microRNA genes are frequently located at fragile sites and genomic regions involved in cancers. Proc Natl Acad Sci U S A 101:2999–3004
23. Ambs S, Prueitt RL, Yi M et al (2008) Genomic profiling of microRNA and messenger RNA reveals deregulated microRNA expression in prostate cancer. Cancer Res 68:6162–6170
24. Calin GA, Ferracin M, Cimmino A et al (2005) A MicroRNA signature associated with prognosis and progression in chronic lymphocytic leukemia. N Engl J Med 353:1793–1801
25. Calin GA, Liu CG, Sevignani C et al (2004) MicroRNA profiling reveals distinct signatures in B cell chronic lymphocytic leukemias. Proc Natl Acad Sci U S A 101:11755–11760
26. Fassan M, Baffa R, Palazzo JP et al (2009) MicroRNA expression profiling of male breast cancer. Breast Cancer Res 11:R58
27. Lorio MV, Visone R, Di Leva G et al (2007) MicroRNA signatures in human ovarian cancer. Cancer Res 67:8699–8707
28. Lu J, Getz G, Miska EA et al (2005) MicroRNA expression profiles classify human cancers. Nature 435:834–838
29. Tong AW, Fulgham P, Jay C et al (2009) MicroRNA profile analysis of human prostate cancers. Cancer Gene Ther 16:206–216
30. Volinia S, Calin GA, Liu CG et al (2006) A microRNA expression signature of human solid tumors defines cancer gene targets. Proc Natl Acad Sci U S A 103:2257–2261
31. Kahlert C, Kalluri R (2013) Exosomes in tumor microenvironment influence cancer progression and metastasis. J Mol Med 91:431–437
32. Mitchell PS, Parkin RK, Kroh EM et al (2008) Circulating microRNAs as stable blood-based markers for cancer detection. Proc Natl Acad Sci U S A 105:10513–10518

Chapter 8

DNA Methylation Analysis of Steroid Hormone Receptor Genes

Ignacio Camacho-Arroyo, Valeria Hansberg-Pastor, and Mauricio Rodríguez-Dorantes

Abstract

Steroid hormone receptors (SHR) are important transcription factors for regulating different physiological and pathological processes. Their altered expression has been strongly associated to cancer progression. Epigenetic marks such as DNA methylation have been proposed as one of the regulatory mechanisms for SHR expression in cancer. DNA methylation occurs at CpG dinucleotides, which form clusters known as CpG islands. These islands are mostly observed at promoter regions of housekeeping genes, and their aberrant methylation in cancer cells is associated with silencing of tumor-suppressor gene expression. SHR genes are characterized for presenting alternative promoters with different CpG island content, which are prone to be methylated. The method of choice for studying DNA methylation is bisulfite sequencing, since it provides information about the methylation pattern at single-nucleotide level. The method is based on the deamination of cytosine residues to uracil after treatment with sodium bisulfite. The converted DNA is amplified by a polymerase chain reaction, cloned, and sequenced. Here, we describe a protocol for bisulfite sequencing suitable for analyzing different CpG regions in SHR genes.

Key words Epigenetic, DNA methylation, Bisulfite sequencing, Steroid hormone receptor, CpG islands

1 Introduction

The steroid hormone receptor (SHR) family of transcription factors is responsible for regulating different biological processes such as development, metabolism, and reproduction. The SHR family includes the estrogen receptor (ERα and ERβ), the glucocorticoid receptor (GR), the mineralocorticoid receptor (MR), the progesterone receptor (PR-A and PR-B), and the androgen receptor (AR) [1]. Differential SHR expression and activity have been associated with cancer susceptibility, progression, and prognosis. Thus, numerous studies have focused on identifying key factors affecting their expression [2–4].

Martha Robles-Flores (ed.), *Cancer Cell Signaling: Methods and Protocols*, Methods in Molecular Biology, vol. 1165, DOI 10.1007/978-1-4939-0856-1_8, © Springer Science+Business Media New York 2014

Epigenetic regulation through DNA methylation is a well-known mechanism affecting gene expression and was the first epigenetic mark related to cancer [5]. DNA methylation consists of the addition of a methyl group to the 5-carbon position of a cytosine (C) forming a 5-methylcytosine residue (5mC). This covalent modification occurs almost exclusively in a cytosine–guanine sequence [6, 7]. The pair of nucleotides is known as CpG dinucleotide, and it is frequently observed in clusters called CpG islands, which are defined as regions of more than 200 bp with at least 50 % of G + C content and a ratio of 0.6 or more between observed and expected CpG dinucleotide frequency. CpG islands are mostly observed in the promoter regions of housekeeping and developmental genes and in half of all tissue-specific genes. Generally, housekeeping genes are unmethylated in normal cells, while tissue-specific genes may be unmethylated or methylated depending on their requirement for lineage commitment [8, 9]. There are also CpG-poor promoter regions found in tissue-specific genes that are also susceptible to be regulated by DNA methylation, and in some cases methylation is required for activation of transcription [10–12]. In cancer cells, hypermethylation of the CpG islands in the promoter regions of tumor-suppressor genes is well documented, while for tissue-specific genes a hyper- or a hypomethylation of the promoter regions has been observed [5, 13]. This aberrant methylation pattern has been studied for different types of cancer cells, and it has been a helpful tool for identifying DNA methylation-dependent biomarkers in clinical oncology [7].

The gold standard method for analyzing DNA methylation at the nucleotide level is bisulfite sequencing. Sodium bisulfite ($NaHSO_3$) is used to convert unmethylated cytosine (C) residues to uracil residues in a single-stranded DNA. The conversion with bisulfite requires fulfilling three chemical reactions: sulfonation of C to cytosine-6-sulfonate, deamination to uracil-6-sulfonate, and desulfonation to uracil. However, sodium bisulfite has no effect on 5mC residues, so they remain unchanged. The bisulfite-treated DNA is then amplified with specific primers designed for converted DNA, and the purified polymerase chain reaction (PCR) products are cloned and sequenced in order to analyze the methylation pattern at the nucleotide level [14–17].

Many SHR have different isoforms as a consequence of the use of alternative promoters, each with its own transcription start site [18–21]. The interplay between CpG methylation and transcript variability in the context of multiple promoters with complex methylation patterns is poorly understood. Here we show a standard bisulfite sequencing protocol for analyzing the methylation pattern of SHR genes with different CpG island content in their promoter regions.

2 Materials

2.1 DNA Extraction

1. DNA extraction kit.
2. Equipment for sample disruption and homogenization.
3. Microcentrifuge.
4. Spectrophotometer.

2.2 Bisulfite Conversion and PCR Amplification and Purification

1. Bisulfite reaction kit for converting C residues into uracil from genomic DNA.
2. Distilled molecular biology-grade water (dH_2O).
3. PCR Mastermix (reaction buffer, Taq DNA polymerase, dNTP mix, $MgCl_2$, and dH_2O) and primers for a converted DNA.
4. Agarose gel, ethidium bromide, and electrophoresis apparatus.
5. DNA recovery kit for purification of target PCR fragment from agarose gels.
6. PCR thermal cycler.
7. UV transilluminator.

2.3 Cloning and Sequencing of Target DNA Fragment

1. A linearized vector system for cloning the target PCR fragment.
2. High-efficiency competent *E. coli* cells.
3. LB medium (for 1 L, pH 7.0: 10 g bactotryptone, 5 g yeast extract, 5 g NaCl, and autoclave; add ampicillin solution to a final concentration of 100 μg/ml; for solid medium add 15 g agar).
4. Isopropyl-b-D-thiogalactoside (IPTG): Dissolve in water to 100 mM, filter-sterilize, and store in aliquots at 4 °C. LB plates are supplemented with 0.5 mM IPTG.
5. X-Gal: Dissolve in dimethyl sulfoxide (DMSO) at 50 mg/ml, filter-sterilize, protect from light, and store in aliquots at −20 °C. LB plates are supplemented with 50 μg/ml of X-Gal.
6. SOC medium (for 100 ml, pH 7.0: 2.0 g bactotryptone, 0.5 g yeast extract, 1 ml 1 M NaCl, 0.25 ml 1 M KCl, autoclave and add 1 ml 1 M $MgCl_2$, 1 ml 1 M $MgSO_4$, 2 ml 1 M glucose, all three filter-sterilized).
7. Bacterial shaker incubator at 37 °C.
8. Bacteria cell culture flasks and Petri dishes.
9. Restriction endonucleases.
10. Plasmid purification kit.
11. Primers that anneal upstream and downstream of clone insertion site in the vector for sequencing analysis.

3 Methods

3.1 Primer Design for DNA Methylation Analysis

In cancer cells, a correlation between loss of expression of a certain gene and methylation of its promoter regions is often observed. When selecting a genomic DNA region for methylation studies, it should first comprise a gene expression study and a database (e.g., NCBI, http://www.ncbi.nlm.nih.gov/) search to locate the promoter sequence and the gene transcription start site (TSS). The selected genomic sequence is then analyzed for potential CpG islands and used for the design and selection of primers (*see* **Note 1**). There are multiple online software available for primer design and in silico analysis of the region of interest. Some examples are the following:

- *CpG island Searcher* (http://cpgislands.usc.edu/): Screens DNA sequences for CpG islands and generates a graphical output [22].
- *MethPrimer* (http://www.urogene.org/methprimer/): Used for designing bisulfite conversion-based PCR primers and generates a graphic view of the location of the primers in the predicted CpG islands of the region of interest [23].
- *BiSearch* (http://bisearch.enzim.hu/): Gives access to PCR primer tests and design and is used for both bisulfite-converted and non-modified sequences.
- *MethGraph* (http://mellfire.ugent.be/public/methgraph/index.php): Used for PCR primer validation and graphic visualization of the position of the primers in the context of the CpG island [24].

The DNA sequence obtained after a bisulfite conversion will be different from the untreated genomic sequence. The regions prone to be methylated will still have Cs at CpGs, while the regions that are unmethylated will have no Cs and thus will be A/T rich. DNA-converted primers are designed to a region without CpGs so that PCR amplification is not dependent on methylation status. Therefore, the primers should not contain any CpG sites and must edge the CpG islands in order to amplify the target internal sequence. For CpG-poor promoters the sequence of interest should include as many CpGs as possible (*see* **Note 2**).

3.2 DNA Extraction

Genomic DNA from tissue samples, cultured cells, and formalin-fixed paraffin-embedded (FFPE) tissues can be purified by using a number of commercially available DNA extraction kits. The purity of genomic DNA is important for a proper bisulfite conversion, and the amount of input DNA for an optimal bisulfite conversion should be between 500 ng and 2 μg of DNA (*see* **Note 3**). FFPE tissues are more difficult to use for bisulfite conversion, as the tissue fixing process causes DNA modifications, including DNA–protein cross-links and DNA fragmentation [25, 26].

3.3 Bisulfite Conversion and PCR Amplification

1. Quantify the DNA sample, and calculate the required volume in order to have 1 μg of input DNA.
2. DNA bisulfite conversion reaction is traditionally done through several steps, including freshly prepared NaOH solution to denaturalize the DNA and a $NaHSO_3$ solution to convert Cs into uracil [27]. The bisulfite reaction can also be performed using different available commercial kits. We highly recommend the kit EZ DNA Methylation-Gold Kit from Zymo Research (*see* **Note 4**) as it integrates temperature DNA denaturation and bisulfite conversion into one step and the obtained converted DNA (~10 μl) is directly used as a template for PCR amplification. According to the kit, for bisulfite conversion perform the following steps in a thermal cycler: 98 °C for 10 min, 64 °C for 2.5 h, and 4 °C storage for up to 20 h (*see* **Note 5**).
3. PCR amplification of bisulfite-treated DNA can be performed as a standard PCR reaction (*see* **Note 6**). 1–3 μl of the bisulfite-converted DNA is used as template for PCR in a 25 μl reaction mixture. However, the conditions should be optimized depending on the CpG content of the region of interest (*see* **Note 7**). It is recommended to use a heat-activated Taq polymerase to prevent amplification of nonspecific fragments (*see* **Note 8**).
4. The extension times for amplifying converted DNA are longer because of the presence of uracil, which decreases the rate of DNA polymerization. Perform the following program for PCR amplification: 10 min at 95 °C, 35–40 cycles of 30 s at 95 °C, 1 min at the annealing temperature, 1 min at 72 °C, and a final extension time of 5 min at 72 °C (*see* **Note 9**).
5. Perform separate PCR reactions, and pool the reactions once amplification conditions have been optimized. Performing and pooling multiple PCR reactions reduce stochastic events that can interfere with the amplification of individual template molecules.
6. Verify the PCR results by 1.5–2 % agarose gel electrophoresis, and a single band of the expected size should be obtained. However, unspecific amplification products are frequently observed.
7. Purify the amplified DNA fragment from the agarose gel with a DNA recovery kit. The purified DNA can be stored at −20 °C until needed (*see* **Note 10**).

3.4 Cloning and Cell Transformation

1. For cloning the PCR product we recommend to use a linearized vector system and high-efficiency competent *E. coli* cells. The linearized vector should have a single 3′-terminal thymidine at both ends in order to improve ligation of PCR products (*see* **Note 11**).

2. Ligate insert and vector at a 1:1 molar ratio using 1 Weiss unit of T4 DNA ligase per 50 ng vector in the buffer supplied by the manufacturer, and incubate the ligation at 4 °C overnight. Include the following controls:
   - Positive control: control insert included in the kit.
   - Negative control: without PCR product.
3. To perform transformation, mix the volume of each ligation reaction with 200 μl of competent *E. coli* cells, incubate for 1 h on ice, then give a heat shock in a water bath for 50 s at 42 °C, and incubate for 2 min at 4 °C.
4. Add 800 μl of room-temperature SOC medium, invert the tubes to mix, and incubate for 1.5 h at 37 °C with shaking (~250 rpm).
5. Prepare LB/ampicillin/IPTG/X-Gal plates for each ligation reaction, and equilibrate the plates at room temperature. It is recommended to freshly prepare the plates before the ligation reaction (*see* **Note 12**).
6. Plate 200 μl of each transformation culture by duplicate in LB/ampicillin/IPTG/X-Gal plates. For a higher number of colonies, cells may be pelleted by centrifugation at 2,800 × *g* for 2 min and resuspended in 200 μl of SOC medium, and 100 μl are plated.
7. Incubate the plates overnight (16–24 h) at 37 °C. Longer incubations or storage of plates at 4 °C (after 37 °C overnight incubation) may be used to ease blue color development. White colonies generally contain inserts; however, inserts may also be present in blue colonies (*see* **Note 13**).

3.5 Vector Insert Analysis and Sequencing

1. Isolate five or more transformant colonies per plate, and suspend each in 5 ml liquid LB medium plus ampicillin. Incubate at 37 °C overnight with shaking, and centrifuge at 1,800 × *g* for 5 min to pellet the bacterial cells.
2. Isolate the plasmid with a purification kit and store at −20 °C until needed.
3. Digest the plasmid with restriction endonucleases depending on the restriction sites present in the vector, incubate for 2–3 h at 37 °C, and verify the presence of the insert by agarose gel electrophoresis (*see* **Note 14**).
4. Sequencing is commonly performed by automated techniques at a core facility or off-site. 10–20 purified plasmid samples are commonly submitted for sequencing (*see* **Note 15**). Standard primers such as T7 or SP6 (depending on the cloning vector used) are required. Nevertheless, the samples should be prepared according to the guidelines supplied by the sequencing facilities (*see* **Note 16**).

5. The DNA methylation pattern is interpreted by comparing the sequencing results with the original DNA sequence. The presence of a C peak in the sequence indicates the occurrence of a 5mC (*see* **Note 17**). If both C and T peaks appear, this indicates a partial methylation status or a potentially incomplete bisulfite conversion. For methylation sequence analysis, there are available software such as BiQ Analyzer (http://biq-analyzer.bioinf.mpi-inf.mpg.de/) [28] or BISMA (http://services.ibc.uni-stuttgart.de/BDPC/BISMA/) [29] that generate publication quality diagrams (*see* **Note 18**).

4 Notes

1. The primers designed for a converted DNA must avoid potential hairpin structures and possible primer dimers, specifically because of the high A/T rate after bisulfite conversion. The melting temperature should be around 50–54 °C. The primer size (25–35 bp) is usually bigger than that for regular PCR, and the length of the PCR product should not exceed 400 bp due to potential DNA degradation during the bisulfite treatment [23].
2. Avoid long poly-T sequences (>9 bases) in the bisulfite-converted template; it can result in nonspecific primer annealing or poor amplification due to polymerase slippage. This is particularly important for CpG-poor regions.
3. The quality and quantity of DNA are important in the bisulfite reaction. Up to 2 μg genomic DNA extracted from cultured cells or fresh tissue may be used, but usually 500 ng to 1 μg of high-quality DNA is recommended. Although a high percent (up to 70 %) of DNA is lost during bisulfite treatment, too much genomic DNA can also induce incomplete bisulfite conversion.
4. Bisulfite conversion with the EZ DNA Methylation-Gold Kit from Zymo Research is a very good option because of the quality of the obtained converted DNA. The bisulfite-transformed DNA is eluted with 10 μl according to the manufacturer's protocol but can be eluted with up to 40 μl. Nevertheless, this could be an impediment if the original DNA sample is very limited. Modifications to the conventional bisulfite protocol for very low amount of sample have been optimized [30–32].
5. For difficult samples to convert with bisulfite, restriction enzyme digestion has been suggested [16].
6. The noncomplementary DNA conformation after bisulfite treatment causes the DNA to be unstable, and repeated freezing–thawing

cycles should be avoided. Converted DNA can be stored at −80 °C for up to 12 months. Nevertheless, freshly made bisulfite-converted DNA is recommended to yield optimal results for subsequent PCR reactions.

7. The optimization of bisulfite PCR conditions can be time consuming and laborious. The bisulfite-treated DNA has reduced specificity due to its high A/T and low G/C composition. To enhance amplification, it is recommended to make an annealing temperature curve or to lower the calculated primer Tm. Still, nonspecific PCR amplification with bisulfite-converted DNA is common.

8. Bisulfite-converted DNA is single stranded, due to loss of complementarity, and prone to form secondary structures. Addition of PCR enhancers such as betaine (1 M) or DMSO (1–10 %) can increase the yield of specific PCR product [33].

9. For low yields of PCR products, the amount of DNA can be modified or the extension time and number of cycles can be increased to improve yield. Also, re-amplification of the PCR product or nested PCR can be helpful for CpG-poor regions or if a limited amount of DNA sample is used.

10. For cloning of bisulfite PCR products, it is recommended to gel-purify the desired fragment in order to avoid unspecific PCR products. The UV exposure in the transilluminator can form pyrimidine dimers that may interfere with the ligation process [34]. Avoid exposure to shortwave UV as much as possible or re-amplify the purified PCR fragment.

11. For efficient ligation of the PCR product to the linearized vector, verify that the used DNA polymerases create a 3′-A overhang. We recommend the HotStarTaq DNA Polymerase from Qiagen or the AmpliTaq Gold DNA Polymerase from Invitrogen. Also, use fresh PCR products since ligation efficiency is reduced when long-stored PCR products are used.

12. The LB plates can be alternatively prepared with antibiotic, and before use, IPTG and X-Gal are spread all over the plate surface and allowed to absorb for 1 h at room temperature.

13. The positive control ligation should yield at least 100 colonies, from which a 60 % should be white. Poor positive control results indicate a failed ligation due to improper ligation conditions, nuclease contamination, or reagent failure.

14. Colony PCR can be done instead of using restriction endonucleases. Colonies are picked after overnight incubation with a sterile pipette tip or toothpick. The toothpick is dipped into a PCR tube containing a PCR reaction mix with the standard primers that anneal to the sites flanking the insertion site (e.g., M13 or T7). The amount of cells should be small, and sufficient

mixing will result in cell lysis. The colony PCR fragments will include the standard forward and reverse primer sequences, which make them suitable for sequencing [27].

15. Cloning PCR products gives the methylation status of individual DNA molecules. PCR products may also be directly sequenced, but consider that the amplified product is a pool of DNA molecules with different methylation patterns. If direct sequencing is chosen, the PCR product has to be cleaned and purified (QIAquick PCR Purification Kit from Qiagen), and the sample must be sequenced several times in order to obtain the methylation percentage of each CpG site.
16. Forward and reverse sequencing reactions are highly recommended to avoid possible misreading.
17. Automated DNA sequencers generate a four-color chromatogram showing the results of the sequencing run. Computer programs that interpret the data can sometimes misread the bases, e.g., at a CpG dinucleotide. It is recommended to check the chromatogram of the corresponding clones and correct the errors if possible.
18. The original DNA sequence should be used without the primer sequence in order to compare only the region of interest with the CpG sites.

References

1. Griekspoor A, Zwart W, Neefjes J et al (2007) Visualizing the action of steroid hormone receptors in living cells. Nucl Recept Signal 5:1–9
2. Ahmad N, Kumar R (2011) Steroid hormone receptors in cancer development: a target for cancer therapeutics. Cancer Lett 300:1–9
3. Green CD, Han JD (2011) Epigenetic regulation by nuclear receptors. Epigenomics 3: 59–72
4. Mani SK, Mermelstein PG, Tetel MJ et al (2012) Convergence of multiple mechanisms of steroid hormone action. Horm Metab Res 44:569–576
5. Berdasco M, Esteller M (2010) Aberrant epigenetic landscape in cancer: how cellular identity goes awry. Dev Cell 19:698–711
6. Korlach J, Turner SW (2012) Going beyond five bases in DNA sequencing. Curr Opin Struct Biol 22:251–261
7. Portela A, Esteller M (2010) Epigenetic modifications and human disease. Nat Biotechnol 28:1057–1068
8. Deaton AM, Bird A (2011) CpG islands and the regulation of transcription. Genes Dev 25:1010–1022
9. Sandoval J, Esteller M (2012) Cancer epigenomics: beyond genomics. Curr Opin Genet Dev 22:50–55
10. Chatterjee R, Vinson C (2012) CpG methylation recruits sequence specific transcription factors essential for tissue specific gene expression. Biochim Biophys Acta 1819:763–770
11. Doi A, Park IH, Wen B et al (2009) Differential methylation of tissue- and cancer-specific CpG island shores distinguishes human induced pluripotent stem cells, embryonic stem cells and fibroblasts. Nat Genet 41:1350–1353
12. Nagae G, Isagawa T, Shiraki N et al (2011) Tissue-specific demethylation in CpG-poor promoters during cellular differentiation. Hum Mol Genet 20:2710–2721
13. Irizarry RA, Ladd-Acosta C, Wen B et al (2009) The human colon cancer methylome shows similar hypo- and hypermethylation at conserved tissue-specific CpG island shores. Nat Genet 41:178–186
14. Brait M, Sidransky D (2011) Cancer epigenetics: above and beyond. Toxicol Mech Methods 21:275–288
15. Hayatsu H (2008) Discovery of bisulfite-mediated cytosine conversion to uracil, the key

reaction for DNA methylation analysis—a personal account. Proc Jpn Acad Ser B Phys Biol Sci 84:321–330

16. Pappas JJ, Toulouse A, Bradley WE (2009) A modified protocol for bisulfite genomic sequencing of difficult samples. Biol Proced Online 11:99–112
17. Zhang Y, Rohde C, Tierling S et al (2009) DNA methylation analysis by bisulfite conversion, cloning, and sequencing of individual clones. Methods Mol Biol 507:177–187
18. Hansberg-Pastor V, Gonzalez-Arenas A, Pena-Ortiz MA et al (2013) The role of DNA methylation and histone acetylation in the regulation of progesterone receptor isoforms expression in human astrocytoma cell lines. Steroids 78: 500–507
19. Sasaki M, Kaneuchi M, Fujimoto S et al (2003) Hypermethylation can selectively silence multiple promoters of steroid receptors in cancers. Mol Cell Endocrinol 202:201–207
20. Turner JD, Pelascini LP, Macedo JA et al (2008) Highly individual methylation patterns of alternative glucocorticoid receptor promoters suggest individualized epigenetic regulatory mechanisms. Nucleic Acids Res 36:7207–7218
21. Breslin MB, Geng CD, Vedeckis WV (2001) Multiple promoters exist in the human GR gene, one of which is activated by glucocorticoids. Mol Endocrinol 15:1381–1395
22. Takai D, Jones PA (2003) The CpG island searcher: a new WWW resource. In Silico Biol 3:235–240
23. Li LC, Dahiya R (2002) MethPrimer: designing primers for methylation PCRs. Bioinformatics 18:1427–1431
24. Lefever S, Hoebeeck J, Pattyn F et al (2010) methGraph: a genome visualization tool for PCR-based methylation assays. Epigenetics 5:159–163
25. Bonin S, Hlubek F, Benhattar J et al (2010) Multicentre validation study of nucleic acids extraction from FFPE tissues. Virchows Arch 457:309–317
26. Bonin S, Stanta G (2013) Nucleic acid extraction methods from fixed and paraffin-embedded tissues in cancer diagnostics. Expert Rev Mol Diagn 13:271–282
27. Darst RP, Pardo CE, Ai L et al (2010) Bisulfite sequencing of DNA. Curr Protoc Mol Biol Chapter 7, Unit 7 9:1–17.
28. Bock C, Reither S, Mikeska T et al (2005) BiQ analyzer: visualization and quality control for DNA methylation data from bisulfite sequencing. Bioinformatics 21:4067–4068
29. Rohde C, Zhang Y, Reinhardt R et al (2010) BISMA-fast and accurate bisulfite sequencing data analysis of individual clones from unique and repetitive sequences. BMC Bioinformatics 11:230. doi:10.1186/1471-2105-11-230
30. Boyd VL, Zon G (2004) Bisulfite conversion of genomic DNA for methylation analysis: protocol simplification with higher recovery applicable to limited samples and increased throughput. Anal Biochem 326:278–280
31. Dallol A, Al-Ali W, Al-Shaibani A et al (2011) Analysis of DNA methylation in FFPE tissues using the MethyLight technology. Methods Mol Biol 724:191–204
32. Pedersen IS, Krarup HB, Thorlacius-Ussing O et al (2012) High recovery of cell-free methylated DNA based on a rapid bisulfite-treatment protocol. BMC Mol Biol 13:12. doi:10.1186/1471-2199-13-12
33. Ralser M, Querfurth R, Warnatz HJ et al (2006) An efficient and economic enhancer mix for PCR. Biochem Biophys Res Commun 347:747–751
34. Alba FJ, Bermudez A, Daban JR (2001) Green-light transilluminator for the detection without photodamage of proteins and DNA labeled with different fluorescent dyes. Electrophoresis 22:399–403

Chapter 9

Control of Oncogenic miRNA Function by Light-Activated miRNA Antagomirs

Colleen M. Connelly and Alexander Deiters

Abstract

MicroRNAs (miRNAs) are single stranded noncoding RNAs of approximately 22 nucleotides that act as posttranscriptional gene regulators by binding partially complementary sequences in the 3′ untranslated region (3′-UTR) of target messenger RNAs (mRNAs). MicroRNAs regulate many biological processes including embryonal development, differentiation, apoptosis, and proliferation and the targets of miRNAs range from signalling proteins and transcription factors to RNA binding proteins. Recently, variations in the expression of certain miRNAs have been linked to a variety of human diseases including cancer and viral infections, validating miRNAs as potential targets for drug discovery. Several tools have been developed to control the function of individual miRNAs and have been applied to study their biological role and therapeutic potential; however, common methods lack a precise level of control that allows for the study of miRNA function with high spatial and temporal resolution. Toward this goal, a light-activated miRNA antagomir for mature miR-21 was developed through the site-specific installation of caging groups on the bases of selected nucleotides. Installation of caged nucleotides led to complete inhibition of the antagomir–miRNA hybridization and inactivation of antagomir function. The miRNA-inhibitory activity of the caged antagomirs was fully restored upon decaging through a brief UV irradiation. The synthesized antagomir was applied to the photochemical regulation of miR-21 function in mammalian cells. Moreover, spatial and temporal control over antagomir activity and thus miR-21 function was obtained in mammalian cells. The presented approach enables the precise regulation of miRNA function with unprecedented spatial and temporal resolution using UV irradiation and can be readily extended to any miRNA of interest.

Key words MicroRNA, Antagomir, Light-activation, Caged oligonucleotides, MicroRNA reporter, Luciferase, EGFP

1 Introduction

MicroRNAs (miRNAs) are a recently discovered class of noncoding RNAs of approximately 22 nucleotides that regulate gene expression in a sequence-specific fashion by binding partially complementary sequences most commonly in the 3′ untranslated regions (3′-UTRs) of target mRNAs. MicroRNAs downregulate gene function by inhibiting translation, accelerating the degradation of their target mRNAs, or mediating deadenylation of the mRNA [1, 2].

Martha Robles-Flores (ed.), *Cancer Cell Signaling: Methods and Protocols*, Methods in Molecular Biology, vol. 1165, DOI 10.1007/978-1-4939-0856-1_9,

After transcription from the genome, miRNAs undergo several posttranscriptional processing steps via a dedicated miRNA pathway to produce mature miRNAs. Since miRNAs can bind to the 3′-UTR of target mRNAs with imperfect complementarity, each miRNA can target many different mRNA transcripts [3], and it is estimated that miRNAs control more than 30 % of all genes and are involved in almost every genetic pathway [4]. Biological processes that are regulated by miRNAs include embryonal development, cell differentiation, apoptosis, and proliferation [5] and the targets of miRNAs range from signalling proteins, metabolic enzymes, and transcription factors to RNA binding proteins [6]. Because the role of miRNAs is to regulate the expression of specific mRNAs, the misregulation of intracellular miRNA levels leads to the misregulation of their mRNA targets, which can have severe implications on cellular homeostasis. The aberrant expression of certain miRNAs has been linked to a wide range of human diseases, including cancer, cardiovascular disease, immune disorders, and viral infections [7–11]. In particular, the misregulation of miRNAs has been associated with the initiation, progression, and metastasis of cancer, and certain miRNAs have been linked to a resistance to apoptosis [12].

Due to the extensive involvement of miRNAs in biological processes and human diseases, several regulatory tools to control the function of individual miRNAs have been developed and applied to study their biogenesis and therapeutic potential [13–16]. Amongst these tools are miRNA antisense oligonucleotides [17, 18], miRNA sponges or decoys [19], miRNA expression vectors [20], and small molecule regulators [21, 22]. The most commonly used tools are miRNA antisense oligonucleotides, or antagomirs, which are chemically modified oligonucleotides that are complementary to a miRNA and act as competitive inhibitors of the target mRNA binding. Several chemical modifications have been introduced into antagomir structures in order to render the oligonucleotides more resistant to degradation, to increase binding affinity, and to improve overall activity. These include 2′ sugar modifications such as 2′-O-methyl (2′OMe), 2′-O-methoxyethyl, 2′-fluoro, and locked nucleic acids (LNAs). In addition, phosphorothioate bonds are typically included in the oligonucleotide backbone [13, 23, 24].

Light-activated antagomirs have potential applications in the dissection of miRNA involvement in spatiotemporally regulated cellular processes, such as cell–cell signalling and the cell cycle. Light has previously been applied to the control of various biological processes since it can be regulated in timing, location, wavelength, and intensity [25–29]. Light-control of oligonucleotide function has been achieved through the installation of caging groups on nucleobases, thereby disrupting hybridization until the caging groups are removed through UV irradiation [30–36]. Here, we are applying the nucleobase-caging methodology to the

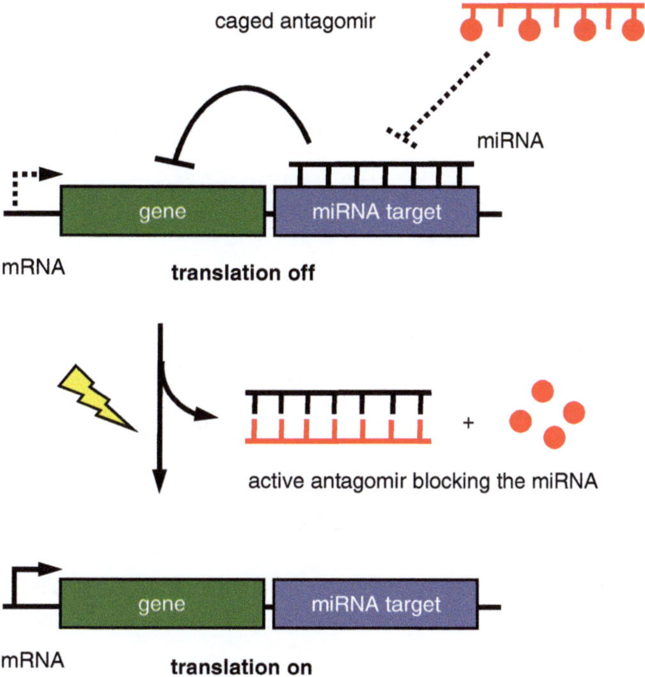

Fig. 1 Regulation of gene expression through photocontrol of endogenous miRNAs with nucleobase-caged antagomirs. A miRNA recognizes a complementary target sequence downstream of a gene and inhibits its expression. A caged antagomir has no effect on miRNA-mediated gene silencing until the caging groups are removed by a brief exposure to UV light. The decaged antagomir then competitively binds the miRNA and blocks its function, leading to the activation of gene expressed. Adapted from *Mol. BioSyst.* **2012**, *8*, 2987 – Reproduced by permission of The Royal Society of Chemistry

photochemical control of antagomir and thus miRNA function. As shown in Fig. 1, a miRNA recognizes a complementary target sequence downstream of a gene and inhibits its expression. A caged antagomir has no effect on miRNA-mediated gene silencing until the caging groups are removed by a brief exposure to UV light. The decaged antagomir then competitively binds the miRNA and blocks its function, leading to the activation of gene expression. The gene of interest can be any endogenous gene that is targeted by a particular miRNA or it can be an exogenous reporter gene, such as luciferase or EGFP, which has been engineered to respond to the miRNA of interest.

In order to demonstrate photochemical control of miRNA function, microRNA miR-21 was selected as a target. MicroRNA miR-21 has been linked to several human malignancies and overexpression of miR-21 has been observed in glioblastomas, breast, pancreatic, cervical, colorectal, ovarian, lung, and hepatocellular cancers [7]. MiR-21 functions as an anti-apoptotic factor in cancer cells and a dependence of tumor growth on miR-21

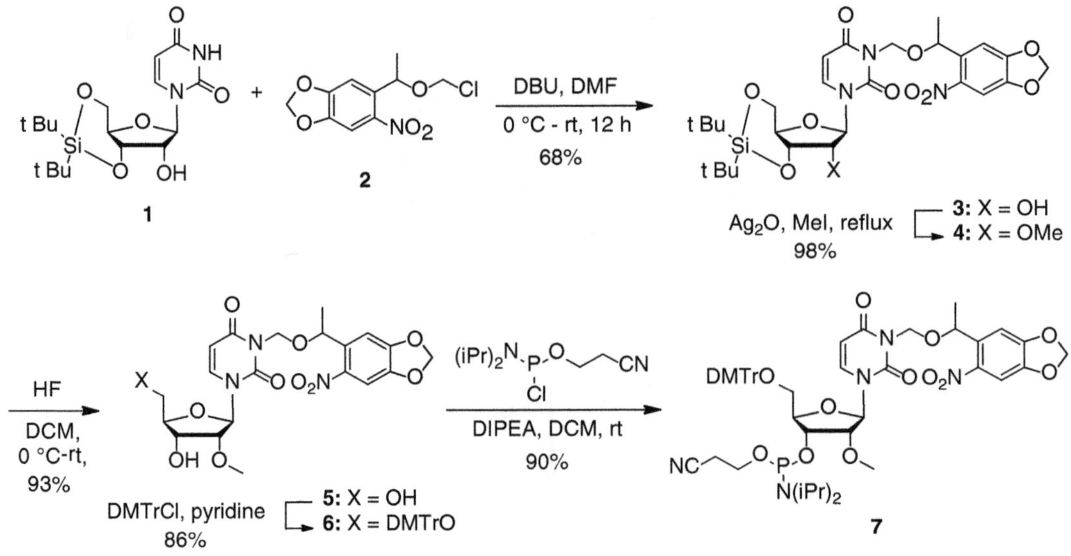

Fig. 2 Synthesis of the NPOM-caged 2′OMe uridine phosphoramidite 7. Adapted from *Mol. BioSyst.* **2012**, *8*, 2987 – Reproduced by permission of The Royal Society of Chemistry

expression has been demonstrated in a mouse model [37]. The involvement of miR-21 in cancer and promising in vivo results make miR-21 a potential target for the development of fundamentally new cancer therapeutics and the development of new photochemical tools to study miR-21 has implications in further studies of its involvement in cancer and precisely controlled therapeutics.

To develop light-activated antagomirs, we designed a perfectly complementary oligonucleotide with 2′OMe modified nucleotides and phosphorothioate backbones targeting mature miR-21 [38]. In order to site-specifically install a light-removable caging group, a 2′-OMe NPOM (6-nitropiperonyloxymethyl)-caged uridine phosphoramidite (7) was synthesized (Fig. 2) and was incorporated at selected locations in the miR-21 antagomir sequence.

The oligonucleotide was designed based on previous investigations regarding the photocontrol of DNA:DNA and DNA:RNA hybridization, which showed that the presence of 3–4 caging groups distributed evenly throughout the oligomer is sufficient to disrupt hybridization [32–34]. Both non-caged and caged 2′OMe phosphorothioate antagomirs were synthesized for mature miR-21 (Table 1). The caged miR-21 antagomir was designed to contain four NPOM-caged 2′OMe uridine nucleotides (Table 1) equally spaced throughout the oligonucleotide sequence including caging groups within the miRNA seed region, which resulted in the complete suppression of antagomir activity prior to UV irradiation. The photochemical control of antagomir activity was investigated in mammalian cell culture using a *Renilla* luciferase sensor for mature miR-21 based on the psiCHECK-2 (Promega) reporter plasmid.

Table 1
Sequences of miR-21 and the synthesized miR-21 antagomirs

miRNA target/antagomir	Sequence 5′ → 3′
Mature miR-21	UAGCUUAUCAGACUGAUGUUGA
miR-21 antagomir	mA*mU*mC*mA*mA*mC*mA*mU*mC*mA*mG*mU*mC *mU*mG*mA*mU*mA*mA*mG*mC*mU*mA
Caged miR-21 antagomir	mA***mU***mC*mA*mA*mC*mA***mU***mC*mA*mG*mU *mC*mU*mG*mA***mU***mA*mA*mG*mC***mU***mA

An *asterisk* indicates a phosphorothioate bond, an m indicates a 2′OMe modified nucleotide, and the *bold* mU denotes a NPOM-caged uridine nucleotide

The psiCHECK-2 vector was selected because it contains both a *Renilla* luciferase as well as an independently transcribed firefly luciferase reporter gene, which can be used for normalization purposes in order to account for variation in transfection efficiency and cell viability. The complementary sequence of miR-21 was inserted downstream of the *Renilla* luciferase gene (Fig. 1) [38]. Thus, the presence of mature miR-21 will lead to a decrease in the *Renilla* luciferase signal enabling the detection of endogenous miR-21 levels. An active antagomir would act as a competitive inhibitor of miRNA target binding and would lead to an increase in *Renilla* luciferase expression. In order to investigate the ability to spatially control antagomir activity using localized UV illumination, a sensor that reports the spatial activity of miR-21 function was developed by inserting the miR-21 complement target sequence downstream of an EGFP gene in the pEGFP-C2 (Clontech) plasmid [38].

The synthesized miR-21 antagomirs were assayed in Huh7 cells transiently transfected with the *Renilla* luciferase sensor for miR-21. In order to determine the best irradiation conditions for light activation through decaging, a time course experiment was performed in a 96-well format and an irradiation time of 2–5 min was identified to efficiently activate miR-21 antagomir function. Huh7 cells were then co-transfected with the psiCHECK-miR21 plasmid and either the non-caged or caged miR-21 antagomir in triplicate and were either kept in the dark or irradiated at 365 nm for 5 min to confirm that decaging fully restores antagomir function to the level of the non-caged antagomir (Fig. 3a). The non-caged miR-21 antagomir produced a threefold increase in relative luciferase units (RLU) compared to a transfection control containing no antagomir. Prior to decaging, the caged miR-21 antagomir showed RLU values equal to the transfection control, indicating that the presence of the four NPOM-caging groups prevents that antagomir from inhibiting miR-21 function. After UV irradiation, the decaged miR-21 antagomir induces a threefold increase in RLU values, equal to that of the non-caged antagomir,

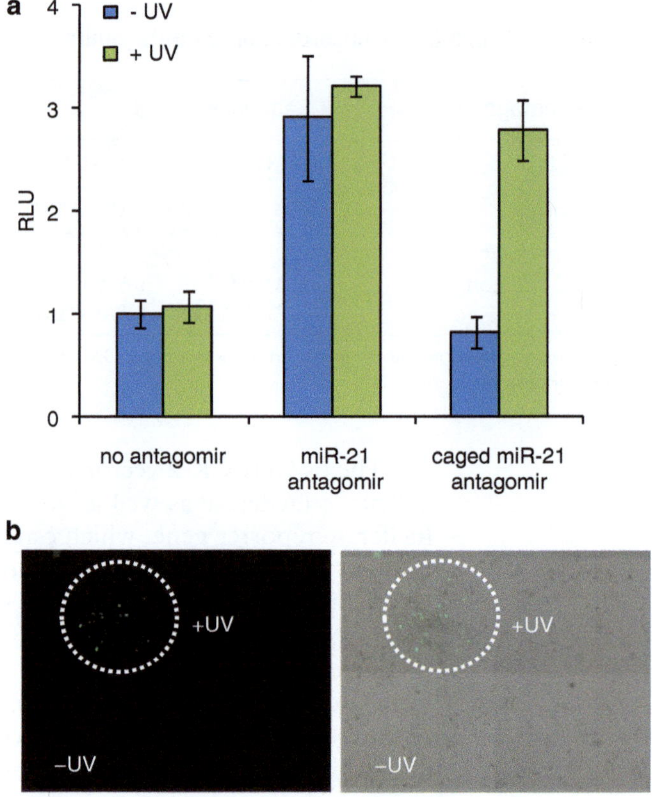

Fig. 3 (**a**) Photochemical activation of miR-21 inhibition and subsequent luciferase expression by decaging of a miR-21 antagomir in Huh7 cells. The cells were co-transfected with a *Renilla* luciferase sensor for miR-21 and either the caged or non-caged miR-21 antagomir, followed by irradiation at 365 nm for 5 min. A Dual Luciferase assay was performed after 48 h. The error bars represent standard deviations from three independent experiments. (**b**) Spatial activation of miR-21 inhibition and subsequent EGFP expression by localized decaging of a miR-21 antagomir. The cells were co-transfected with an EGFP sensor for miR-21 function and the caged miR-21 antagomir, followed by irradiation at 365 nm using a LED fiber-optics probe. The cells were imaged after 48 h and the EGFP channel and corresponding bright-field images are shown. Adapted from *Mol. BioSyst.* **2012**, *8*, 2987 – Reproduced by permission of The Royal Society of Chemistry

demonstrating that the miR-21 inhibitory activity is fully restored. Temporal control over miRNA function through UV irradiation was further demonstrated by using the caged miR-21 antagomir and the *Renilla* luciferase reporter assay [38]. Spatial control over miR-21 function was achieved in Huh7 cells that were co-transfected with the EGFP sensor for miR-21 (pEGFP-C2-miR21) and the caged miR-21 antagomir. The cells were exposed to UV light in a localized fashion using a UV LED fiber-optics probe (365 nm, 5 min). After a 48 h incubation, fluorescence was imaged using a 10× objective and a 2×2 tile scan (Fig. 3b). EGFP expression was only observed within the irradiated area and no EGFP expression was visible in the non-irradiated cells, demonstrating spatial control over antagomir activity. The presented approach enables the precise deactivation of miRNA function using UV irradiation and can be extended to any miRNA of interest. These caged antagomirs can be used as tools to further explore the spatial and temporal role of specific miRNAs in cellular networks, embryonal development, and human diseases including cancer.

The steps to create these light-activated miRNA antagomirs include the synthesis of NPOM caged 2′OMe uridine, the synthesis and purification of the antagomir oligonucleotides, construction of the reporter plasmids for miRNA function, determination of the optimal decaging time through a UV irradiation time course, the photochemical activation of antagomir function in mammalian cells, and the spatial control of antagomir function. This approach can be applied not only to the photochemical control of miR-21 function, but to any other miRNA of interest.

2 Materials

2.1 Synthesis of NPOM-Caged 2′OMe Uridine

All solvents and reagents can be purchased from commercial sources (such as Sigma Aldrich, St. Louis, MO or Fisher Scientific, Pittsburgh, PA, USA) or synthesized from commercially available starting materials.

1. *N*,*N*-dimethylformamide (DMF), ethyl acetate (EtOAc), hexane, triethylamine (TEA), pyridine, dichloromethane (DCM), and methanol (MeOH).
2. 3′,5′-*O*-(di-*t*-butylsilanediyl)uridine [39], NPOM chloride [40], 1,8-diazabicyclo[5.4.0]undec-7-ene (DBU), 6-nitropiperonal, saturated aqueous solution of sodium bicarbonate, saturated aqueous solution of sodium chloride (brine), anhydrous sodium sulfate, methyl iodide, silver oxide, 70 % HF in pyridine, 4,4′-dimethoxytrityl chloride (DMTrCl), 1 M hydrochloric acid (HCl), 5 % aqueous citric acid, *N,N-diisopropylethylamine* (DIPEA), 2-cyanoethyl *N,N-diisopropylchlorophosphoramidite.*

2.2 Synthesis and Purification of Antagomirs

1. Applied Biosystems Model 394 automated DNA/RNA synthesizer or equivalent instrument.
2. Standard solid-phase supports and standard reagents for automated DNA synthesis (Glen Research, Sterling, VA).
3. Anhydrous acetonitrile.
4. Beaucage sulfurizing reagent: 3H-1,2-benzodithiole-3-one-1,1-dioxide [41].
5. Ammonium hydroxide.
6. Illustra NAP-10 columns (GE Healthcare, Pittsburg, PA).
7. Sephadex G-25 DNA grade resin (GE Healthcare, Pittsburg, PA).

2.3 Construction of psiCHECK-miR21 and pEGFP-C2- miR21

1. psiCHECK-2 plasmid (Promega, Madison, WI, USA).
2. pEGFP-C2 plasmid (Clontech, Mountain View, CA).
3. Antarctic phosphatase (New England Biolabs, Ipswich, MA).
4. Restriction enzymes: SgfI, PmeI, XhoI, EcoRI-HF.
5. 5′-Phosphorylated sense and antisense oligonucleotides containing the miR-21 binding site and designed to have overhangs that generate (cut) SfgI/PmeI (psiCHECK-miR21) or XhoI/EcoRI (pEGFP-C2-miR21) restriction sites: psiCHECK-miR21 insert: 5′-CGCAGTAGAGCTCTAGTTCAACATCAG TCTGATAAGCTAGTTT-3′ and 5′-AAACTAGCTTATCAG ACTGATGTTGAACTAGAGCTCTACTGCGAT-3′, pEGFP-C2-miR21 insert: 5′-TCGACTCAACATCAGTCTG ATAAGCTACTCGAGTAG-3′ and 5′-AATTCTACTCGAGT AGCTTATCAGACTGATGTTGAG-3′). Dissolved to 100 μM in ultrapure water.
6. 1 % Agarose gel.
7. 1× TBE buffer: 89 mM Tris base, 89 mM boric acid, 2 mM EDTA.
8. QIAquick Gel Extraction Kit (Qiagen, Valencia, CA, USA).
9. T4 DNA ligase (New England Biolabs, Ipswich, MA, USA).
10. NovaBlue competent *E. coli* cells.
11. LB Agar.
12. Ampicillin (50 mg/mL stock) and kanamycin (50 mg/mL stock).
13. psiCHECK-miR21 sequencing primer: 5′-GCTAAGAAGTTC CCT-3′; pEGFP-C2-miR21 sequencing primer: 5′-TTCATC TGCACCACCGGCAAG-3′.

2.4 General Cell Culture

1. Huh7 cells (Japanese Collection of Research Bioresources, Ibaraki-City Osaka, Japan).
2. Dulbecco's Modified Eagle's Medium (DMEM), pH 7.4, supplemented with 10 % fetal bovine serum and 2 % penicillin/streptomycin, and filter sterilized (*see* **Note 1**).

2.5 Photochemical Activation of miR-21 Antagomir Function in Mammalian Cells

1. White clear-bottom 96-well cell culture plates.
2. psiCHECK-miR21 reporter plasmid (or a similar reporter for the miRNA of interest).
3. X-tremGENE siRNA transfection reagent (Roche Applied Science, Indianapolis, IN).
4. Opti-MEM Reduced Serum Medium, pH to 7.3, filter sterilized (*see* **Note 2**).
5. High Performance UV transilluminator (UVP, Upland, CA).
6. Dual Luciferase Reporter Assay Kit (Promega, Madison, WI): Prepare 1× Passive Lysis Buffer, Luciferase Assay Reagent II,

and Stop & Glo Reagent according to the manufacturer's protocol.

7. Biotek Synergy 4 microplate reader (BioTek, Winooski, VT) or an equivalent microplate reader capable of reading luminescence (*see* **Note 3**).
8. 1× PBS: 137 mM NaCl, 2.7 mM KCl, 10 mM Na_2HPO_4, 2 mM KH_2PO_4, pH 7.4, sterilize.

2.6 Spatial Control of miR-21 Function in Mammalian Cells

1. 6-well cell culture plate treated with poly-D-lysine.
2. pEGFP-C2-miR21 reporter plasmid (or a similar reporter for the miRNA of interest).
3. LED fiber-optics probe (Prizmatix, Givat Shmuel, Israel).
4. DMEM without phenol red.
5. Zeiss Axio Observer inverted microscope with filter set 38 (excitation 470/40, emission 525/50).

3 Methods

3.1 Synthesis of NPOM-Caged 2′OMe Uridine

The light-activated antagomir presented here was designed to contain 2′OMe NPOM-caged uridine groups, however, this approach can be extended to other caging groups and nucleobase combinations [32, 42, 43]. Reactions should be performed using anhydrous or dry solvents and under inert atmosphere.

1. Synthesis of 1-((4a*R*,6*R*,7*R*,7a*S*)-2,2-di-*tert*-butyl-7-hydroxy-tetrahydro-4H-furo[3,2-d][1–3]dioxasilin-6-yl)-3-((1-(6-nitrobenzo[d][1,3]dioxol-5-yl)ethoxy)methyl)pyrimidine-2,4(1H,3H)-dione (**3**): 3′,5′-*O*-(di-*t*-butylsilanediyl)uridine (**1**) can be synthesized in one step from uracil and di-*tert*-butylsilyl bis(trifluoromethanesulfonate) [39]. The NPOM chloride **2** is synthesized in three steps from commercially available 6-nitropiperonal [40]. Add DBU (697 μL, 4.5 mmol) to a solution of 3′,5′-*O*-(di-*t*-butylsilanediyl)uridine (**1**) (0.88 g, 2.3 mmol) in DMF (5 mL) and stir under inert atmosphere at room temperature. After 30 min, add freshly prepared **2** (2.76 mmol) dissolved in DMF (0.5 mL) to the reaction mixture and stir for 12 h under an argon atmosphere. Pour the reaction mixture into a saturated solution of $NaHCO_3$ (20 mL). Extract the product using EtOAc (2×20 mL) and then wash the combined organic layers with a saturated aqueous $NaHCO_3$ (3×20 mL), followed by brine (15 mL), and dry over anhydrous sodium sulfate. Filter the mixture, concentrate the filtrate, and purify by silica gel column chromatography using EtOAc–hexane (1:1) with 1 % TEA to obtain the NPOM-caged uridine **3** as a yellow solid.

2. Synthesis of 1-((2*R*,3*R*,4*R*,5*R*)-4-hydroxy-5-(hydroxymethyl)-3-methoxytetrahydrofuran-2-yl)-3-((1-(6-nitrobenzo[d][1,3]dioxol-5-yl)ethoxy)methyl)pyrimidine-2,4(1H,3H)-dione (**4**): Add methyl iodide (20 mL) to neat NPOM-caged uridine **3** (2.1 g, 3.4 mmol) under an argon atmosphere. Add Ag_2O [44] (2.3 g, 10.3 mmol) to the stirred solution and heat the reaction mixture under reflux (50 °C) for 5 h. Cool the reaction mixture to room temperature and add EtOAc (20 mL). Filter the mixture and wash the residue with EtOAc. Combine the organic layers and wash with saturated aqueous $NaHCO_3$ (3×20 mL), followed by brine (20 mL), and dry over anhydrous sodium sulfate. After filtration, concentrate the filtrate and purify the product by silica gel column chromatography using EtOAc–hexane (2:3) with 1 % TEA to obtain the 2′OMe uridine **4** as a white solid.
3. Synthesis of 1-((2*R*,3*R*,4*R*,5*R*)-4-hydroxy-5-(hydroxymethyl)-3-methoxytetrahydrofuran-2-yl)-3-((1-(6-nitrobenzo[d][1,3]dioxol-5-yl)ethoxy)methyl)pyrimidine-2,4(1H,3H)-dione (**5**): Add diluted HF-pyridine (330 μL, 70 % HF-pyridine) in dry pyridine (2 mL) to a solution of 2′OMe uridine **4** (2.0 g, 3.21 mmol) in DCM at 0 °C in a polyethylene reaction vessel and stir under a nitrogen atmosphere for 2 h (*see* **Note 4**). Add DCM (40 mL) to the reaction mixture followed by saturated aqueous $NaHCO_3$ (30 mL) to neutralize residual amounts of HF. Separate the organic layer and wash with saturated aqueous $NaHCO_3$ (30 mL), followed by 1 M HCl (2×20 mL), water (20 mL), and brine (15 mL), and dry over anhydrous sodium sulfate. After filtration, reduce the volume of filtrate and purify the crude product by silica gel column chromatography using DCM–MeOH (92:8) with 1 % TEA to obtain the caged 2′OMe uridine **5** as a yellow solid.
4. Synthesis of 1-((2*R*,3*R*,4*R*,5*R*)-5-((*bis*(4-methoxyphenyl)(phenyl)methoxy)methyl)-4-hydroxy-3-methoxytetrahydrofuran-2-yl)-3-((1-(6-nitrobenzo[d][1,3]dioxol-5-yl)ethoxy)methyl)pyrimidine-2,4(1H,3H)-dione (**6**): Add DMTrCl (1.5 g, 4.6 mmol) to a solution of the alcohol **5** (1.4 g, 2.9 mmol) in dry pyridine (15 mL) and stir under an argon atmosphere at room temperature for 24 h. Add MeOH (3 mL) to the reaction mixture in order to quench the unreacted DMTrCl and stir the mixture for 20 min. Remove the solvent, dissolve the residue in EtOAc (40 mL), and wash with a 5 % aqueous citric acid solution (2×20 mL), followed by saturated aqueous $NaHCO_3$ (2×20 mL), and brine (15 mL), and dry over anhydrous sodium sulfate. After filtration, concentrate the filtrate and purify the crude product by silica gel column chromatography using DCM–EtOAc (1:5) with 1 % TEA to furnish the DMT protected caged 2′OMe uridine **6** as a white solid.

5. Synthesis of (2*R*,3*R*,4*R*,5*R*)-2-((*bis*(4-methoxyphenyl)(phenyl)methoxy)methyl)-4-methoxy-5-(3-((1-(6-nitrobenzo[d][1,3]dioxol-5-yl)ethoxy)methyl)-2,4-dioxo-3,4-dihydropyrimidin-1(2H)-yl)tetrahydrofuran-3-yl (2-cyanoethyl) diisopropylphosphoramidite (7): Add DIPEA (0.55 mL, 3.2 mmol) and 2-cyanoethyl *N,N*-diisopropylchlorophosphoramidite (0.284 mL, 1.27 mmol) to a solution of the alcohol **6** (0.5 g, 0.64 mmol) in DCM (5.0 mL) and stir under an argon atmosphere at room temperature for 2 h or until TLC shows the complete consumption of the starting material. Concentrate the mixture and directly purify the crude product by silica gel chromatography using EtOAc–hexanes (4:6) with 1 % TEA to obtain the caged phosphoramidite 7 as a white solid (*see* **Note 5**).

3.2 Synthesis and Purification of Antagomirs

Both non-caged and caged 2′OMe phosphorothioate antagomirs for mature miR-21 were synthesized using standard DNA synthesis protocols (Table 1). The miR-21 caged antagomir was designed to contain four NPOM-caged 2′OMe uridines equally distributed throughout the oligonucleotide. The position and number of the caging groups present in the antagomir may need to be optimized for other sequences in order to produce an inactive antagomir prior to UV irradiation. The oligonucleotide synthesis was performed using an Applied Biosystems Model 394 automated DNA/RNA synthesizer and standard β-cyanoethyl phosphoramidite chemistry on a 0.2 μM scale. The solid-phase supports and reagents for automated DNA synthesis were obtained from Glen Research.

1. Resuspend the NPOM-2′OMe uridine phosphoramidite 7 in anhydrous acetonitrile to a concentration of 0.1 M.
2. Use standard synthesis cycles provided by Applied Biosystems with 15 min coupling times for all bases to assemble the desired oligonucleotide.
3. Perform the sulfurization step using the Beaucage sulfurizing reagent 3H-1,2-benzodithiole-3-one-1,1-dioxide [41] at 0.05 M in acetonitrile.
4. Elute the oligonucleotides from the solid-phase supports with ammonium hydroxide and deprotect at 65 °C for 16 h.
5. Purify the deprotection reactions using Illustra NAP-10 columns by gravity flow through Sephadex G-25 DNA-grade resin and elute the antagomirs in water.
6. Perform polyacrylamide gel analysis to confirm the purity of the final antagomirs.

3.3 Construction of psiCHECK-miR21 and pEGFP-C2-miR21

1. Sequentially digest the psiCHECK-2 plasmid (1 μg) with SgfI (10 units, 50 μL reaction) followed by PmeI (10 units, 50 μL reaction) at 37 °C for 2 h and heat-inactivate at 75 °C for 20 min. Similarly, digest the pEGFP-C2 plasmid (5 μg) with

XhoI (20 units, 100 μL reaction) and EcoRI-HF (20 units) at 37 °C for 2 h and heat-inactivate at 75 °C for 20 min. Treat the digested plasmids with Antarctic phosphatase at 37 °C for 1 h.

2. Separate the digested backbone on a 1 % agarose gel in TBE buffer at 90 V for 30 min. Excise the digested backbone and purify using a QIAquick Gel Extraction Kit (see manufacturer's protocol).
3. Hybridize the insert DNA containing the miR-21 binding site by combining 50 μL of both the sense and the antisense oligonucleotides (1 μM) (psiCHECK-miR21 insert: 5′-CGCAGT AGAGCTCTAGTTCAACATCAGTCTGATAAGCTAG TTT-3′ and 5′-AAACTAGCTTATCAGACTGATGTTGAACT AGAGCTCTACTGCGAT-3′, pEGFP-C2-miR21 insert: 5′-TC GACTCAACATCAGTCTGATAAGCTACTCGAGTAG-3′ and 5′-AATTCTACTCGAGTAGCTTATCAGACTGATGTTG AG-3′) and incubate at 72 °C for 15 min followed by cooling to 20 °C over 5 min.
4. Ligate the annealed inserts with T4 ligase (200 units, 10 μL reaction, 1:10 vector–insert ratio) into either the digested psiCHECK-2 or pEGFP-C2 vector at 4 °C overnight.
5. Transform the ligation reaction into NovaBlue competent cells and plate on a LB agar plate containing either ampicillin (50 μg/mL) or kanamycin (50 μg/mL) for psiCHECK-miR21 or pEGFP-C2-miR21, respectively. Incubate at 37 °C overnight.
6. The construction of the psiCHECK-miR21 and pEGFP-C2-miR21 vectors is confirmed by sequencing (psiCHECK-miR21 sequencing primer: 5′-GCTAAGAAGTTCCCT-3′; pEGFP-C2-miR21 sequencing primer: 5′-TTCATCTGCACCACCGG CAAG-3′).

3.4 General Cell Culture

All cell culture experiments using the Huh7 human hepatoma cell line are performed in standard Dulbecco's Modified Eagle's Medium (DMEM) supplemented with 10 % fetal bovine serum (FBS) and 1 % penicillin/streptomycin and maintained at 37 °C in a 5 % CO_2 atmosphere.

3.5 Photochemical Activation of miR-21 Antagomir Function in Mammalian Cells

The optimal irradiation conditions for decaging are determined using a UV irradiation time course. Non-irradiated cells transfected with the caged antagomir should exhibit *Renilla* luciferase expression similar to that of the cells transfected with psiCHECK-miR21 alone. With increasing irradiation times, an increase in *Renilla* luciferase expression is observed as the caging groups are removed and antagomir function is restored, enabling the fine tuning of miRNA inhibition. The optimal irradiation conditions will provide maximum antagomir function following decaging and produce minimal phototoxicity to the cells. For the caged miR-21 antagomir, an irradiation time of 2–5 min efficiently activated antagomir function [38]. In order to confirm that removal of the caging groups fully restores antagomir function to the level of the non-caged antagomir, Huh7 cells are co-transfected with the psiCHECK-miR21 plasmid and non-caged or caged miR-21 antagomir and either kept in the dark or irradiated at 365 nm for 5 min. Following irradiation, the decaged antagomir should inhibit miR-21 function and lead to an increase in *Renilla* luciferase expression similar to that of the non-caged antagomir.

1. Passage Huh7 cells at 10,000 cells per well into a white clear-bottom 96-well plate.
2. After an overnight incubation, transfect the Huh7 cells with the psiCHECK-miR21 plasmid (250 ng/well) and either the non-caged or caged miR-21 antagomir (10 pmol, 50 nM) using X-tremeGENE siRNA transfection reagent (1.0 μL/well) in Opti-MEM media.
3. Incubate the cells at 37 °C for 4 h. Remove the media and replace with PBS (50 μL/well).
4. Keep the cells in the dark or irradiate the cells on a transilluminator (365 nm, 25 W) for 5 min (or the optimal time determined by an irradiation time course) in triplicate.
5. Replace the PBS with standard DMEM growth media (supplemented with 10 % FBS and 1 % penicillin/streptomycin) and incubate the cells at 37 °C for 48 h.
6. Remove the media, wash the cells with 1× PBS, then lyse the cells by adding 1× Passive Lysis Buffer (25 μL; supplied in Dual Luciferase Reporter Assay Kit) and shake for 15 min at room temperature.
7. Assay the lysed cells using a Dual Luciferase Reporter Assay Kit according to the manufacturer's protocol and record the luminescence on a microplate reader. In short, for each well, dispense the Luciferase Assay Reagent II (100 μL), read firefly luciferase activity with a measurement time of 10 s and a delay time of 2 s, dispense the Stop & Glo Reagent (100 μL), read *Renilla* luciferase activity with a measurement time of 10 s and a delay time of 2 s.
8. Calculate the ratio of *Renilla* to firefly luciferase expression for each well to provide relative luciferase units (RLU). The average and standard deviation for each of the triplicates is calculated from the RLU values.

3.6 Spatial Control of miR-21 Function in Mammalian Cells

To control miR-21 function in a spatial fashion, light-activation of the caged miR-21 antagomir is performed through localized UV irradiation. Huh7 cells are co-transfected with the pEGFP-C2-miR21 reporter and the caged miR-21 antagomir and are irradiated following transfection using a UV LED fiber-optics probe (365 nm, 5 min). The fluorescence is imaged using a 10× objective and a 2×2 tile scan 48 h after irradiation. EGFP expression is only observed within the irradiated area, thus demonstrating spatial control over antagomir activity (Fig. 3b).

1. Passage Huh7 cells at 100,000 cells per well into a 6-well plate treated with poly-D-lysine.
2. After a 48 h incubation, transfect the Huh7 cells with the pEGFP-C2-miR21 plasmid (2.5 μg/well) and the caged miR-21 antagomir (100 pmol, 500 nM) using X-tremeGENE siRNA transfection reagent (10 μL/well) in Opti-MEM media.
3. After a 4 h incubation at 37 °C, remove the media and replace with PBS (1 mL/well).
4. Irradiate the cells from the top using a LED fiber-optics probe at 365 nm for 5 min (*see* **Note 6**).
5. Replace the PBS with standard DMEM growth media (supplemented with 10 % FBS and 1 % penicillin/streptomycin) and incubate the cells at 37 °C for 48 h.
6. Remove the media and replace with DMEM without phenol red and image the cells on a Zeiss Axio Observer inverted microscope using a 10× objective, a 2×2 tile scan (*see* **Note 7**), and filter set 38 (excitation 470/40, emission 525/50).

4 Notes

1. The Dulbecco's Modified Eagle's Medium (DMEM) described here is made from DMEM/HIGH with L-glutamine powder (Hyclone): Add DMEM/HIGH with L-glutamine powder (11.8 g) and sodium bicarbonate (3.26 g) to ultrapure water (880 mL). Mix and adjust the pH to 7.4. Add fetal bovine serum (100 mL) and penicillin/streptomycin (50× solution, 20 mL) and filter sterilize. Store at 4 °C. The DMEM growth media can also be purchased as pre-sterilized liquid media to simplify the media preparation. In this case only the FBS and penicillin/streptomycin need to be added.
2. The Opti-MEM used here is made from Opti-MEM Reduced Serum Medium powder (Invitrogen): Add Opti-MEM Reduced Serum Medium powder (6.8 g) and sodium bicarbonate (1.2 g) to ultrapure water (450 mL). Mix and adjust the pH to 7.3. Adjust the volume to 500 mL with ultrapure water and filter sterilize. Store at 4 °C.
3. For the Dual Luciferase Reporter Assay, a microplate reader with dispensers is preferable for delivering the luciferase substrates; however, the procedure can be adjusted for microplate readers without dispensers (see manufacturer's protocol).

4. This reaction should be carried out in an appropriate container for HF, such as a polyethylene reaction vessel, and the use of glassware should be avoided. Other appropriate safety measures should be taken including use of personal protective equipment.
5. Phosphoramidites are moisture sensitive and should be handled as such and stored under argon.
6. In the spatial control experiment, the cells were irradiated from the top using a LED fiber-optics probe for 5 min, which was sufficient irradiation time to observe restored antagomir activity. The irradiation time will need to be optimized for the particular light source and caged antagomir. In addition, the irradiated area was approximately 800 μm, which can be controlled by both the distance between the fiber-optics probe and the cell monolayer and the specifications of the fiber-optics probe diameter.
7. The 10× objective and 2 × 2 tile scan were used in order to view the irradiated area of cells and the surrounding non-irradiated cells. Depending on the distance of the fiber-optics probe from the cell monolayer during irradiation and the diameter of the probe, the fluorescence imaging may need to be adjusted to view a larger area.

References

1. Carthew R (2006) Gene regulation by microRNAs. Curr Opin Genet Dev 16:203–208
2. Djuranovic S, Nahvi A, Green R (2012) miRNA-mediated gene silencing by translational repression followed by mRNA deadenylation and decay. Science 336:237–240
3. Pillai RS (2005) MicroRNA function: multiple mechanisms for a tiny RNA? RNA 11: 1753–1761
4. Lewis BP, Burge CB, Bartel DP (2005) Conserved seed pairing, often flanked by adenosines, indicates that thousands of human genes are microRNA targets. Cell 120:15–20
5. Appasani K (2008) MicroRNAs: from basic science to disease biology. Cambridge University Press, Cambridge
6. Janga SC, Vallabhaneni S (2011) MicroRNAs as post-transcriptional machines and their interplay with cellular networks. Adv Exp Med Biol 722:59–74
7. Tong AW, Nemunaitis J (2008) Modulation of miRNA activity in human cancer: a new paradigm for cancer gene therapy? Cancer Gene Ther 15:341–355
8. Sevignani C, Calin G, Siracusa L, Croce C (2006) Mammalian microRNAs: a small world for fine-tuning gene expression. Mamm Genome 17:189–202
9. Port JD, Sucharov C (2010) Role of microRNAs in cardiovascular disease: therapeutic challenges and potentials. J Cardiovasc Pharmacol 56:444–453
10. Lindsay MA (2008) microRNAs and the immune response. Trends Immunol 29: 343–351
11. Cullen BR (2011) Viruses and microRNAs: RISCy interactions with serious consequences. Genes Dev 25:1881–1894
12. Esquela-Kerscher A, Slack F (2006) Oncomirs – microRNAs with a role in cancer. Nat Rev Cancer 6:259–269
13. Esau C (2008) Inhibition of microRNA with antisense oligonucleotides. Methods 44: 55–60
14. Veedu R, Wengel J (2010) Locked nucleic acids: promising nucleic acid analogs for therapeutic applications. Chem Biodiver 7: 536–542

15. Brown B, Naldini L (2009) Exploiting and antagonizing microRNA regulation for therapeutic and experimental applications. Nat Rev Genet 10:578–585
16. Deiters A (2010) Small molecule modifiers of the microRNA and RNA interference pathway. AAPS J 12:51–60
17. Meister G, Landthaler M, Dorsett Y, Tuschl T (2004) Sequence-specific inhibition of microRNA- and siRNA-induced RNA silencing. RNA 10:544–550
18. Krützfeldt J, Rajewsky N, Braich R, Rajeev K, Tuschl T, Manoharan M et al (2005) Silencing of microRNAs in vivo with 'antagomirs'. Nature 438:685–689
19. Ebert MS, Neilson JR, Sharp PA (2007) MicroRNA sponges: competitive inhibitors of small RNAs in mammalian cells. Nat Methods 4:721–726
20. Kota J, Chivukula RR, O'Donnell KA, Wentzel EA, Montgomery CL, Hwang HW et al (2009) Therapeutic microRNA delivery suppresses tumorigenesis in a murine liver cancer model. Cell 137:1005–1017
21. Gumireddy K, Young D, Xiong X, Hogenesch J, Huang Q, Deiters A (2008) Small-molecule inhibitors of microrna miR-21 function. Angew Chem Int Ed Engl 47:7482–7484
22. Young D, Connelly C, Grohmann C, Deiters A (2010) Small molecule modifiers of microRNA miR-122 function for the treatment of hepatitis C virus infection and hepatocellular carcinoma. J Am Chem Soc 132:7976–7981
23. Liu Z, Sall A, Yang D (2008) MicroRNA: an emerging therapeutic target and intervention tool. Int J Mol Sci 9:978–999
24. Grünweller A, Hartmann R (2007) Locked nucleic acid oligonucleotides: the next generation of antisense agents? BioDrugs 21:235–243
25. Riggsbee CW, Deiters A (2010) Recent advances in the photochemical control of protein function. Trends Biotechnol 28:468–475
26. Deiters A (2010) Principles and applications of the photochemical control of cellular processes. Chembiochem 11:47–53
27. Deiters A (2009) Light activation as a method of regulating and studying gene expression. Curr Opin Chem Biol 13:678–686
28. Young DD, Deiters A (2007) Photochemical control of biological processes. Org Biomol Chem 5:999–1005
29. Fenno L, Yizhar O, Deisseroth K (2011) The development and application of optogenetics. Annu Rev Neurosci 34:389–412
30. Prokup A, Hemphill J, Deiters A (2012) DNA computation: a photochemically controlled AND gate. J Am Chem Soc 134:3810–3815
31. Govan JM, Lively MO, Deiters A (2011) Photochemical control of DNA decoy function enables precise regulation of nuclear factor κB activity. J Am Chem Soc 133:13176–13182
32. Young D, Lively M, Deiters A (2010) Activation and deactivation of DNAzyme and antisense function with light for the photochemical regulation of gene expression in mammalian cells. J Am Chem Soc 132:6183–6193
33. Young DD, Lusic H, Lively MO, Yoder JA, Deiters A (2008) Gene silencing in mammalian cells with light-activated antisense agents. Chembiochem 9:2937–2940
34. Young DD, Edwards WF, Lusic H, Lively MO, Deiters A (2008) Light-triggered polymerase chain reaction. Chem Commun (Camb) 462–4
35. Joshi KB, Vlachos A, Mikat V, Deller T, Heckel A (2012) Light-activatable molecular beacons with a caged loop sequence. Chem Commun (Camb) 48:2746–2748
36. Mikat V, Heckel A (2007) Light-dependent RNA interference with nucleobase-caged siRNAs. RNA 13:2341–2347
37. Medina PP, Nolde M, Slack FJ (2010) OncomiR addiction in an in vivo model of microRNA-21-induced pre-B-cell lymphoma. Nature 467:86–90
38. Connelly CM, Uprety R, Hemphill J, Deiters A (2012) Spatiotemporal control of microRNA function using light-activated antagomirs. Mol Biosyst 8:2987–2993
39. Furusawa K, Ueno K, Katsura T (1990) Synthesis and restricted conformation of 3′,5′-O-(di-tert-butylsilanediyl) ribonucleosides. Chem Lett 97–100
40. Lusic H, Deiters A (2006) A new photocaging group for aromatic N-heterocycles. Synth Stutt 2147–50
41. Iyer R, Egan W, Regan J, Beaucage S (1990) 3H-1,2-benzodithiole-3-one 1,1-dioxide as an improved sulfurizing reagent in the solid-phase synthesis of oligodeoxyribonucleoside phosphorothioates. J Am Chem Soc 112:1253–1254
42. Govan JM, Uprety R, Hemphill J, Lively MO, Deiters A (2012) Regulation of transcription through light-activation and light-deactivation of triplex-forming oligonucleotides in mammalian cells. ACS Chem Biol 7:1247–1256
43. Lusic H, Lively MO, Deiters A (2008) Light-activated deoxyguanosine: photochemical regulation of peroxidase activity. Mol Biosyst 4:508–511
44. Akiyama T, Nishimoto H, Ozaki S (1990) The selective protection of uridine with a para-methoxybenzyl-chloride- a synthesis of 2′-O-methyluridine. Bull Chem Soc Jpn 63: 3356–3357

Chapter 10

Methods for the Study of Long Noncoding RNA in Cancer Cell Signaling

Yi Feng, Xiaowen Hu, Youyou Zhang, Dongmei Zhang, Chunsheng Li, and Lin Zhang

Abstract

With the advances in sequencing technology and transcriptome analysis, it is estimated that up to 75 % of the human genome is transcribed into RNAs. This finding prompted intensive investigations on the biological functions of noncoding RNAs and led to very exciting discoveries of microRNAs as important players in disease pathogenesis and therapeutic applications. Research on long noncoding RNAs (lncRNAs) is in its infancy, yet a broad spectrum of biological regulations has been attributed to lncRNAs. Here, we provide a collection of detailed experimental protocols for lncRNA studies, including lncRNA immunoprecipitation, lncRNA pull-down, lncRNA northern blot analysis, lncRNA in situ hybridization, and lncRNA knockdown. We hope that the information included in this chapter can speed up research on lncRNAs biology and eventually lead to the development of clinical applications with lncRNA as novel prognostic markers and therapeutic targets.

Key words Long noncoding RNA, RNA immunoprecipitation, RNA pull-down, In situ hybridization, Northern blot, Short hairpin RNA

1 Introduction

Cancer is a genetic disease involving multistep changes in the genome [1]. While up to 75 % of the human genome is transcribed to RNA, only less than 2 % of the genome encodes protein-coding transcripts, leaving most of the genome to noncoding RNA transcripts [2, 3]. The recent discovery of the noncoding RNA genes has dramatically altered our understanding of cancer genetics. In the last decade, the functional significance of small noncoding RNA and microRNA in tumorigenesis and progression has been extensively documented; yet research on long noncoding RNA (lncRNA) is still in its infancy [4–8]. lncRNAs are operationally defined as RNA genes larger than 200 bp that do not appear to have protein coding potential [4–8]. Recent studies demonstrated that lncRNAs act as key regulators of development, differentiation, apoptosis, and cell proliferation, all of

Martha Robles-Flores (ed.), *Cancer Cell Signaling: Methods and Protocols*, Methods in Molecular Biology, vol. 1165,
DOI 10.1007/978-1-4939-0856-1_10,

which have been implicated in tumor initiation and progression. In addition, the expression of lncRNAs has been found to be remarkably deregulated by epigenetic and genomic alterations in tumors. In various experimental systems, lncRNAs have been reported to have tumor suppressor or oncogene activity. Therefore, it is reasoned that lncRNAs may play important roles in the development of cancer, hence represent the leading edge of cancer research. Investigations on lncRNA functions in cancer will lead to a greater understanding of molecular mechanisms of this disease, and eventually lead to the development of lncRNA-based novel applications in cancer diagnosis and therapeutic management.

1.1 The Human Genome Contains Many Thousands of Unexplored lncRNAs

lncRNAs are operationally defined as RNA transcripts larger than 200 bp that do not appear to have coding potential [4–8]. Given that up to 75 % of the human genome is transcribed to RNA, while only a small portion of the transcripts encodes proteins [3], the number of lncRNA genes can be large. After the initial cloning of functional lncRNAs such as H19 [9, 10] and XIST [11] from cDNA libraries, two independent studies using high-density tiling array reported that the number of lncRNA genes is at least comparable to that of protein-coding genes [12, 13]. Recent advances in tiling array [12–15], chromatin signature [16, 17], computational analysis of cDNA libraries [18, 19], and next-generation sequencing (RNA-seq) [20–23] have revealed that thousands of lncRNA genes are abundantly expressed with exquisite cell-type and tissue specificity in human. In fact, the GENCODE consortium within the framework of the ENCODE project recently reported 14,880 manually annotated and evidence-based lncRNA transcripts originating from 9,277 gene loci in human [3, 23], including 9,518 intergenic lncRNAs (also called lincRNAs) and 5,362 genic lncRNAs (Fig. 1a) [16, 17, 22]. These studies indicate that (1) lncRNAs are independent transcriptional units; (2) lncRNAs are spliced with fewer exons than protein-coding transcripts and utilize the canonical splice sites; (3) lncRNAs are under weaker selective constraints during evolution and many are primate specific; (4) lncRNA transcripts are subjected to typical histone modifications as protein-coding mRNAs; and (5) the expression of lncRNAs is relatively low and strikingly tissue-specific or cell-type-specific.

1.2 lncRNAs Regulate Gene Expression and Protein Functions via Various Mechanisms

The discovery of lncRNA has provided an important new perspective on the centrality of RNA in gene expression regulation. lncRNAs can regulate the transcriptional activity of a chromosomal region or a particular gene by recruiting epigenetic modification complexes in either *cis-* or *trans-*regulatory manner. For example, Xist, a 17 kb X-chromosome specific noncoding transcript, initiates X chromosome inactivation by targeting and tethering Polycomb-repressive complexes (PRC) to X chromosome in *cis* [24–26]. HOTAIR regulates the HoxD cluster genes in *trans* by serving as a scaffold, which enables RNA-mediated assembly of PRC2 and LSD1 and coordi-

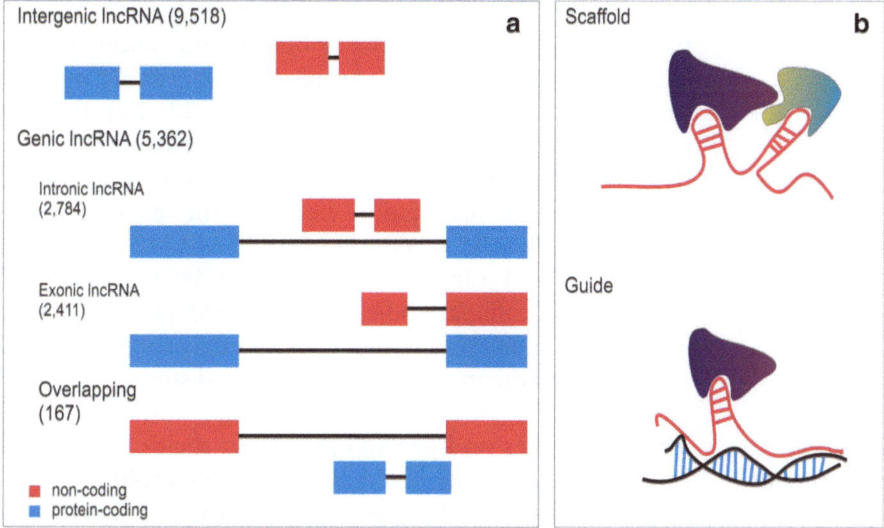

Fig. 1 Schematic diagram of lncRNA gene organization and mechanistic themes. (**a**) lncRNA gene organization in human genome. Two types of lncRNA genes exist in the genome. Intergenic lncRNA genes (*red bar, upper panel*) have no overlap with other protein-coding genes (*blue bar, upper panel*). Genic lncRNA genes (*red bar, lower panel*) are overlapping with protein coding genes (*blue bar, upper panel*) in various fashions. (**b**) lncRNAs can either function as scaffolds via RNA–protein interaction (*upper panel*), or as guides via RNA–protein and RNA–DNA interactions (*lower panel*)

nates the binding of PRC2 and LSD1 to chromatin [14, 27]. Based on the knowledge obtained from studies on a limited number of lncRNAs, at least two working models have been proposed (Fig. 1b). First, lncRNAs can function as scaffolds. lncRNAs contain discrete protein-interacting domains that can bring specific protein components into the proximity of each other, resulting in the formation of unique functional complexes [27–29]. These RNA mediated complexes can also extend to RNA–DNA and RNA–RNA interactions. Second, lncRNAs can act as guides to recruit proteins [26, 30, 31], such as chromatin modification complexes, to chromosome [26, 31]. This may occur through RNA–DNA interactions [31] or through RNA interaction with a DNA-binding protein [26]. In addition, lncRNAs have been proposed to serve as decoys that bind to DNA-binding proteins [32], transcription factors [33], splicing factors [34–36], or miRNAs [37]. Some studies have also identified lncRNAs transcribed from the enhancer regions [38–40] or a neighbor loci [20, 41] of certain genes. Given that their expressions correlated with the activities of the corresponding enhancers, it was proposed that these RNAs (termed enhancer RNA/eRNA [38–40] or ncRNA-activating/ncRNA-a [20, 41]) may regulate gene transcription.

1.3 lncRNA Expression Is Deregulated in Human Cancer

The advances in high-throughput RNA quantification technologies unveiled a profound deregulation of the lncRNome in human cancer. First, lncRNA expression profiles are dramatically different between tumors and their adjacent normal tissues. A large-scale RNA-seq analysis in prostate cancer identified 121 lncRNAs whose expressions pattern can distinguish between benign and cancer specimens [21]. Second, given that lncRNA expression patterns are more tissue-specific than those of protein coding genes [22, 23], it has been proposed that lncRNA expression signatures may be able to accurately determine the developmental lineage and tissue origin of human cancers. Third, the association between the expressions of several lncRNAs, such as MALAT-1 [42], HOTAIR [15], PCAT-1 [21] and LET [43], and cancer metastasis have been identified by high-throughput profiling studies and validated by further independent investigations, suggesting that lncRNAs may also serve as robust biomarkers in predicting cancer prognosis and survival.

1.4 lncRNAs Serve as Tumor Suppressor Genes or Oncogenes

Though studies on lncRNAs are still in their early stage, it is clear that lncRNAs are involved in regulating proliferation [33, 36], differentiation [16, 31, 44–46], migration [15, 20], and apoptosis [30, 47]. Therefore, it is reasoned that deregulation of lncRNA expression may contribute to the development and progression of cancer. In fact, some lncRNAs have been shown to function as oncogenes or tumor suppressors. For example, HOTAIR [14] can induce metastasis [15] by operating as a tether that links EZH2/PRC2 and LSD1, and therefore coordinating their epigenetic regulatory functions [27]. ANRIL, an antisense lncRNA of the CDKN2A/CDKN2B gene, represses INK4A/INK4B expression [48] by binding to CBX7/PRC1 [28] and SUZ12/PRC2 [29]. On the other hand, deleting Xist resulted in the development of highly aggressive myeloproliferative neoplasm and myelodysplastic syndrome with 100 % penetrance in female mice [49].

1.5 lncRNAs Represent Promising Biomarker and Therapeutic Candidates for Cancer Diagnosis and Treatment

The ability to fully characterize cancer genome contributed significantly to the development of biomarkers and therapeutic applications for cancer diagnoses and treatments. PCA3, a prostate cancer-associated lncRNA [50], is the first prominent example of lncRNA as a novel biomarker. The noninvasive method to detect PCA3 transcript in urine has been developed and used clinically to detect prostate malignancy [51]. The transition from lncRNA-based diagnostics to lncRNA-based therapies is also under intensive investigations. The rapid advances in oligonucleotide/nanoparticle therapy create realistic optimism for developing lncRNA-based therapeutic tools for cancer treatment. Although the majority of cancer-related studies still focus on the protein-coding genes, given that almost 75 % of the genome is transcribed to RNAs and initial studies on a handful of lncRNAs clearly demonstrated their functional significance in cancer development and high potential in clinical applications, we argue that investigations on lncRNAs is the leading edge of cancer research.

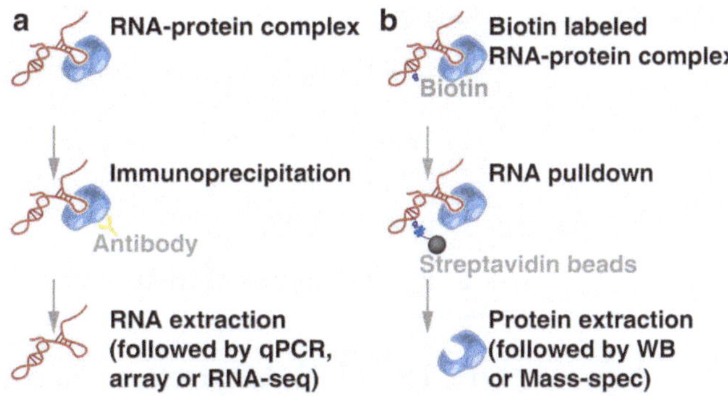

Fig. 2 Schematic diagram of lncRNA IP and lncRNA pull-down. (**a**) lncRNA IP, to identify lncRNA molecules interacting with a protein of interest. (**b**) lncRNA pull-down, to identify proteins interacting with a specific lncRNA

1.6 Methods in lncRNAs Research

In the following sections, we provide detailed protocols on characterizing lncRNA expression and functions. They are RNA-immunoprecipitation (RNA-IP), RNA pull-down, northern blot analysis on lncRNA expression, In situ hybridization (ISH) of lncRNA and lncRNA knockdown, respectively. As one major mechanism for lncRNA to exert its function is to serve as a scaffold via RNA–protein interaction, it is important to investigate which lncRNAs are binding to a protein of interest. RNA-IP has been developed to identify lncRNA species that bind to a protein of interest. On the other hand, if the research focus is to identify the proteins that are bound to a given lncRNA, lncRNA pull-down will help to identify the protein molecules that interact with a specific lncRNA (Fig. 2). Moreover, as a novel class of RNA transcripts, it is important to characterize the expression of lncRNAs in various systems. While the northern blot can be used to determine lncRNA abundance and to identify different splicing variants of a given lncRNA, lncRNA in situ hybridization can also provide information regarding the expression level of a given lncRNA. More importantly, it can reveal the cellular or tissue localization of the lncRNA of interest. Knocking down the expression of a target gene has been a gold standard assay to elucidate its endogenous function. In this regard, we also included two lncRNA knockdown protocols in this chapter. We hope this chapter would help the readers to develop assays for their lncRNA research, which will lead to a better understanding of the roles of lncRNAs in carcinogenesis and other pathological conditions.

2 Materials

Prepare all solutions using ultrapure RNase-free water and analytical grade reagents. Contamination of the solutions with RNase can result in RNA degradation. Use filtration or/and autoclave sterilization to ensure that all reagents and supplies used in this section are RNase free. Use RNase ZAP to clean all equipment and work surface.

2.1 lncRNA-IP

1. 1 M Dithiothreitol (DTT).
2. 10× Phosphate-Buffered Saline (PBS): 1.37 μM NaCl, 27 mM KCl, 100 mM Na_2HPO_4, 20 mM KH_2PO_4, pH 7.4, sterilize. To make 1× PBS, mix one part of 10× PBS with nine parts RNase-free water. Store at 4 °C.
3. Protein A/G beads.
4. RNase inhibitor (Invitrogen).
5. Protease inhibitor cocktail (Sigma).
6. TRIzol RNA extraction reagent (Invitrogen).
7. 1 mL Dounce homogenizer.
8. Nuclear Isolation Buffer: 1.28 M sucrose, 40 mM Tris–HCl (pH 7.4), 20 mM $MgCl_2$, 4 % Triton X-100. Put 40 mL RNase-free water in a beaker with a stir bar and dissolve 21.9 g sucrose in the beaker. Add 2 mL 1 M Tris–HCl (pH 7.5), 1 mL 1 M $MgCl_2$, and 2 mL Triton X-100 and mix well. Make up to a final volume of 50 mL with RNase-free water, store at 4 °C.
9. RNA Immunoprecipitation (RIP) Buffer: 150 mM KCl, 25 mM Tris (pH 7.4), 5 mM EDTA, 0.5 % NP-40. Mix 7.5 mL 1 M KCl, 1.25 mL 1 M Tris–HCl (pH 7.4), 500 μL 0.5 M EDTA, and 250 μL NP-40 and make up to a final volume of 48 mL with RNase-free water. Store at 4 °C. Right before use, add DTT (0.5 mM final concentration), RNase inhibitor (100 U/mL final concentration), and protease inhibitor cocktail (1× final concentration).

2.2 Materials for lncRNA Pull-Down

1. 1 M Dithiothreitol (DTT).
2. 10× Phosphate-Buffered Saline (10× PBS: 1.37 μM NaCl, 27 mM KCl, 100 mM Na_2HPO_4, 20 mM KH_2PO_4, pH 7.4). To make 1× PBS, mix one part of 10× PBS with nine parts RNase-free water. Store at 4 °C.
3. DNA template (*see* **Note 1**).
4. Restriction Enzyme.
5. Vanadyl-ribonucleoside complex (VRC, NEB).
6. 10× Biotin RNA labeling mix (Roche).
7. T7 RNA polymerase (10 U/μL) and 5× transcription buffer (Agilent).
8. RNase inhibitor (Invitrogen).
9. Protease inhibitor cocktail (Sigma).
10. Dnase I, 2,000 U/mL (NEB).
11. Streptavidin Agarose Beads.
12. 0.5 M EDTA (pH 8.0).
13. Yeast tRNA (Ambion).

14. Nuclear Isolation Buffer: 1.28 M sucrose, 40 mM Tris–HCl (pH 7.4), 20 mM $MgCl_2$, 4 % Triton X-100. Put 40 mL RNase-free water in a beaker with a stir bar and dissolve 21.9 g sucrose in the beaker. Add 2 mL 1 M Tris–HCl (pH 7.5), 1 mL 1 M $MgCl_2$, and 2 mL Triton X-100 and mix well. Make up to a final volume of 50 mL with RNase-free water, store at 4 °C.
15. RNA Immunoprecipitation (RIP) Buffer: 150 mM KCl, 25 mM Tris (pH 7.4), 5 mM EDTA, 0.5 % NP-40. Mix 7.5 mL 1 M KCl, 1.25 mL 1 M Tris–HCl (pH 7.4), 500 μL 0.5 M EDTA, and 250 μL NP-40 and make up to a final volume of 48 mL with RNase-free water. Store at 4 °C. Right before use, add DTT (0.5 mM final concentration), RNase inhibitor (100 U/mL final concentration), and protease inhibitor cocktail (1× final concentration).
16. NT2 Buffer: 50 mM Tris–HCl (pH 7.4), 150 mM NaCl, 1 mM $MgCl_2$, 0.05 % NP-40. Store at 4 °C. For 50 mL NT2 buffer, mix 2.5 mL 1 M Tris–HCl, 5 mL 150 nM NaCl, 1 mL 1 mM $MgCl_2$, 2.5 mL 1 % NP-40, add 39 mL RNase-free water to a final volume of 50 mL, filter stock solution, store at 4 °C. Right before use, add RNase inhibitor (100 U/mL final concentration), Vanadyl-ribonucleoside complex (VRC, 400 nM final concentration), DTT (1 mM final concentration), EDTA (20 mM final concentration), and protease inhibitor cocktail (1× final concentration).
17. RNA structure buffer: 10 mM Tris (pH 7.0), 0.1 M KCl, 10 mM $MgCl_2$.
18. 1 mL Dounce homogenizer.
19. Agarose gel electrophoresis supplies for DNA fragment purification.
20. Quick Spin Columns for radiolabeled RNA purification Sephadex G-50 (Roche).
21. Gel Extraction Kit (Qiagen).
22. BCA protein assay kit (Pierce).
23. 2× Laemmli loading buffer: 4 % SDS, 120 mM Tris–HCl (pH 6.8), 0.02 % bromophenol blue, and 0.2 M DTT. Mix 4 mL 10 % SDS, 1.2 mL 1 M Tris–HCL (pH 6.8), 200 μL 1 % bromophenol blue, and 2 mL 1 M DTT and add Milli-Q water to make the final volume to 10 mL. Make 500 μL aliquots to minimize the freeze–thaw cycles.

2.3 Materials for lncRNA Northern Blot Analysis

2.3.1 Digoxigenin (DIG)-Labeled RNA Probe Synthesis

1. DNA template (*see* **Note 1**).
2. Restriction Enzyme.
3. 10× DIG RNA labeling mix (Roche).
4. T7 RNA polymerase (10 U/μL) and 5× transcription buffer (Agilent).

5. Dnase I 2,000 U/mL (NEB).
6. Agarose gel electrophoresis supplies for DNA fragment purification.
7. Quick Spin Columns for radiolabeled RNA purification Sephadex G-50 (Roche).
8. Gel Extraction Kit (Qiagen).

2.3.2 Separating RNA by Electrophoresis

1. Nucleic acid agarose.
2. 55 °C water bath.
3. 10× Denaturing gel buffer (Invitrogen).
4. Heat block.
5. Gel electrophoresis apparatus.
6. 10× MOPS buffer: 200 mM MOPS, 50 mM Sodium acetate, 20 mM EDTA, adjust pH to 7.0. To make 1× MOPS gel running buffer, mix one part of 10× MOPS buffer with nine parts of RNase-free water.
7. RNA loading buffer (Invitrogen).
8. Ethidium bromide (only if RNA visualization is needed).
9. DIG-labeled RNA marker (Roche).

2.3.3 Transferring RNA to the Membrane

1. 20× SSC: 3 M NaCl (1.75 g/L), 0.3 M Sodium citrate dihydrate (88 g/L). Adjust pH to 7.0 with 1 M HCl.
2. Razor blade.
3. 3 M Filter paper.
4. Positively Charged Nylon Membrane (Roche).
5. Blunt end forceps.
6. Paper towel.
7. RNase-free flat-bottom container as buffer reservoir.
8. Clean glass Pasteur pipet as roller.
9. Light weight (150–200 g) object serving as weight during transfer.
10. Supports for the reservoir (i.e., a stack of books).
11. Stratalinker® UV Crosslinker.

2.3.4 Probe–RNA Hybridization

1. 20× SSC: 3 M NaCl (1.75 g/L), 0.3 M Sodium citrate dihydrate (88 g/L). Adjust pH to 7.0 with 1 M HCl.
2. 10 % SDS.
3. DIG Easy Hyb Granules (Roche).
4. 68 °C shaking water bath.
5. Heat block.
6. Hybridization oven.
7. Hybridization bags.

8. Low Stringency Buffer: 2× SSC with 0.1 % SDS.
9. High Stringency Buffer: 0.1× SSC with 0.1 % SDS.

2.3.5 Detection of Probe–RNA Hybrids

1. Washing and blocking buffer set (Roche).
2. Anti-DIG-alkaline phosphatase antibody (Roche).
3. NBT/BCIP Stock Solution (Roche).
4. CDP-Star, Ready-to-Use (Roche).
5. TE buffer: 10 mM Tris–HCL, 1 mM EDTA, adjust pH to approximately 8.

2.4 Materials for lncRNA In Situ Hybridization

1. 8-well chamber slide.
2. 10× Phosphate-Buffered Saline (10× PBS: 1.37 μM NaCl, 27 mM KCl, 100 mM Na_2HPO_4, 20 mM KH_2PO_4, pH 7.4). To make 1× PBS, mix one part of 10× PBS with nine parts RNase-free water. Store at 4 °C.
3. 4 % paraformaldehyde (in RNase-free PBS).
4. 0.1 M triethanolamine.
5. Acetic anhydride.
6. 0.2 M HCl in RNase-free water.
7. 20× SSC: 3 M NaCl (1.75 g/L), 0.3 M Sodium citrate dihydrate (88 g/L). Adjust pH to 7.0 with 1 M HCl.
8. 50× Denhardt solution (Sigma).
9. 50 % Dextran sulfate solution: dissolve 5 g Dextran sulfate in 10 mL RNase-free water, stir at room temperature until completely dissolved.
10. Yeast tRNA.
11. Pre-hybridization buffer: 2× SSC, 50 % formamide, 1× Denhardt solution, and 1 mg/mL yeast tRNA. For 50 mL hybridization buffer, add 1 mL 20× SSC, 5 mL formamide, 200 μL 50× Denhardt solution, and 50 mg yeast tRNA, and make up a final volume of 50 mL with RNase-free water.
12. Hybridization buffer: 2× SSC, 50 % formamide, 1× Denhardt solution, 10 % dextran sulfate, 1 mg/mL yeast tRNA. For 50 mL hybridization buffer, add 1 mL 20× SSC, 5 mL formamide, 200 μL 50× Denhardt solution, 10 mL 50 % dextran sulfate, and 50 mg yeast tRNA, and make up a final volume of 50 mL with RNase-free water.
13. DIG Wash and Block Buffer Set (Roche).
14. Anti-DIG-alkaline phosphatase antibody (Roche).
15. NBT/BCIP stock solution (Roche).
16. TE buffer: 10 mM Tris–HCL, 1 mM EDTA, adjust pH to approximately 8.
17. Humidity oven.

2.5 Materials for lncRNA Knockdown

2.5.1 lncRNA Knockdown Using siRNAs

1. siRNA targeting lncRNA of interest: sources of siRNAs include (1) ready-to-use Lincode RNA from http://www.thermoscientificbio.com/; (2) custom-designed siRNAs (*see* **Note 2** for a list of siRNA design Web sites and **Note 3** for a list of siRNA synthesis vendors).
2. Control siRNA (Thermo Scientific).
3. siRNA transfection reagents. We use Lipofectamine RNAiMAX from Invitrogen, but there are a handful of similar reagents that one can choose from. *See* **Note 4** for a list of siRNA transfection reagents.
4. Opti-MEM.

2.5.2 lncRNA Knockdown Using shRNAs

1. pLKO.1 (Addgene).
2. shRNA (designed at http://www.broadinstitute.org/rnai/public/seq/search).
3. MISSION® pLKO.1-puro Non-Mammalian shRNA Control Plasmid DNA (Sigma).
4. pRSV-Rev (Addgene).
5. pMDLg/pRRE (Addgene).
6. pMD2G (Addgene).
7. HEK293T cells.
8. RPMI 1640 media
9. Fetal Bovine serum.
10. Penicillin–Streptomycin solution with 10,000 U penicillin and 10 mg streptomycin/mL.
11. FuGENE6.
12. 0.45 μm disk filter.
13. Hexadimethrine bromide (polybrene) (Sigma): to make 4 mg/mL stock, dissolve 40 mg powder in 10 mL Milli-Q water and filter through 0.22 μm disk filter, make 500 μL aliquots and store at −20 °C.

3 Methods

The procedures must be performed in an RNase-free environment. Use filter tips and RNase-free tubes and clean all equipment and work surface with RNase ZAP before starting the experiment.

3.1 lncRNA-IP

lncRNA-IP aims to identify lncRNA species that bind to a protein of interest. The protocol includes two parts: (1) preparing protein lysate from target cells and (2) immunoprecipitating the protein of interest and extract protein-bound RNAs. It is up to the readers to decide the subsequent analysis on the isolated RNAs.

Before harvesting the cells, precool 1× PBS, RNase-free water, nuclear isolation buffer, and RIP buffer on ice; estimate the amount of RIP buffer needed and add RNase inhibitor and protease inhibitor cocktail to the buffer accordingly (*see* **Note 5**).

*3.1.1 Whole-Cell Lysate Preparation (See **Note 6**)*

If nuclear RNA–protein interaction is the focus of the research, skip this step and go directly to Subheading 3.1.2 for nuclear lysate preparation.

1. Harvest cells using regular trypsinization technique and count the cell number.
2. Wash the cells in ice-cold 1× PBS once and resuspend the cell pellet (1×10^7 cells) in 1 mL ice-cold RIP buffer containing RNase and protease inhibitors.
3. Shear the cells on ice using a Dounce homogenizer with 15–20 strokes.
4. Centrifuge at 15,000 ×*g* for 15 min at 4 °C and transfer the supernatant into a clean tube. This supernatant is the whole-cell lysate.

*3.1.2 Cell Harvest and Nuclear Lysate Preparation (See **Note 6**)*

1. Harvest cells using regular trypsinization technique and count the cell number.
2. Wash the cells in ice-cold 1× PBS three times and resuspend 1×10^7 cells in 2 mL ice-cold PBS (*see* **Note 7**).
3. Put cell suspension in 1× PBS on ice, add 2 mL ice-cold nuclear isolation buffer and 6 mL ice-cold RNase-free water into the tube and mix well. Incubate the cells on ice for 20 min with intermittent mixing (4–5 times).
4. Harvest nuclei by spinning the tube at 2,500 ×*g* for 15 min at 4 °C. The pellet contains the purified nuclei.
5. Resuspend nuclei pellet in 1 mL freshly prepared ice-cold RIP buffer containing DTT, RNase, and protease inhibitors.
6. Shear the nuclei on ice with 15–20 strokes using a Dounce homogenizer.
7. Pellet nuclear membrane and debris by centrifugation at 16,000 ×*g* for 10 min at 4 °C.
8. Carefully transfer the clear supernatant (nuclear lysate) into a new tube. The supernatant is nuclear lysate.

3.2 RNA Immune-Precipitation and Purification

1. Wash 40 μL protein A/G beads with 500 μL ice-cold RIP buffer three times. After the wash, spin down the beads at 600 ×*g* for 30 s at 4 °C, take off the RIP buffer and add 40 μL RIP buffer to resuspend the beads.
2. Add the prewashed beads and 5–10 μg IgG into the whole-cell lysate from Subheading 3.1.1 or nuclear lysate from Subheading 3.1.2.

3. Incubate the lysate with IgG and beads at 4 °C with gentle rotation for 1 h. Pellet the IgG with beads by centrifugation at 16,000 × *g* for 5 min.
4. Carefully transfer the supernatant (pre-cleared nuclear lysate) into a new tube. At this point, the lysate can be divided into multiple portions of equal volume for different antibodies and corresponding controls. Take 50 μL lysate and set aside on ice as input control.
5. Add antibody of interest into nuclear lysate (*see* **Note 8**). Incubate the lysate and antibody overnight at 4 °C with gentle rotation.
6. The next day, add 40 μL prewashed protein A/G beads and incubate at 4 °C for 1 h with gentle rotation.
7. Pellet the beads by spinning at 600 × *g* for 30 s at 4 °C. Remove supernatant.
8. Wash the beads with 500 μL ice-cold RIP buffer three times, invert 5–10 times during each wash and pellet the beads by spinning at 600 × *g* for 30 s at 4 °C.
9. Wash the beads with 500 μL ice-cold PBS and pellet the beads by spinning at 600 × *g* for 30 s at 4 °C. Use a fine needle or tip to remove as much PBS as possible without disturbing the beads.
10. Resuspend the beads in 1 mL TRIzol RNA extraction reagent and isolate co-precipitated RNA according to manufacturer's instructions.
11. Dissolve RNA in nuclease-free water and store the RNA at −80 °C for further application (*see* **Note 9**).

3.3 lncRNA Pull-Down

lncRNA pull-down aims to identify proteins that bind to a lncRNA of interest. The protocol includes three parts: (1) synthesis and labeling the lncRNA of interest; (2) preparing protein lysate from target cells; (3) pull-down labeled lncRNA with its interacting proteins. The readers can decide the subsequent analyses on the lncRNA-bound proteins.

Before harvesting cells, precool PBS, water, RIP buffer, and NT2 buffer on ice; estimate the amount of buffers needed and add VRC, EDTA, DTT, RNase inhibitor, and protease inhibitor to the buffers accordingly (*see* **Note 5**).

3.3.1 Biotinylated RNA Synthesis by In Vitro Transcription

1. DNA template preparation: Linearize 3–4 μg of the plasmid containing the desired template DNA (*see* **Note 1**) with a suitable restriction enzyme at the 3′ end of the insert (*see* **Note 10**).
2. DNA template purification: run the digested DNA by DNA gel electrophoresis, excise the band of correct size, and extract DNA from agarose using Gel Extraction Kit.

Table 1
Components for in vitro transcription of template DNA (20 μL system)

Linearized plasmid DNA (1 μg) or PCR product (100–200 ng)	
Biotin RNA labeling mix (10×)	2 μL
5× Transcription buffer	4 μL
T7 RNA polymerase (20 U/μL)	2 μL
RNase-free sterile water	Up to 20 μL

3. In vitro synthesis of biotinylated RNA using T7 RNA polymerase: add the components listed in Table 1 into an RNase-free tube on ice and mix thoroughly. Centrifuge briefly and incubate at 37 °C for 2 h. After incubation, add 2 μL Dnase I (RNase-free) into the reaction and incubate at 37 °C for 15 min to remove DNA template. Stop the reaction by adding 0.8 μL 0.5 M EDTA (pH 8.0).
4. Biotinylated RNA purification of using G-50 Sephadex Columns: Before use, gently invert the column several times to resuspend the medium. Remove the top cap and then remove the bottom tip (*see* **Note 11**). Drain the buffer in the column by gravity and then centrifuge at 1,100×*g* for 2 min to eliminate residual buffer. Place the column in an upright position (*see* **Note 12**) with a new collection tube. Apply the RNA sample (up to 100 μL) to the center of the column carefully (*see* **Note 13**) and centrifuge for 4 min at 1,100×*g* at 4 °C. The elution contains the purified biotinylated RNA. Determine the RNA concentration and store at −80 °C.

3.3.2 Whole-Cell Lysate Preparation (See **Note 6**)

If nuclear RNA–protein interaction is the focus of the research, skip this step and go directly to Subheading 3.2.3 for nuclear lysate preparation.

1. Harvest cell by regular trypsinization (~10^7 cells) and wash cells with ice-cold 1× PBS once.
2. Resuspend cell pellet in 1 mL ice-cold RIP buffer containing RNase and protease inhibitors.
3. Shear the cell pellet on ice using a Dounce homogenizer with 15–20 strokes.
4. Centrifuge at 15,000×*g* for 15 min at 4 °C to clear the cell lysate. The supernatant contains whole-cell lysate.

3.3.3 Nuclear Lysate Preparation (See **Note 6**)

1. Cell harvest and nuclei isolation: harvest cells using regular trypsinization technique, wash cell pellet in ice-cold 1× PBS once and resuspend 1×10^7 cells in 2 mL ice-cold 1× PBS. Put cell suspension on ice, add 2 mL ice-cold nuclear isolation buffer and 6 mL ice-cold RNase-free water into the tube and mix well. Incubate the cells on ice for 20 min with intermittent mixing. Harvest nuclei by spinning the tube at 2,500 × *g* for 15 min at 4 °C. The pellet contains the purified nuclei.
2. Nuclei lysis: resuspend nuclei pellet in 1 mL freshly prepared ice-cold RIP buffer containing RNase and protease inhibitors. Shear the nuclei on ice with 15–20 strokes using a Dounce homogenizer. Pellet nuclear membrane and debris by centrifugation at 15,000 × *g* for 15 min at 4 °C. Carefully transfer the clear supernatant into a new tube. The supernatant contains nuclear lysate (*see* **Note 7**).
3. Pre-clear lysate: take 60 μL Streptavidin agarose beads slurry and wash the beads with pre-cold NT2 buffer three times. After wash, spin down the beads at 12,000 × *g* for 1 min and resuspend the beads in 60 μL pre-cold NT2 buffer. Add the prewashed Streptavidin agarose beads into the whole-cell lysate (Subheading 3.2.2) or nuclear lysate (Subheading 3.2.3) and incubate at 4 °C for at least 1 h with gentle rotation. Centrifuge the lysate briefly. Carefully transfer the supernatant into a new tube, and determine the protein concentration using BCA protein assay. Save 3–5 % of the lysate as input.

3.3.4 RNA Pull-Down (See **Note 14**)

1. Dilute ~10 pmol biotinylated RNA into 40 μL of RNA structure buffer and heat the tube at 90 °C for 2 min. Immediately transfer the tube on ice and incubate for another 2 min. Then let the tube sit at room temperature for 20 min to allow proper RNA secondary structure formation.
2. Add the properly folded RNA from previous step into 200 μg of pre-cleared lysate from Subheading 3.2.4, and supplement with tRNA to a final concentration of 0.1 μg/μL (*see* **Note 15**). Incubate at 4 °C for 2 h with gentle rotation.
3. Add 60 μL prewashed Streptavidin agarose beads and incubate for 1 h at 4 °C.
4. At the end of the incubation, centrifuge at 12,000 × *g* 4 °C for 1 min and take off the supernatant. Wash the beads with 1 mL ice-cold NT2 buffer at 4 °C five times (*see* **Note 16**).
5. After the last wash, carefully remove any residual buffer without disturbing the beads.
6. Add 40 μL 2× Laemmli loading buffer and boil the beads in loading buffer for 5–10 min. Centrifuge at 12,000 × *g* for 1 min at room temperature and transfer the supernatant, which contain the lncRNA interacting proteins, into a new tube and store at −80 °C for further analysis.

3.4 lncRNA Northern Blot Analysis (See Note 17)

lncRNA northern blot analysis aims to characterize lncRNA expression. The protocol includes five parts: (1) RNA probe synthesis and labeling; (2) RNA sample electrophoresis; (3) RNA transfer; (4) RNA–probe hybridization; and (5) RNA–probe hybrid detection.

3.4.1 DIG-Labeled RNA Probe Synthesis by In Vitro Transcription

The DIG-labeled RNA probe synthesis is very similar to the biotinylated RNA synthesis described in Subheading 3.2.1. The differences between the two procedures are as follows:

1. Since the probe needs to be complement to the target sequence, the probe RNA is transcribed from the 3′ end of the target sequence. We clone the gene of interest in reverse orientation to make the in vitro transcription template for northern probes.
2. Use DIG labeling mix in place of the biotin-labeling mix.

3.4.2 Separating RNA Samples by Electrophoresis

1. Wipe the gel rack, tray and combs with RNAZap, rinse with water and let dry. Weight 100 g agarose in a clean glass flask and mix with 90 mL RNase-free water. Melt the agarose completely by heating with a microwave. Put the flask with melted agarose in a 55 °C water bath.
2. In a fume hood (*see* **Note 18**), add 10 mL 10× denaturing gel buffer to the gel mix that is equilibrated to 55 °C. Mix the gel solution by gentle swirling to avoid generating bubbles. Slowly pour the gel mix into the gel tray. Pop any bubbles or push them to the edges of the gel with a clean pipet tip. The thickness of the gel should be about 6 mm. Slowly place the comb in the gel. Allow the gel to solidify before removing the comb.
3. Right before RNA electrophoresis, place the gel tray in the electrophoresis chamber with the wells near the negative lead and add 1× MOPS gel running buffer in the chamber until it is 0.5–1 cm over the top of the gel (*see* **Note 19**).
4. Mix no more than 30 μg sample RNA with 3 volumes of RNA loading buffer (*see* **Notes 20** and **21**). To destruct any secondary structure of the RNAs, incubate the RNA with loading buffer at 65 °C for 15 min using a heat block. Spin briefly to collect samples to the bottom of the tube and put the tubes on ice (*see* **Note 22**).
5. Carefully draw the RNAs in the tip without trapping any bubbles at the end of pipet tip. Place the pipet tip inside of the top of the well, slowly push samples into the well and exit the tip without disturbing the loaded samples. If markers are needed, load one lane with DIG-labeled RNA marker.
6. Run the gel at 5 V/cm (*see* **Note 23**).
7. (Optional) Stain the gel with ethidium bromide and visualize the RNA under UV (*see* **Note 24**).

3.4.3 Transfer RNA from Agarose Gel to the Membrane (Fig. 3)

1. Use a razor blade to trim the gel by cutting through the wells and discard the unused gel above the wells. For marking the orientation, make a notch at a corner.
2. Cut the membrane to the size slightly larger than the gel. Make a notch at a corner to align the membrane with gel in the same orientation. Handle the membrane with care only touching the edges with gloved hands or blunt tip forceps.
3. Cut eight pieces of filter paper the same size of the membrane.
4. Cut a stack of paper towels that are 3 cm in height and 1–2 cm wider than the gel.
5. Pour 20× SSC into a flat-bottom container that has bigger dimension of the agarose gel. This serves as the buffer reservoir and can also be used to wet the paper and membrane. Put the reservoir on a support (i.e., a stack of books) so that its bottom is higher than the paper towel stack.
6. Cut three pieces of filter paper large enough to cover the gel and long enough to reach over to the reservoir. These papers serve as the bridge to transfer buffer from the reservoir to the gel.
7. Stack paper towel on a clean bench and put three pieces dry filter paper on top.
8. Wet two more pieces of filter paper and put on top of the dry filter paper. Gently roll out any bubbles between the filter paper layers.

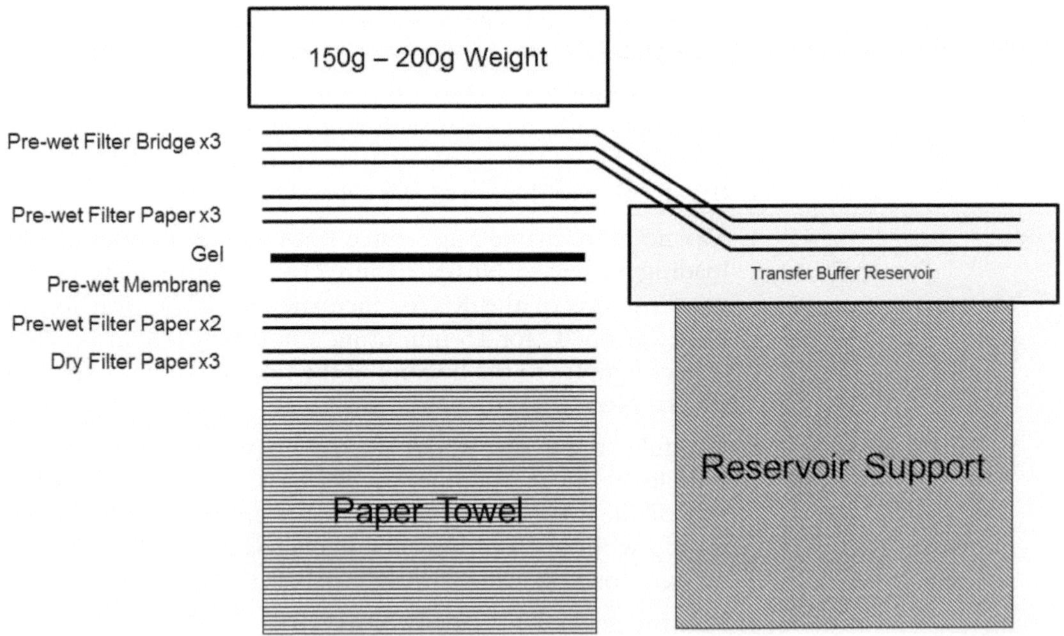

Fig. 3 Schematic diagram of downward transfer of northern blot analysis (adapted from NorthernMax kit instruction, Invitrogen)

9. Carefully put the membrane on top of the wet filter paper. Gently roll out bubbles between the membrane and the filter papers.
10. Put the trimmed gel onto the center of the membrane with the bottom of the gel touching the membrane (i.e., the gel plane that faces down during electrophoresis will be in contact with the membrane). Align the notches of the gel and the membrane. Roll out bubbles between the membrane and the gel.
11. Place three more pieces pre-wet filter paper on top of the gel and roll out bubbles between filter paper layers.
12. Wet the three pieces of paper bridge and place them with one end on top of the stack and the other end in buffer reservoir. Make sure there is no bubble between any layers of paper (*see* **Note 25**).
13. Place a 150–200 g object with the size similar to the gel on top of the stack.
14. Transfer the gel for 15–20 min/mm of gel thickness. It usually takes about 2 h (*see* **Note 26**).
15. RNA cross-link: disassemble the transfer stack carefully and rinse the member with 1× MOPS gel running buffer to remove residual agarose. Blot off excessive liquid and immediately subject the membrane to cross-link treatment. Cross-link the RNA to the membrane with Stratalinker® UV Crosslinker using the autocrosslink setting (*see* **Note 27**). Air-dry the membrane at room temperature. At this point, the membrane can be subjected to hybridization immediately or stored in a sealed bag between two pieces filter paper at 4 °C for several months before hybridization.

3.4.4 Hybridization of DIG-Labeled Probes to the Membrane (See **Note 28**)

1. Pre-hybridization: Reconstitute the DIG Easy Hyb Granules: add 64 mL RNase-free water into one bottle of the DIG Easy Hyb Granules, stir for 5 min at 37 °C to complete dissolve the granules. DIG Easy Hyb buffer will be used in pre-hybridization and hybridization. The reconstituted DIG Easy Hyb buffer is stable at room temperature for up to 1 month. For every 100 cm^2 membrane, 10–15 mL Hyb buffer should be used for pre-hybridization. Measure the appropriate amount of Hyb buffer for pre-hybridization and place it in a clean tube and pre-warm it in a 68 °C water bath (*see* **Note 29**). Put the membrane in a hybridization bag, add the pre-warmed Hyb buffer from the previous step, seal the bag properly and incubate the membrane in Hyb buffer at 68 °C for at least 30 min with gentle agitation (*see* **Note 30**). Pre-hybridization can be up to several hours as far as the membrane remains wet.
2. Hybridization: For every 100 cm^2 membrane, 3.5 mL Hyb buffer is needed for hybridization. Measure the appropriate

amount of Hyb buffer for hybridization and place it in a clean tube and pre-warm it in a 68 °C water bath (*see* **Note 29**). Determine the amount of RNA probe needed (*see* **Note 31**) and place it into a microcentrifuge tube with 50 μL RNase-free water. Denature the probe by heating the tube at 85 °C for 5 min and chill on ice immediately. Mix the denatured probe with pre-warmed Hyb buffer by inversion. Remove pre-hybridization buffer from the membrane and immediately replace with the pre-warmed hybridization buffer containing the probe. Seal the bag properly and incubate the membrane in probe-containing Hyb buffer at 68 °C overnight with gentle agitation (*see* **Note 30**).

3. The next day, pre-warm the High Stringency Buffer to 68 °C and pour Low Stringency Buffer in an RNase-free container at room temperature and make sure it is enough to cover the membrane.
4. Cut open the hybridization bag, remove the Hyb buffer, and immediately submerge the membrane in the Low Stringency Buffer.
5. Wash the membrane twice in Low Stringency Buffer at room temperature for 5 min each time with shaking.
6. Wash the membrane twice in High Stringency Buffer at 68 °C for 5 min each time with shaking (*see* **Note 32**).

3.4.5 Detection of DIG-Probe–Target RNA Hybrids

1. Localizing the probe–target hybrid with anti-DIG antibody. Transfer the membrane from the last wash in High stringency buffer to a plastic container with 100 mL washing buffer. Incubate for 2 min at room temperature and discard the washing buffer. Add 100 mL Blocking buffer onto the membrane and incubate for more than 30 min (up to 3 h) with shaking at room temperature. Dilute anti-DIG-alkaline phosphatase antibody at the ratio of 1:5,000 in Blocking buffer and incubate the membrane in 20 mL diluted antibody for 30 min at room temperature with shaking. Wash the membrane twice with 100 mL of washing buffer for 15 min each time at room temperature.
2. Visualizing probe–target hybrids using chromogenic or chemiluminescent method (*see* **Note 33**): Equilibrate the membrane in 20 mL Detection buffer for 3 min at room temperature. If using the chromogenic method, prepare the color substrate solution while equilibrating the membrane. Put the membrane with the RNA side facing up in a container and incubate in 10 mL color substrate solution in the dark without shaking. When the desired intensity for the band is observed, discard the color substrate solution and rinse the membrane in 50 mL of TE buffer for 5 min (*see* **Note 34**). Document the result by photographing the membrane (*see* **Note 35**).

For chemiluminescent detection: put the membrane with the RNA side facing up on a plastic sheet (i.e., cut out of a hybridization bag) and add 20 drops of CDP-Star, Ready-to-Use reagent. Immediately cover the membrane with another sheet to evenly distribute the reagent without creating any bubbles. Incubate for 5 min at room temperature. Squeeze out excess reagent and seal the bag. Develop the membrane with an X-ray film in a dark room (*see* **Note 35**).

3.5 lncRNA In Situ Hybridization

lncRNA in situ hybridization aims to characterize and quantify lncRNA expression in cells. The protocol includes three parts: (1) RNA probe synthesis and labeling; (2) cell preparation and pretreatment; (3) RNA–probe hybridization and detection.

3.5.1 DIG-Labeled RNA Probe Synthesis by In Vitro Transcription

The DIG-labeled RNA probe synthesis is very similar to the biotinylated RNA synthesis described in Subheading 3.2.1. The differences between the two procedures are as follows:

1. Both sense and antisense probes need to be synthesized (*see* **Note 36**). The antisense probe contains the complementary sequence to the target gene and the sense probe contains the target gene sequence. The hybridization signal from the antisense probe represents target gene expression while the sense probe is used as negative control.
2. To generate antisense probes, we clone the gene of interest in reverse orientation and transcribe it using in vitro transcription as described in Subheading 3.2.1.
3. DIG labeling mix is used in place of the biotin-label mix during in vitro transcription.

3.5.2 Cell Preparation and Pretreatment

1. Culture the cells of interest on multi-well chamber glass slides. At the day of experiment, the cells should be 60–80 % confluent. We use the 8-chamber glass slide to reduce the amount of hybridization buffer and probes used. The amount of reagents used in the following protocol is for 8-well chamber slides. Scale up or down according to the cell culture device used in each specific experiment.
2. On the day of the experiment, wash the cells with 400 μL 1× RNase-free-PBS per well, then fix the cells using 200 μL 4 % PFA per well at room temperature for 10 min. Wash the fixated cells three times with 1× PBS at room temperature.
3. Add acetic anhydride into 0.1 M triethanolamine to make the final acetic anhydride concentration 0.25 % (v/v, *see* **Note 37**). Pre-treat cells with 0.1 M triethanolamine containing 0.25 % acetic anhydride for 10 min at room temperature. Wash the cells with 1× PBS for 5 min afterwards.

4. Permeablize the cells with 0.2 M HCl for 10 min and wash the cells twice with 1× PBS, 5 min each time, at room temperature.
5. The slides can be either subject to hybridization immediately or stored in 1× PBS at 4 °C for a couple of days before proceeding to the next step.

3.5.3 In Situ Hybridization and Detection of Probe–Target Hybrid

1. Add 200 μL pre-hybridization buffer (hybridization buffer without probe and dextran sulfate, *see* Subheading 2 for details) into each well. Incubate at 60 °C for 2 h (*see* **Note 38**).
2. During pre-hybridization, take the amount of probe needed for the experiment (*see* **Note 39**) and denature the probes by heating them at 85 °C for 10 min and immediately cool down on ice for at least 5 min.
3. Add denatured probes into the hybridization buffer. To ensure assay specificity, always use a control probe for each target probe, *see* **Note 36** for details.
4. Remove the pre-hybridization buffer from chambers and add 200 μL hybridization buffer containing denatured probes on to the slide and incubate at 60 °C in humidity oven with lid on overnight.
5. The next day, wash the cells by adding 500 μL 0.1× SSC with 50 % formamide in each well and incubate the slides with lid at 60 °C for 30 min twice (*see* **Note 40**).
6. Then wash the cells by adding 500 μL 2× SSC to each well and incubate for 5 min at room temperature twice (*see* **Note 40**).
7. Wash the cells with 500 μL washing buffer per well at room temperature for 5 min.
8. Add 200 μL blocking buffer per well and incubate for 1 h at room temperature.
9. While the cells are incubated with blocking buffer, dilute the anti-DIG-alkaline phosphatase antibody in blocking buffer (*see* **Note 41**).
10. Discard the blocking buffer. Add 100 μL diluted antibody per well and incubate for 1 h at room temperature.
11. Wash the slides with 500 μL wash buffer per well for 15 min at room temperature twice.
12. Incubate the slides in 500 μL detection buffer per well for 10 min at room temperature. Meanwhile, dilute 200 μL NBT/BCIP stock solution in 10 mL detection buffer.
13. Discard the detection buffer and add 500 μL diluted NBT/BCIP solution per well and incubate cells in the dark (up to 16 h).
14. When the desired color intensity is observed (*see* **Note 42**), stop the color reaction by discarding the NBT-BCIP solution and incubate with TE buffer for 15 min.

Table 2
siRNA transfection components and composition

Reagent per well for 24-well plate	Amount of reagent
Opti-MEM for siRNA dilution	50 μL
siRNA	0.6–30 pmol
Opti-MEM for RNAiMAX dilution	50 μL
RNAiMAX	0.5–1.5 μL

3.6 shRNA Knockdown

Like mRNAs, the endogenous lncRNA expression can be down-regulated using two RNA silencing-mediated approaches. One is transfecting small-interfering RNAs (siRNAs) targeting the gene of interest into cells. The other is stably expressing gene-specific RNA hairpins (shRNAs) in target cells. The siRNA approach provides acute gene down-regulation and allows gradient knock-down of target gene expression by adding different amounts of siRNAs; whereas the shRNA-approach provides sustained down-regulation of the target lncRNA, making it more suitable for experiments that take a long period to get the end-point readout. The following section will provide protocols for both methods.

3.6.1 lncRNA Knockdown Using siRNAs

The following protocol provides siRNA concentration and the amount of transfection reagents based on Lipofectamine RNAiMAX transfection protocol. If you are using different transfection reagents, please follow the specific instructions from the manufacturers.

1. Make 20 μM siRNA stock solution with RNase-free water, aliquot into 20 μL per tube and store at −80 °C. It is highly recommended to reduce the freeze–thaw cycles of siRNA stocks.
2. The day before transfection, plate cells in 24-well plate so that it will be 60–80 % confluent the next day.
3. On the day of transfection, calculate the amount of control and targeting siRNAs needed for each well according to Table 2 (*see* **Note 43**). Dilute the siRNA and transfection reagents in Opti-MEM in separate tubes. Mix together and incubate at room temperature for at least 20 min.
4. Without removing any media from the wells, add the siRNA and transfection reagent mixture to the wells.
5. Incubate the cells with siRNA/transfection reagents for at least 6 h (*see* **Note 44**).
6. Remove the media containing siRNA/transfection reagents and feed the cells with fresh media.

Table 3
Plasmid composition for pLKO.1 virus packaging

Plasmid	μg
pRSV-Rev	0.65
pMDLg/pRRE	1.3
pMD2G	0.65
pLKO.1	1.4

7. Harvest the infected cells in Trizol, 48–72 h post-transfection.
8. Check target gene knockdown efficiency by quantitative RT-PCR of RNA from control and targeting siRNA transfected cells.

3.6.2 lncRNA Knockdown Using shRNAs

1. To generate shRNA knockdown constructs: design shRNA sequence targeting specific lncRNA at http://www.broadinstitute.org/rnai/public/seq/search (*see* **Note 45**). Synthesize the DNA containing the shRNA sequence of choice and clone it into lentiviral vector pLKO.1. Purify pLKO.1 constructs (containing specific shRNA or scramble controls) and other packing constructs (pRSV-Rev and pMDLg/pRRE) using QIAGEN Plasmid *Plus* Maxi Kit (*see* **Note 46**).
2. Packing virus using 293T cells: plate HEK293T cells in 6-well plate the day before transfection so it will be 80 % confluent the next day. Re-feed cells with antibiotic-free media before transfection. Mix 4 μg plasmid cocktail and 12 μL FuGENE6 transfection reagent according to the FuGENE protocol. The composition of the plasmid cocktail is shown in Table 3: add the FuGENE:DNA mixture onto cells. Eight hours after transfection, change into regular media with antibiotics. Forty-eight hours after transfection, harvest virus by collecting the culture media and filter through 0.45 μm disk filter to get rid of cell debris. The virus stock can be aliquoted and stored at −80 °C for future use.
3. Cell infection: plate target cells in 6-well plate the day before infection so that it will be 50–60 % confluency the next day. Mix three parts of virus with one part of culture media (i.e., 3 mL virus with 1 mL medium) and add polybrene to final concentration of 8 μg/mL. Add virus mixture directly onto cells and incubate overnight.
4. After 24 h of incubation, change media or split cells depending on the cell confluency.

5. Harvest the infected cells in Trizol 72–96 h postinfection.
6. Check target gene knockdown efficiency by quantitative RT-PCR of RNA from scrambled and shRNA-infected cells.

4 Notes

1. The desired DNA template should be a plasmid containing a promoter for in vitro transcription (i.e., T7 or T3) and a target sequence whose 5' end is placed as close as possible to the 3' end of the promoter. We usually use pBluescript SK(+) and transcribe the target sequence using T7 polymerase. Minimize any unnecessary addition of non-lncRNA sequence into the plasmid to avoid inappropriate RNA folding.
2. List of siRNA design Web sites:
   (a) siDESIGN center at http://www.thermoscientificbio.com/.
   (b) BLOCK-iT™ RNAi Designer at http://rnaidesigner.invitrogen. com/rnaiexpress/.
   (c) http://www.broadinstitute.org/rnai/public/seq/search.
3. List of siRNA synthesis vendors:
   (a) http://www.idtdna.com/.
   (b) http://www.invitrogen.com.
   (c) http://www.genscript.com/.
   (d) http://www.thermoscientificbio.com/.
   (e) http://www.sigmaaldrich.com/.
4. List of siRNA transfection reagents:
   (a) Lipofectamine RNAiMAX (Invitrogen).
   (b) DharmaFECT Transfection Reagents (Thermo Scientific).
   (c) X-tremeGENE siRNA transfection reagent (Roche).
5. It is utterly important that the experiment described above be conducted with extra precaution to avoid RNA degradation. All materials have to be RNase-free and the buffers need to be precooled on ice.
6. To ensure result reproducibility, the cells need to be maintained consistently.
7. The abundance of different target proteins and lncRNAs may vary from cell line to cell line. Therefore, the amount of lysate input needs to be empirically determined for each assay. We found 1×10^7 cells is a good starting point. In some cases more cells are needed. Scale up the amount of buffer used to ensure high nuclear lysing efficiency.

8. The amount of antibody used for each experiment need to be empirically determined. Our suggestion is to start at around 1–2 μg antibody per million cells.
9. The amount of nuclease-free water used to dissolve the RNAs is determined by several factors including the type of downstream analysis, the amount of lncRNA bound to the target protein and the cell type. We recommend the researchers start at 20 μL and adjust according to their specific situations.
10. Make sure the restriction enzyme digests efficiently and generate a 5′ overhang.
11. The top-to-bottom sequence is necessary to avoid creating vacuum and uneven flow of buffer.
12. Maintaining the column in an upright position is very important, especially after centrifugation. Tipping the column can cause backflow of the RNA sample and reduce the yield after purification.
13. Avoid to apply the sample to the side of the column or to overload the column, since it will reduce the yield and purity of the RNA.
14. The amount of total protein used in each assay need to be empirically determined based on the specific questions the researchers try to address. If the interaction between a specific lncRNA and a target protein is to be tested, the abundance of the target protein in the nuclear lysate needs to be taken into account to determine the amount of total protein used in the assay.
15. The purpose of adding tRNA in the pull-down assay is to reduce nonspecific binding. Therefore, the amount of tRNA added to each reaction can be optimized according to specific conditions.
16. We found that sometimes increasing the number of washes can greatly reduce the background. Therefore, it is recommended to optimize the wash condition for each specific assay.
17. Northern blot analysis is a golden standard in RNA detection and analysis. There are many protocols developed by laboratories specialized in RNA research or companies. The protocol described here is adapted from the NorthernMax procedure from Invitrogen and DIG application manual for filter hybridization from Roche. In our hands, this protocol is time efficient and gives satisfying results without using radioactivity.
18. Always cast the gel in a fume hood as the denaturing solution contains formaldehyde. Solidified gels can be wrapped up and stored at 4 °C overnight.
19. Do not let gel soaked in running buffer for more than 1 h before loading.

20. Load no more than 30 μg total RNA in each lane. As the binding capacity of the membrane is limited, more RNA loaded does not guarantee a stronger signal. Overloading can lead to the detection of minor degradation of targeted RNAs.
21. If the total volume of sample and dye exceed the capacity of the wells, it is necessary to concentrate the RNA by precipitation and to resuspend the pellet in smaller volume of water before adding the loading dye.
22. Use a heat block instead of a water bath to avoid contaminating the samples with water.
23. The voltage is decided by the distance between the two electrodes (not by the size of the gel). Usually, the run takes about 2 h. If the run is longer than 3 h, change the buffer at the two end chambers to avoid the pH gradient.
24. RNA gels stained with ethidium bromide are not suitable for northern blot analysis. Therefore, if a visual examination or photograph of total RNA samples is needed as a reference for the northern blot, we suggest the researchers to either run the same set of samples on a separate gel or stain with ethidium bromide the gel just for visualization; or de-stain the gel before continuing northern blot analysis. If a gel will be subjected to northern analysis after UV visualization, avoid prolonged exposure of the gel to UV light.
25. It is essential to ensure that the only way for the transfer buffer to run from reservoir to the dry paper stack is through the gel. Therefore, extra care is needed to assemble the stack properly to avoid shortcut. The most common shortcut happens between the bridge and the paper beneath the gel. One can cover the edges of the gel with Parafilm to prevent this from happening.
26. Transfers longer than 4 h may cause small RNA hydrolysis and reduce yield.
27. The autocrosslink Mode of Stratalinker® UV Crosslinker delivers a preset exposure of 1,200 μJ to the membrane and takes about 40 s. Other methods of cross-linking RNA to membrane are available and can be used at this step as well.
28. Once the membrane is wet during pre-hybridization, it is important to avoid it getting dry during the hybridization and detection process. Dried membrane will have high background. Only if the membrane will not be stripped and reprobed, it can be dried after the last high stringency wash and stored at 4 °C for future analysis.
29. For most northern blot hybridization using DIG Easy Hyb buffer, 68 °C is appropriate for both pre-hybridization and hybridization. In cases of more heterologous RNA probes

being used, the pre-hybridization and hybridization temperatures need to be optimized.

30. Pre-hybridization/hybridization can be performed in containers other than bags, as far as it can be tightly sealed. Sealing the hybridization container can prevent the release of NH_4, which changes the pH of the buffer during incubation.
31. For RNA probe synthesized by in vitro transcription, it is recommended that the probe concentration should be 100 ng/mL Hyb buffer.
32. If the probe is less than 80 % homologous to the target RNA, the high stringency wash should be performed at a lower temperature, which needs to be empirically determined.
33. The DIG probe–target RNA hybrids can be detected in two ways. One uses chemiluminescent method, whereas the other uses chromogenic method. The chemiluminescent method is sensitive and fast, but it requires the usage of the films and the accessibility of a darkroom. The chromogenic method requires no film or dark room and different targets can be detected simultaneously using different colored substrate. However, the chromogenic method may not be sensitive enough for low-abundant targets.
34. At this step, if there are multiple membranes, process one at a time. Depending on the abundance of target RNAs, the band may appear as quickly as a few minutes after adding the chromogenic agents. The reaction can be stopped when the band reaches a desired intensity.
35. If reprobing is needed, photograph the result while the membrane is wet and proceed to stripping and reprobing. If no reprobing is needed, dry the membrane, document the result by photograph and store the dried membrane in a clean bag at room temperature.
36. During assay development for each specific gene target, it is critical to know whether the hybridization is specific to the gene of interest. One way to ensure hybridization specificity is to include samples hybridizing with sense probe (containing the target sequence, serving as negative controls) at the same concentration as those with the antisense probe (containing the complementary sequence). If the control probe gives comparable signal as the target probe does, the hybridization may not be specific enough to the gene of interest. In this case, optimization of the probe sequence, probe concentration, hybridization condition, and wash stringency will be needed.
37. The acetic anhydride needs to be added freshly each time and discard any leftover 0.1 M triethanolamine with acetic anhydride.

38. To avoid evaporation of the buffer, we put on the lid for the chamber slide, and use the humidity oven with temperature set at 60 °C for the incubation period.
39. The concentrations of different probes need to be empirically determined. We found that 100–400 ng probes per mL hybridization buffer are a good starting point.
40. The wash conditions can be optimized by adjusting the salt concentration and the temperature.
41. The recommended range for anti-DIG antibody is from 1:500 to 1:2,000. The optimal concentration needs to be empirically determined.
42. The process usually takes from 5 min to 2 h. Stop the reaction once the desired signal is visible.
43. The amounts of siRNA and transfection reagents for each cell line need to be empirically determined.
44. Leaving the transfection reagents on cells for extended period of time may cause cell toxicity.
45. It is common to clone five or more shRNA and choose the two that gives the highest knockdown efficiency for future analysis.
46. We prefer to use the Qiagen Plasmid *Plus* Maxi Kit. However, any plasmid isolation kit that gives high quality DNA for efficient transfection can be used. A good mini-prep kit can be also used to isolate small amount plasmid during pilot experiments.

Acknowledgements

This work was supported, in whole or in part, by National Institutes of Health Grant R01CA142776 (L. Zhang), Ovarian Cancer SPORE P50-CA83638-7951 Project 3 (L. Zhang), Department of Defense Grant W81XWH-10-1-0082 (L. Zhang), the Ovarian Cancer Research Fund Tilberis Scholar Award the Basser Research Center grant for BRCA (L. Zhang), and Marsha Rivkin Center for Ovarian Cancer Research (L. Zhang). D. Zhang was supported by the China Scholarship Council.

References

1. Hanahan D, Weinberg RA (2011) Hallmarks of cancer: the next generation. Cell 144:646–674
2. International Human Genome Sequencing, C (2004) Finishing the euchromatic sequence of the human genome. Nature 431:931–945
3. Djebali S, Davis CA, Merkel A et al (2012) Landscape of transcription in human cells. Nature 489:101–108
4. Rinn JL, Chang HY (2012) Genome regulation by long noncoding RNAs. Annu Rev Biochem 81:145–166
5. Lee JT (2012) Epigenetic regulation by long noncoding RNAs. Science 338:1435–1439
6. Prensner JR, Chinnaiyan AM (2011) The emergence of lncRNAs in cancer biology. Cancer Discov 1:391–407

7. Guttman M, Rinn JL (2012) Modular regulatory principles of large non-coding RNAs. Nature 482:339–346
8. Spizzo R, Almeida MI, Colombatti A et al (2012) Long non-coding RNAs and cancer: a new frontier of translational research? Oncogene 31:4577–4587
9. Pachnis V, Brannan CI, Tilghman SM (1988) The structure and expression of a novel gene activated in early mouse embryogenesis. EMBO J 7:673–681
10. Bartolomei MS, Zemel S, Tilghman SM (1991) Parental imprinting of the mouse H19 gene. Nature 351:153–155
11. Brown CJ, Ballabio A, Rupert JL et al (1991) A gene from the region of the human X inactivation centre is expressed exclusively from the inactive X chromosome. Nature 349: 38–44
12. Kapranov P, Cawley SE, Drenkow J et al (2002) Large-scale transcriptional activity in chromosomes 21 and 22. Science 296:916–919
13. Rinn JL, Euskirchen G, Bertone P et al (2003) The transcriptional activity of human chromosome 22. Genes Dev 17:529–540
14. Rinn JL, Kertesz M, Wang JK et al (2007) Functional demarcation of active and silent chromatin domains in human HOX loci by noncoding RNAs. Cell 129:1311–1323
15. Gupta RA, Shah N, Wang KC et al (2010) Long non-coding RNA HOTAIR reprograms chromatin state to promote cancer metastasis. Nature 464:1071–1076
16. Guttman M, Amit I, Garber M et al (2009) Chromatin signature reveals over a thousand highly conserved large non-coding RNAs in mammals. Nature 458:223–227
17. Khalil AM, Guttman M, Huarte M et al (2009) Many human large intergenic noncoding RNAs associate with chromatin-modifying complexes and affect gene expression. Proc Natl Acad Sci U S A 106:11667–11672
18. Maeda N, Kasukawa T, Oyama R et al (2006) Transcript annotation in FANTOM3: mouse gene catalog based on physical cDNAs. PLoS Genet 2:e62
19. Jia H, Osak M, Bogu GK et al (2010) Genome-wide computational identification and manual annotation of human long noncoding RNA genes. RNA 16:1478–1487
20. Orom UA, Derrien T, Beringer M et al (2010) Long noncoding RNAs with enhancer-like function in human cells. Cell 143:46–58
21. Prensner JR, Iyer MK, Balbin OA et al (2011) Transcriptome sequencing across a prostate cancer cohort identifies PCAT-1, an unannotated lincRNA implicated in disease progression. Nat Biotechnol 29:742–749
22. Cabili MN, Trapnell C, Goff L et al (2011) Integrative annotation of human large intergenic noncoding RNAs reveals global properties and specific subclasses. Genes Dev 25: 1915–1927
23. Derrien T, Johnson R, Bussotti G et al (2012) The GENCODE v7 catalog of human long noncoding RNAs: analysis of their gene structure, evolution, and expression. Genome Res 22:1775–1789
24. Brown CJ, Hendrich BD, Rupert JL et al (1992) The human XIST gene: analysis of a 17 kb inactive X-specific RNA that contains conserved repeats and is highly localized within the nucleus. Cell 71:527–542
25. Zhao J, Sun BK, Erwin JA et al (2008) Polycomb proteins targeted by a short repeat RNA to the mouse X chromosome. Science 322:750–756
26. Jeon Y, Lee JT (2011) YY1 tethers Xist RNA to the inactive X nucleation center. Cell 146: 119–133
27. Tsai MC, Manor O, Wan Y et al (2010) Long noncoding RNA as modular scaffold of histone modification complexes. Science 329:689–693
28. Yap KL, Li S, Munoz-Cabello AM et al (2010) Molecular interplay of the noncoding RNA ANRIL and methylated histone H3 lysine 27 by polycomb CBX7 in transcriptional silencing of INK4a. Mol Cell 38:662–674
29. Kotake Y, Nakagawa T, Kitagawa K et al (2011) Long non-coding RNA ANRIL is required for the PRC2 recruitment to and silencing of p15(INK4B) tumor suppressor gene. Oncogene 30:1956–1962
30. Huarte M, Guttman M, Feldser D et al (2010) A large intergenic noncoding RNA induced by p53 mediates global gene repression in the p53 response. Cell 142:409–419
31. Grote P, Wittler L, Hendrix D et al (2013) The tissue-specific lncRNA Fendrr is an essential regulator of heart and body wall development in the mouse. Dev Cell 24:206–214
32. Kino T, Hurt DE, Ichijo T et al (2010) Noncoding RNA gas5 is a growth arrest- and starvation-associated repressor of the glucocorticoid receptor. Sci Signal 3:8
33. Hung T, Wang Y, Lin MF et al (2011) Extensive and coordinated transcription of noncoding RNAs within cell-cycle promoters. Nat Genet 43:621–629
34. Tripathi V, Ellis JD, Shen Z et al (2010) The nuclear-retained noncoding RNA MALAT1 regulates alternative splicing by modulating SR splicing factor phosphorylation. Mol Cell 39: 925–938
35. Bernard D, Prasanth KV, Tripathi V et al (2010) A long nuclear-retained non-coding

RNA regulates synaptogenesis by modulating gene expression. EMBO J 29:3082–3093

36. Tripathi V, Shen Z, Chakraborty A et al (2013) Long noncoding RNA MALAT1 controls cell cycle progression by regulating the expression of oncogenic transcription factor B-MYB. PLoS Genet 9:e1003368
37. Salmena L, Poliseno L, Tay Y et al (2011) A ceRNA hypothesis: the Rosetta Stone of a hidden RNA language? Cell 146:353–358
38. Kim TK, Hemberg M, Gray JM et al (2010) Widespread transcription at neuronal activity-regulated enhancers. Nature 465:182–187
39. Wang D, Garcia-Bassets I, Benner C et al (2011) Reprogramming transcription by distinct classes of enhancers functionally defined by eRNA. Nature 474:390–394
40. Melo CA, Drost J, Wijchers PJ et al (2013) eRNAs are required for p53-dependent enhancer activity and gene transcription. Mol Cell 49:524–535
41. Lai F, Orom UA, Cesaroni M et al (2013) Activating RNAs associate with mediator to enhance chromatin architecture and transcription. Nature 494:497–501
42. Ji P, Diederichs S, Wang W et al (2003) MALAT-1, a novel noncoding RNA, and thymosin beta4 predict metastasis and survival in early-stage non-small cell lung cancer. Oncogene 22:8031–8041
43. Yang F, Huo XS, Yuan SX et al (2013) Repression of the long noncoding RNA-LET by histone deacetylase 3 contributes to hypoxia-mediated metastasis. Mol Cell 49:1083–1096
44. Loewer S, Cabili MN, Guttman M et al (2010) Large intergenic non-coding RNA-RoR modulates reprogramming of human induced pluripotent stem cells. Nat Genet 42: 1113–1117
45. Guttman M, Donaghey J, Carey BW et al (2011) lincRNAs act in the circuitry controlling pluripotency and differentiation. Nature 477:295–300
46. Kretz M, Siprashvili Z, Chu C et al (2013) Control of somatic tissue differentiation by the long non-coding RNA TINCR. Nature 493:231–235
47. Zhang A, Zhou N, Huang J et al (2013) The human long non-coding RNA-RoR is a p53 repressor in response to DNA damage. Cell Res 23:340–350
48. Pasmant E, Laurendeau I, Heron D et al (2007) Characterization of a germ-line deletion, including the entire INK4/ARF locus, in a melanoma-neural system tumor family: identification of ANRIL, an antisense noncoding RNA whose expression coclusters with ARF. Cancer Res 67:3963–3969
49. Yildirim E, Kirby JE, Brown DE et al (2013) Xist RNA is a potent suppressor of hematologic cancer in mice. Cell 152:727–742
50. Bussemakers MJ, van Bokhoven A, Verhaegh GW et al (1999) DD3: a new prostate-specific gene, highly overexpressed in prostate cancer. Cancer Res 59:5975–5979
51. Lee GL, Dobi A, Srivastava S (2011) Prostate cancer: diagnostic performance of the PCA3 urine test. Nat Rev Urol 8:123–124

Part III

Metastasis Promotion

Chapter 11

Microvesicles as Mediators of Intercellular Communication in Cancer

Marc A. Antonyak and Richard A. Cerione

Abstract

The discovery that cancer cells generate large membrane-enclosed packets of epigenetic information, known as microvesicles (MVs), that can be transferred to other cells and influence their behavior (Antonyak et al., Small GTPases 3:219–224, 2012; Cocucci et al., Trends Cell Biol 19:43–51, 2009; Rak, Semin Thromb Hemost 36:888–906, 2010; Skog et al., Nat Cell Biol 10:1470–1476, 2008) has added a unique perspective to the classical paracrine signaling paradigm. This is largely because, in addition to growth factors and cytokines, MVs contain a variety of components that are not usually thought to be released into the extracellular environment by viable cells including plasma membrane-associated proteins, cytosolic- and nuclear-localized proteins, as well as nucleic acids, particularly RNA transcripts and micro-RNAs (Skog et al., Nat Cell Biol 10:1470–1476, 2008; Al-Nedawi et al., Nat Cell Biol 10:619–624, 2008; Antonyak et al., Proc Natl Acad Sci U S A 108:4852–4857, 2011; Balaj et al., Nat Commun 2:180, 2011; Choi et al., J Proteome Res 6:4646–4655, 2007; Del Conde et al., Blood 106:1604–1611, 2005; Gallo et al., PLoS One 7:e30679, 2012; Graner et al., FASEB J 23:1541–1557, 2009; Grange et al., Cancer Res 71:5346–5356, 2011; Hosseini-Beheshti et al., Mol Cell Proteomics 11:863–885, 2012; Martins et al., Curr Opin Oncol 25:66–75, 2013; Noerholm et al., BMC Cancer 12:22, 2012; Zhuang et al., EMBO J 31:3513–3523, 2012). When transferred between cancer cells, MVs have been shown to stimulate signaling events that promote cell growth and survival (Al-Nedawi et al., Nat Cell Biol 10:619–624, 2008). Cancer cell-derived MVs can also be taken up by normal cell types that surround the tumor, an outcome that helps shape the tumor microenvironment, trigger tumor vascularization, and even confer upon normal recipient cells the transformed characteristics of a cancer cell (Antonyak et al., Proc Natl Acad Sci U S A 108:4852–4857, 2011; Martins et al., Curr Opin Oncol 25:66–75, 2013; Al-Nedawi et al., Proc Natl Acad Sci U S A 106:3794–3799, 2009; Ge et al., Cancer Microenviron 5:323–332, 2012). Thus, the production of MVs by cancer cells plays crucial roles in driving the expansion of the primary tumor. However, it is now becoming increasingly clear that MVs are also stable in the circulation of cancer patients, where they can mediate long-range effects and contribute to the formation of the pre-metastatic niche, an essential step in metastasis (Skog et al., Nat Cell Biol 10:1470–1476, 2008; Noerholm et al., BMC Cancer 12:22, 2012; Peinado et al., Nat Med 18:883–891, 2012; Piccin et al., Blood Rev 21:157–171, 2007; van der Vos et al., Cell Mol Neurobiol 31:949–959, 2011). These findings, when taken together with the fact that MVs are being aggressively pursued as diagnostic markers, as well as being considered as potential targets for intervention against cancer (Antonyak et al., Small GTPases 3:219–224, 2012; Hosseini-Beheshti et al., Mol Cell Proteomics 11:863–885, 2012; Martins et al., Curr Opin Oncol 25:66–75, 2013; Ge et al., Cancer Microenviron 5:323–332, 2012; Peinado et al., Nat Med 18:883–891, 2012; Piccin et al., Blood Rev 21:157–171, 2007; Al-Nedawi et al., Cell Cycle 8:2014–2018, 2009; Cocucci and Meldolesi, Curr Biol 21:R940–R941, 2011; D'Souza-Schorey and Clancy, Genes Dev

Martha Robles-Flores (ed.), *Cancer Cell Signaling: Methods and Protocols*, Methods in Molecular Biology, vol. 1165, DOI 10.1007/978-1-4939-0856-1_11,

26:1287–1299, 2012; Shao et al., Nat Med 18:1835–1840, 2012), point to critically important roles for MVs in human cancer progression that can potentially be exploited to develop new targeted approaches for treating this disease.

Key words Microvesicles, Shedding vesicles, Microparticles, Oncosomes, Exosomes, Intercellular communication, ARF6, RhoA, EGF-receptor, Extracellular matrix, Cytoskeleton, Tumor microenvironment, Angiogenesis, Metastasis

1 Introduction

1.1 General Cell Communication

Cell-to-cell communication is a fundamental cellular process that is important for embryogenesis and maintaining tissue homeostasis in the adult organism. One of the best studied forms of intercellular signaling is paracrine signaling, which is initiated when one cell secretes soluble forms of growth factors or pro-inflammatory cytokines into its local environment [1–3]. These ligands, after traveling short distances, are then capable of binding their corresponding receptors that are expressed on the plasma membrane of a neighboring (recipient) cell and triggering their activation. Depending on the particular type of receptor that is activated, a specific set of intracellular signaling events is initiated in the recipient cell that impacts its ultimate fate. A good example of this form of communication involves epidermal growth factor (EGF)-mediated signaling [3–5]. As the name implies, EGF was initially shown to be produced by skin cells, but it is now known that several additional cells types produce this growth factor as well. The EGF-receptor, also referred to as ErbB1 or HER1, is a transmembrane receptor tyrosine kinase that is widely expressed in embryonic and adult tissues, suggesting that it exerts pleiotropic effects in development and tissue maintenance and function. Consistent with this idea, when cultures of normal cells that express this receptor are stimulated with EGF, the kinase domain of the ligand-bound EGF-receptor becomes transiently activated and mediates signaling events that cause changes in gene expression, giving rise to controlled cell growth, differentiation, and survival responses [6–9]. Moreover, mouse models where the gene encoding the EGF-receptor has been deleted exhibit a range of severe phenotypes including early embryonic lethality, central nervous system defects, and craniofacial malformations [10–12]. Signaling paradigms analogous to that described for EGF are a recurring theme in cell biology and can be used to describe the mechanism of action for a wide range of growth factors and pro-inflammatory cytokines [2].

Interestingly, cancer cells hijack and de-regulate these same communication processes to promote tumor growth and the development of the metastatic state. A case in point, once again, involves EGF-coupled signaling. The gene encoding the EGF-receptor is frequently amplified or mutated in human cancer cells of brain, breast, lung, bladder, and ovarian origin [13–16]. The resulting overexpression and/or excessive activation of the EGF-receptor in these cancer cells cause aberrant signaling events that have deleterious effects on the health of the cells. These include promoting unregulated cellular growth and chemoresistance, as well as stimulating the migration and metastatic activity of highly aggressive forms of cancer cells [8, 15, 17, 18]. Thus, these findings highlight the important roles that paracrine signaling mechanisms play in both physiological processes and the progression of disease states like cancer.

1.2 Microvesicles: An Unusual Form of Paracrine Signaling

As already mentioned, growth factors and pro-inflammatory cytokines are major mediators of intercellular communication. However, it is becoming increasing clear that cancer cells are also capable of carrying-out a distinct form of paracrine signaling that involves the generation of relatively large extracellular vesicular structures known as microvesicles (MVs). Although initially met with skepticism and thought to be nothing more than cellular debris, the number of studies showing that MVs are actively produced by viable cells and mediate a vast array of critical biological processes has increased exponentially in just the past few years [19–22]. MVs have been referred to in the literature by many different names, with some of the more common examples being ectosomes, microparticles, shedding vesicles, oncosomes, and tumor-derived microvesicles (TMVs). The different names used to describe MVs have created some confusion in the field, although this is often merely a reflection of the biological context in which these vesicles are being studied (i.e., oncosomes and TMVs refer to MVs produced exclusively by cancer cells) or in some cases refer to specific features or physical traits that a type of extracellular vesicle exhibits based on the isolation and characterization approaches used [19, 21–24]. Throughout the remainder of this chapter, we will refer to these vesicular structures as MVs.

Like soluble growth factors (e.g., EGF), MVs are released by cancer cells into their surroundings where they can function as “satellites of intercellular communication” and signal to other nearby cells. It has been proposed that MV-mediated signaling can occur as a result of MVs rupturing and releasing cargo into the extracellular milieu where it comes in contact with and activates cells [25]. In this context, MVs may provide a mechanism for paracrine signals such as growth factors to travel a significant distance from the donor cell, as well as confer a protective effect against their degradation, and/or help to deliver these factors at higher localized concentrations to their target recipient cells, than could otherwise be achieved by classical secretion. More often, however, it seems that the mode of action of MVs involves their surprising uptake by cells [26–30]. This horizontal transfer of MVs between cells has been shown to alter signaling events and change the expression of certain genes in the recipient cells that can influence their fate or function, sometimes in unexpected ways. For example, the sharing of MVs between two cells has been frequently shown to stimulate the signaling activities of the mitogen activated protein (MAP)-kinase family member, ERK (for extracellular-regulated protein kinase) and AKT/protein kinase B (PKB) to promote cell growth and chemoresistance in cell culture and animal tumor models [26, 27, 30–32]. Thus, it is easy to imagine how MVs might contribute to the overall growth of a tumor. Cancer cell-derived MVs have also been implicated in regulating a host of additional pathophysiological processes that are intimately linked to the progression of the malignant state. These include reshaping of the tumor microenvironment, promoting the formation of new blood vessels, evading immunosurveillance responses, eliminating cytotoxic agents from cells, and stimulating the invasive and metastatic activities of highly aggressive forms of cancer cells [20–22, 33]. Many of these topics will be discussed further below.

MVs generally range in size from 0.1 to 2 μm in diameter [21, 27, 34], although MVs as large as 10 μm in diameter have recently been detected in certain human pancreatic cancer cell lines [35]. They are released from cells as an outcome of a poorly understood mechanism that is characterized by the outward budding and fission of the plasma membrane, and do not require the components of the classical secretory pathway, such as the endoplasmic reticulum and Golgi apparatus, for their formation [27, 31, 36]. Moreover, MVs are to be distinguished from exosomes, which are significantly smaller (approximately 40–100 nm in diameter) vesicular structures that are formed through a well-established mechanism that involves the rerouting of multi-vesicular bodies containing endosomes to the cell surface in an ESCRT (for endosomal sorting complex required for transport)-dependent manner [37, 38]. The multi-vesicular bodies then fuse with the plasma membrane and exocytose their contents (i.e., endosomes) into the extracellular milieu. Upon their release from cells, these endocytic vesicles (exosomes) can mediate cell-to-cell communication events that are similar to those stimulated by MVs. Figures 1a, b highlight the general processes of MV and exosome formation and release. Reviews providing an in-depth description of exosome biogenesis and function have recently been published [34, 37–40].

It is generally believed that essentially all human cancer cells, and at least a few types of specialized normal cells (i.e., macrophages, and embryonic stem cells), are capable of generating MVs, but they do so to different extents [27, 31, 33, 41, 42]. Interestingly, higher grade and more aggressive forms of cancer cells consistently generate larger, as well as greater numbers of MVs than lower grade and less aggressive cancer cells. This was demonstrated by two separate studies where comparisons were made between the amounts of MVs generated by human cell lines

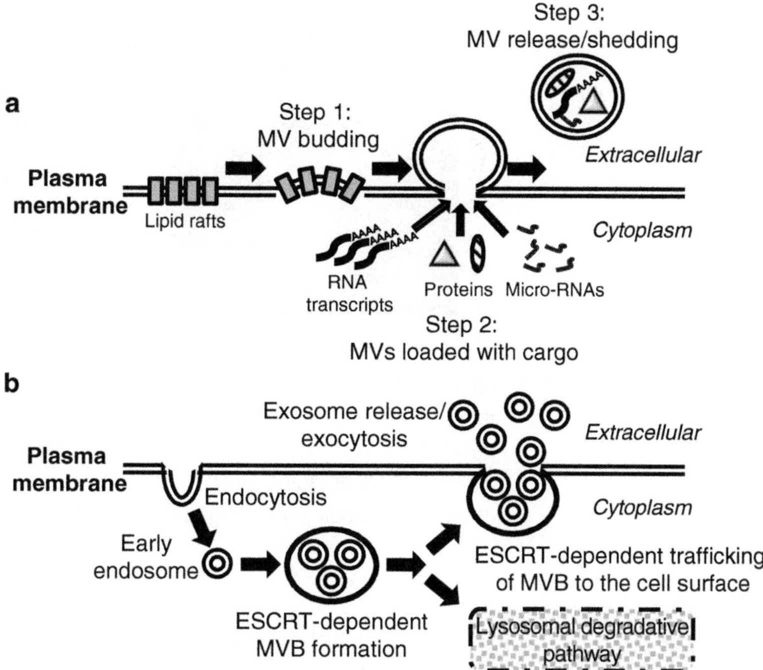

Fig. 1 MVs and exosomes represent two forms of extracellular vesicles that are generated via distinct mechanisms. (**a**) MVs are formed and shed directly from the surface of a cell. During the first step of this process, MVs bud from distinct micro-domains within the plasma membrane known as lipid rafts. The vesicles are then loaded with specific cargo (i.e., proteins, RNA transcripts, and micro-RNAs), while in the third and final step of the process, the MVs are released or shed into the extracellular environment. (**b**) Exosomes are formed as an outcome of exocytosis. Early endosomes are accumulated in multi-vesicular bodies (MVBs) through an ESCRT-dependent mechanism. The MVBs are then either targeted for degradation in the lysosome or are redirected to the cell surface where they fuse with the plasma membrane and release their contents (exosomes) into the extracellular milieu

of increasing transforming activity including MCF10A cells, an immortalized normal (non-transformed) mammary epithelial cell line, SKBR3 cells, a low grade breast cancer cell line, and MDAMB231 cells, a highly aggressive breast cancer cell line [31, 42]. Cultures of the MCF10A cells were shown to generate barely detectable levels of MVs, while MV production by SKBR3 cells was detectable but low. In contrast, the highly aggressive MDAMB231 breast cancer cells produced considerably more MVs than either normal MCF10A epithelial cells or the lower grade SKBR3 cancer cells, with the MVs shed by the MDAMB231 cells frequently reaching sizes of 1–2 μm in diameter.

Because of their relatively large size, MVs can often be detected along the surfaces of cancer cells using electron microscopy, or even

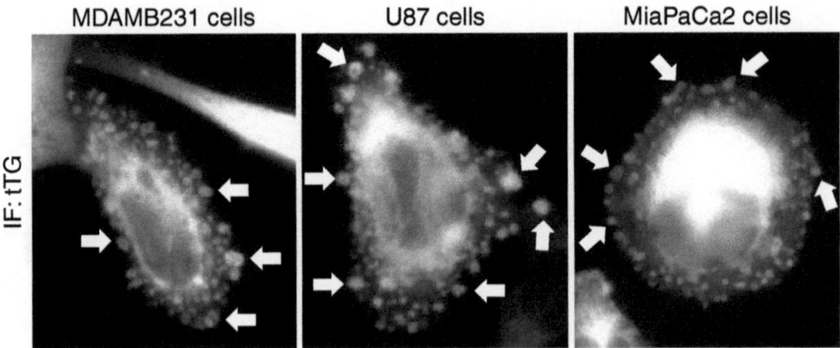

Fig. 2 Highly aggressive human cancer cells generate MVs. MDAMB231 breast cancer cells, U87 glioblastoma cells, and MiaPaCa2 pancreatic cancer cells were immuno-stained with a tTG antibody and then visualized using fluorescent microscopy. A representative image of each cell type is shown. Note that MVs can be readily detected along the surfaces of these cells, with some of the larger and more obvious MVs being indicated with *arrows*

by standard light or fluorescent microscopy [19, 24, 27, 35, 43]. Fluorescent-tagged forms of plasma membrane-targeting sequences, fluorescent dyes that intercalate into lipid bilayers, and immuno-staining of MV cargo have all been successfully used to study various aspects of MV biogenesis and function [24, 27, 28, 31, 32, 35, 43–47]. Examples of three different highly aggressive human cancer cell lines generating MVs, as detected by immuno-fluorescent microscopy, are shown in Fig. 2. Specifically, the images are of a MDAMB231 breast cancer cell, a U87 glioblastoma cell, and a MiaPaCa2 pancreatic cancer cell stained for tissue transglutaminase (tTG), a major protein component of cancer cell-derived MVs [31, 45]. The tTG-labeled MVs of various sizes can be readily detected decorating the surfaces of each of these cells, with some of the largest and most pronounced MVs being indicated with arrows.

In addition to their unusually large size, there are two additional important and interrelated reasons that MVs have been receiving a good deal of attention from the basic and translational cancer research and the pharmaceutical communities. The first has to do with their contents. In addition to containing conventional paracrine signaling molecules, such as growth factors and pro-inflammatory cytokines, MVs also have been shown to contain a variety of cargo not normally thought to be released by viable cells, including cell surface receptor tyrosine kinases (e.g., the EGF-receptor), cytosolic signaling proteins, metabolic enzymes, extra-cellular matrix proteins, metalloproteases (MPPs), nuclear proteins, molecular chaperones (e.g., heat shock protein family members), and RNA transcripts and micro-RNAs [27, 29–31, 36, 41, 44, 48–58]. Figure 3 provides specific examples of some of the more frequently detected proteins in cancer cell-derived MVs

Proteins commonly found in cancer cell-derived MVs:

Growth factors/cytokines		Metabolic enzymes
VEGF	TNFα	PKM isoform 2
EGF	FasL	A-enolase
bFGF	IL-1	Glyceraldehyde-3-phosphate dehygrogenase
TGF	IL-3	
Survivin	IL-8	

Cell surface proteins		Apoptotic proteins
EGF receptor	Flotillin	Caspase-1
EGFRvIII	ErbB2/HER2	Caspase-3
Met	β1-integrins	
TNF receptor		

Extracellular matrix components/modifiers		Signaling/trafficking/ Chaperonin proteins	
Fibronectin	MMP-2	Rab5	Arf6
Collagen	MMP-9	Rab27	AKT
Actin	uPA	Tsg101	PTEN
Tubulin	EMMPRIN	Ran	HSP70
tTG		Ras	HSP90

Fig. 3 A list of proteins frequently contained within MVs derived from human cancer cells. The proteins are separated based upon their primary function

in each of these categories, while databases containing comprehensive lists of the lipids, proteins, RNA transcripts, and microRNAs identified to date in extracellular vesicles, including both MVs and exosomes, can be found on the Web at either Vesiclepedia (www.microvesicles.org) or Exocarta (www.exocarta.com). Although some of the cargo identified in MVs collected from different cancer cells are fairly well conserved, many are unique to their cancer cell type of origin and often include the oncogenic components that are responsible for the transformed characteristics exhibited by the donor cancer cell (the cell that produced the MVs) [26–28, 31, 59, 60].

This then leads to the second important reason that MVs are being intensely studied, namely, because they are stable in the circulation of cancer patients. Many groups have shown that MVs can be isolated from blood samples taken from cancer patients and their contents determined [26, 27, 35, 51, 53, 55, 61]. Because cancer cell-derived MVs often contain oncogenic proteins that reflect their cell of origin, and the extent of their production is correlated to tumor grade/aggressiveness, the idea that MVs can be used as a potential source for cancer biomarkers and/or for monitoring tumor progression is being intensely pursued [22]. The use of MVs in such a way could prove to be especially useful in situations where tumor biopsies are difficult to obtain.

MV biology and their development as potential diagnostic tools have emerged as key areas of cancer research in just the past few years.

Here we will summarize some of the recent advances that have been made regarding the mechanisms that underlie MV biogenesis. Moreover, the critical roles played by MVs in promoting several different aspects of human cancer progression will also be discussed in the following sections.

2 The Biogenesis of MVs

Although the production of MVs is still in many ways poorly understood, sufficient information has been obtained on the subject to provide a general description of this cellular process. MV biogenesis can be broken-down into three distinct steps (*see* Fig. 1a). The first step is the outward protrusion or budding of the MV from the plasma membrane, while the second involves the loading of specific cargo into these vesicles. The third and final step is the scission or shedding of the mature MV from the cell surface.

2.1 Step 1: MV Budding

There are several lines of evidence suggesting that the initial step of MV biogenesis, MV budding, occurs at specific micro-domains (20–200 nm) within the plasma membrane that are enriched in specific lipids and cytoskeletal and signaling proteins (Fig. 1a). These micro-domains, often referred to as "lipid rafts," were originally identified back in the late 1980s and function as hubs of cell signaling, endocytosis, and plasma membrane sorting and trafficking events [62, 63]. One of the strongest indications that MVs originate from lipid rafts stems from the observation that several resident lipid raft proteins, such as flotillin, tissue factor, and stomatin, are also consistently detected in MVs [64–66]. In fact, one of these proteins, flotillin, is now commonly used as a marker of MVs [27, 31, 45]. Another indication that MVs bud from lipid rafts is their dependence on cholesterol. Cholesterol is a major component of lipid rafts and pharmacological depletion of cholesterol from the membranes of cells using compounds like Methyl-β-cyclodextrins (MβCD) and 2-hydroxyl-β-CD (2OHpβCD) potently inhibits MV budding [65, 67, 68].

Another intriguing connection between MV formation and lipids exists. This stems from the unique organization of phospholipids in the cell membrane and how their redistribution can contribute to the membrane curvature that accompanies MV budding. Normally, the outer leaflet of the plasma membrane contains disproportionately high concentrations of phosphatidylcholine and sphingomyelin, while the inner leaflet contains high levels of phosphatidylserine and phosphatidylethanolamine [69, 70]. The maintenance of the asymmetric distribution of these lipids across the surface of the cell is a tightly controlled, active process that is mediated by a group of membrane-associated and calcium-regulated enzymes, including flippases, floppases, and scramblases [69, 71].

Under resting conditions (low calcium concentrations) the flippases and floppases are active and directly assist in lipid transfer, or "flip-flopping" as it is often referred to, from one leaflet of the plasma membrane to the other. For MV budding to occur, cytosolic calcium stores within the cell must be released, an event that reduces or inhibits the activities of certain flippases and floppases, while stimulating the activities of scramblases. The changes in the activation states of these enzymes allow for the accumulation of the negatively charged phosphatidylserine on the outer leaflet of the plasma membrane. This change in net charge at specific sites along the outer surface of a cell promotes the membrane curvature that underlies one of the earliest steps of MV budding, a notion that is supported by the finding that the outer surfaces of shed MVs are enriched with phosphatidylserine [28, 68, 72, 73].

While alterations in the lipid composition of the plasma membrane help contribute to MV budding, they alone are likely not sufficient to drive the process. Rather, the actin cytoskeleton also provides a physical connection with the plasma membrane that is important for MV formation [31, 43, 74, 75]. One of the first suggestions that this was likely going to be the case came from the identification of actin, as well as several other cytoskeletal components, as major constituents of MVs isolated from both normal and cancer cell lines [31, 43, 45, 74, 75]. Moreover, staining cells with fluorescently labeled forms of phalloidin to detect filamentous actin (F-actin) was found to label MVs budding from the surfaces of several different cancer cell lines, including LOX melanoma cells, MDAMB231 breast cancer cells, U87 glioblastoma cells, and HeLa cervical carcinoma cells [31, 43]. It is worth emphasizing that the presence of F-actin in MVs distinguishes them from the more conventional membrane blebbing events that frequently occur in cells, where the plasma membrane extends out from the surface of a cell as an outcome of a localized separation of the plasma membrane from its underlying cytoskeleton [76, 77]. Thus, blebs do not contain cytoskeletal components, including F-actin, except at a very specific and late stage of their life cycle when they are retracted back into the cell. Since blebs also do not normally detach from the surfaces of cells like MVs [76], the formation and shedding of MVs likely represent two distinct cellular processes from membrane blebs.

The suggestion that actin-cytoskeletal rearrangements underlie MV budding has recently been further corroborated by findings that show specific signaling events known to regulate actin dynamics and cell morphology also influence MV formation. RhoA is a member of the Rho subfamily of the Ras superfamily of small GTP-binding proteins (GTPases) and functions as a molecular switch in signaling pathways [78]. Similar to nearly all other GTPases, RhoA cycles between an inactive GDP-bound state and an active GTP-bound state. When bound to GTP, RhoA signals to

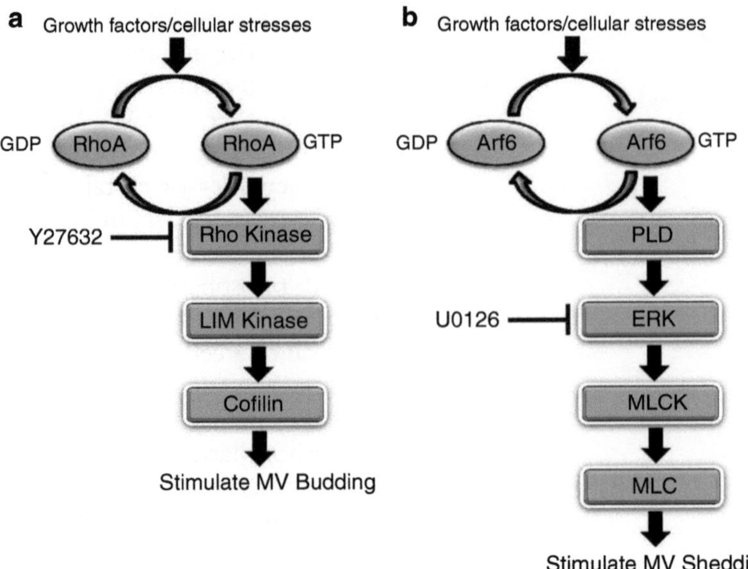

Fig. 4 The signaling pathways that regulate MV budding and shedding in human cancer cells. (**a**) RhoA-dependent signaling induces MV budding. RhoA signaling through Rho kinase to LIM kinase, results in the phosphorylation of cofilin, which inhibits its actin severing capabilities. This results in increased rates of actin polymerization and the budding of MVs from the surfaces of cells. Blocking the activities of components within this pathway (i.e., Rho kinase using Y27632) can effectively inhibit RhoA-mediated MV budding in cancer cells. (**b**) MVs are shed from cells via an ARF6-dependent signaling event that culminates in the phosphorylation of myosin light chain. This results in the "tightening" of the myosin cables at the base of the MV to the point where the vesicle detaches from the surface of the cell

downstream effectors and influences specific cellular processes. One of the best known roles for RhoA is to promote rearrangements of the actin cytoskeleton that induce cell spreading and migration [59, 78, 79]. However, it has recently been shown that RhoA signaling is also capable of stimulating MV budding in a number of different cancer cell lines [31]. The pathway through which RhoA mediates this effect was determined (Fig. 4a) and includes the sequential activation of Rho-associated, coiled-coil containing protein kinase (Rho kinase) and LIM kinase, two proteins that have been heavily implicated in cell migration and especially in the invasive activity exhibited by highly aggressive cancer cells [80–84]. LIM kinase then phosphorylates the actin severing enzyme cofilin on serine 3 and functionally inactivates it. Thus, the normal balance struck in cells between the polymerization of actin filaments and their disassembly is tipped towards actin polymerization and the localized buildup of actin fibers at discrete sites along the surfaces of cells, such as lipid rafts [62, 63, 70], thereby triggering MV budding. The targeted disruption of any component within this pathway, using a combination of small molecule inhibitors, siRNA-mediated knockdown of protein expression, as well as the ectopic expression of dominant-negative forms of various proteins (*see* Fig. 4a), can potently block MV budding. For example, treatment of MDAMB231 breast cancer cells or U87 glioblastoma cells, two cell lines that constitutively produce MVs at a high rate, with the Rho kinase inhibitor Y-27632 [84] markedly reduces the amount of MVs detected along the surfaces of these cells or that are released into their culturing medium [45]. These findings, when combined with others that show that DRF-3 (for Diaphanous-related formin-3), which functions as an effector of RhoA and regulates actin dynamics [83], is frequently deleted in prostate cancer cells that generate large amounts of MVs [24], underscores the importance of RhoA signaling to the cytoskeleton for MV budding.

RhoA and other closely related family members (i.e., Cdc42 and Rac) are frequently activated by growth factor stimulation [78, 79]. Thus, it is not surprising to find that the signals triggered by excessive activation of growth factor receptors, like the EGF-receptor, in cancer cells is functionally coupled to MV biogenesis. EGF stimulation of the human HeLa cervical carcinoma cell line or the prostate cancer cell lines, PC-3 and DU145, dramatically increased the amount of MVs that were detected on the surfaces of these cells or shed into their conditioned medium, compared to their untreated control counterparts [24, 31]. Moreover, the ectopic expression of a naturally occurring in-frame deletion mutant of the EGF-receptor that is constitutively active and highly oncogenic, called EGF-receptor variant type III (EGFRvIII) [85], in human U373 brain tumor cells, similarly enhances MV formation [27]. Importantly, the effects of EGF-receptor activation on MV formation were shown to be blocked by inhibiting the activities of various components of the RhoA signaling pathway in HeLa cervical carcinoma cells, further supporting the idea that an important connection exists between growth factor receptor signaling and the biogenesis of MVs [19, 31].

In addition to growth factor stimulation, challenging cancer cells with stresses that they would typically encounter during tumor progression, or as a part of a therapeutic regiment to treat or manage the disease, including chemotherapeutic agents, gamma irradiation, and hypoxic and serum-limiting conditions, has also recently been shown to increase the rate of MV formation and shedding by cells [30, 31, 39, 86–89]. Moreover, connections between MV biogenesis and alterations to the metabolic programming of cancer cells that often accompany malignant transformation, such as changes in glycolysis (the "Warburg effect") and acquiring an addiction to glutamine [90–92], are just beginning to emerge as well [19, 92]. It will be interesting to determine whether the ability of these different cellular stresses and metabolic changes to trigger MV formation will have a similar requirement for RhoA-signaling, or whether they mediate this effect through a distinct mechanism.

2.2 Step 2: The Loading of Cargo into MVs

The trafficking of cargo into MVs (*see* Fig. 1a) is by far the least understood aspect of the MV life cycle. One thing that is certain regarding this step of MV biogenesis is that the inclusion of proteins, RNA transcripts, and micro-RNAs, in MVs is not random, but rather is the result of a highly selective process. This is perhaps exemplified by findings showing that RhoA localizes to the necks of MVs that are emanating from the surfaces of cells, as detected by immunofluorescent microscopy, where it presumably promotes the formation of F-actin that drives MV formation [45]. Surprisingly, however, despite its physical and functional association with budding MVs, RhoA is conspicuously absent from the MVs once they are released or shed from cells, even under conditions where it is ectopically over-expressed in the donor cells (i.e., the cells from which the MVs are shed) [31, 45]. This is in direct contrast to survivin, a protein whose expression is frequently induced in cancer cells to promote cell survival and chemoresistance [93, 94]. While survivin predominantly resides in the nucleus and is sometimes found associated with mitochondria, several recent studies have shown that it is enriched in extracellular vesicles derived from a host of cancer cells [95–98]. How a protein that is primarily expressed in the nucleus, like survivin, is trafficked into MVs, while another protein (RhoA) that co-localizes with the MVs as they are forming on the surfaces of cells is excluded from these vesicles, speaks to the high degree of regulation that must underlie the process of sorting and loading cargo into MVs.

There have been a few recent suggestions from the literature that proteins and RNA transcripts are more likely to be trafficked into MVs based on a specific modification that they may undergo, or because of unique sequences that they contain. An example of such an MV-targeting signature that has been proposed for proteins is the glycosylphosphatidylinositol (GPI) anchor that is added posttranslationally to the C-terminus of many plasma membrane-associated proteins [99]. Several GPI-anchored proteins, such as Gce1 and CD73, have been identified in MVs derived from normal and cancer cells [100], raising the interesting possibility that post-translational modifications of proteins, particularly those that direct proteins to the cell surface, may help target them to MVs. Indeed, this appears to be true, as the addition of acyl, myristoyl, and palmitoyl moieties to proteins causes them to be more efficiently targeted to extracellular vesicles [47]. However, hundreds of cytosolic and nuclear proteins that lack such protein modifications are also present in MVs. While it has been suggested that these proteins might be recruited into MVs by forming complexes with membrane-targeted proteins inside cells and then, in effect, "piggyback" their way into MVs [47], it is also conceivable that additional mechanisms that regulate the trafficking of cytosolic and nuclear proteins into MVs exist.

Similar to proteins, RNA transcripts and micro-RNAs are also found in MVs [26, 44, 48, 50, 53, 54, 56, 101–103]. Much of the original work showing that RNA transcripts and micro-RNAs are MV cargo was performed using MVs isolated from either cultures of primary brain tumor cells or from blood samples taken from patients afflicted with brain tumors [26, 53]. Microarray analyses performed on donor cancer cells, as well as on the MVs that they produced, revealed that more than 25,000 different RNA transcripts could be detected in the MVs, with approximately 5,000 transcripts being preferentially or exclusively contained within the MVs, suggesting that certain transcripts are selectively enriched in these vesicles [26]. But how are RNA transcripts, especially those that have specifically accumulated in MVs, loaded into the vesicles? Although it should be stressed that the mechanisms that regulate this process remain largely unknown, recent evidence points to the noncoding regions of RNA transcripts as playing an important role. When aligning the sequences of the different RNA transcripts expressed in MVs isolated from two human primary brain tumor cell cultures and two human primary melanoma cultures, it was discovered that several of the most abundant transcripts contained a conserved 25 base-pair sequence in their 3′-untranslated region [104]. The addition of this sequence to the 3′-untranslated region of a reporter RNA transcript and then introducing this construct into cells, led to the enrichment of the reporter RNA transcript in the MVs produced by the transfectants. Thus, these findings provide some of the first insights into how RNA transcripts might be trafficked into MVs.

2.3 Step 3: MV Shedding

The final step of MV biogenesis involves the release or shedding of the mature vesicle from the cell into its surrounding environment (Fig. 1a). Physically, this involves a scission event where the base or neck of the MV continuously gets smaller and smaller until the two plasma membrane bilayers become less than 10 nm apart, at which time the membranes fuse together and the vesicle pinches off from the surface of the cell [105]. At the molecular level, MV shedding, like MV budding, appears to be highly dependent on rearrangement of the actin cytoskeleton. In this case, signaling through another member of the Ras superfamily of small GTPases, ADP-ribosylation factor 6 (ARF6) [106], plays a central role in coordinating MV shedding by triggering signaling events that promote the generation of actin-myosin-based contractile forces at the base or neck of the MVs as they extend outward from the surface of a cell. The connection between ARF6 and MV shedding was established using mouse LOX melanoma cells that were genetically engineered to express either a dominant-negative, inactive form of ARF6 (ARF6 T27N), or a constitutively active form of ARF6 (ARF6 Q67L) [43]. The expression of ARF6 T27N, but not ARF6 Q67L, in these cells resulted in a buildup of MVs on their surfaces, suggesting that interfering with the ability of ARF6 to signal blocks the release of MVs from cells. This ARF6-dependent signaling pathway has been delineated (Fig. 4b) and involves the coordinated activation of several downstream effectors that include phospholipase D (PLD), the mitogen activated protein kinase family member ERK, as well as myosin light chain kinase (MLCK). Once recruited to the neck of the MV where it is activated by ERK, MLCK phosphorylates the regulatory light chain of myosin (MLC) on serine 19, inducing myosin's ability to bind actin filaments and begin the contraction process that underlies MV scission [105].

3 The Consequences of Producing MVs for Cancer Progression

3.1 MVs and Tumor Growth

The release of MVs by cancer cells into their local environment can dramatically alter tumor dynamics in several different ways to drive the growth of the primary tumor. Perhaps the most straightforward mode of action of MVs is through their transfer between two neighboring cancer cells. This was originally demonstrated in a landmark study that in many ways launched much of the current interest in the MV field [27]. In this work, the authors determined that the human U373 brain tumor cell line, engineered to stably express the highly oncogenic form of the EGF-receptor, EGFRvIII [85], generated significantly more MVs compared to the parental U373 cell line, providing some of the earliest evidence suggesting that growth factor receptor-signaling regulates MV biogenesis. When MVs shed by U373 cells expressing EGFRvIII were collected and then added back to cultures of parental U373 cells, it was shown that not only were the MVs taken up by the recipient cells, but they also triggered signaling events that stimulated their growth. The transfer of MVs between cancer cells was also monitored using fluorescence microscopy. Here, a small number of GFP-labeled U373 cells expressing the EGFRvIII were plated or injected into animals together with a large excess of unlabeled parental U373 cells. Within a limited period of time too short to be accounted for by cell growth alone, the number of cells expressing GFP increased exponentially, as detected by fluorescence microscopy, suggesting that the GFP was being transferred between cells via MVs.

However, the authors then went on to make one more unexpected discovery, namely, that EGFRvIII was part of the cargo in the MVs shed by these cells and was responsible for mediating their growth promoting capabilities [27]. These findings were the first of many studies showing that oncogenic proteins (i.e., EGFRvIII), RNA transcripts, and micro-RNAs can be transferred between cancer cells in MVs to elicit growth responses [26, 86, 103]. Subsequent studies performed on patients with brain tumors that express EGFRvIII revealed that the mutant receptor itself, or the

RNA transcript encoding the EGFRvIII, could be routinely detected in MVs isolated from these patients, setting the stage for the potential use of MVs as potential biomarkers [26, 51]. This is an aspect of MV-based research that is now receiving a good deal of attention from translational cancer researchers and the pharmaceutical community [22, 44, 56, 107].

Another way that MVs participate in the growth of the primary tumor involves the effects they have upon the normal cell lineages that are located adjacent to the tumor. This population of cells makes up what is collectively referred to as the tumor microenvironment and plays important roles in cancer progression, largely by producing growth factors and other soluble factors that cancer cells need to sustain their growth [108]. However, reciprocal mechanisms of communication also exist, as we have recently shown that MVs derived from MDAMB231 breast cancer cells can be transferred to fibroblasts or normal (non-transformed) mammary epithelial cells [31, 45], two of the principle cell types found in the microenvironment of mammary tumors [109, 110]. Surprisingly, the uptake of MVs by these recipient cells causes them to acquire many of the transformed characteristics exhibited by the donor MDAMB231 cells, including survival advantages and the ability to form colonies when grown under anchorage-independent conditions, an in vitro measure of tumorgenicity [31, 45]. Likewise, the subcutaneous injection of MDAMB231 cells that were mitotically disabled, but still capable of producing MVs, together with normal fibroblasts into nude mice, resulted in tumors that were primarily composed of cells of fibroblastic origin, suggesting that MVs are capable of transferring oncogenic material to normal cell types that are commonly found in the microenvironment of a tumor (i.e., fibroblasts and mammary epithelial cells) and inducing their transformation.

These findings also revealed some important mechanistic insights into how MVs help to promote cancer progression. For example, the ability of MVs derived from MDAMB231 breast cancer cells, or U87 brain tumor cells, to transform normal recipient cells appears to be transient in nature and requires the continuous addition of freshly prepared MVs to the cells in order to maintain the transformed phenotype [31]. This implies that MVs transfer epigenetic information that has a relatively short half-life (proteins, RNA transcripts, and microRNAs), rather than genetic information (genes or chromosomal material) that can be stably integrated into the genome of the recipient cells and permanently alter their genetic makeup, in order to mediate biological activities. Consistent with this idea, it was shown that the protein crosslinking enzyme, tissue transglutaminase (tTG), and the extracellular matrix protein, fibronectin, are major constituents of cancer cell-derived MVs and that they work together in a unique way to promote the transforming capabilities of MVs derived from MDAMB231 and U87 cells [31].

At least a portion of the tTG expressed in MVs is associated with the outer surfaces of the vesicles where it crosslinks fibronectin, producing a ligand for β1-integrins expressed on the surfaces of recipient fibroblasts and mammary epithelial cells that is capable of triggering excessive signaling and inducing cellular transformation. Importantly, however, if the MVs are not routinely added to the cells to maintain β1-intergrin signaling, then they revert back to a normal phenotype. When considered within the context of a tumor, this could mean that the continuous transfer of epigenetic information in MVs derived from cancer cells to normal cell lineages comprising the tumor microenvironment could have important consequences for tumor development. Instead of a patient's tumor being simply an outcome of the clonal expansion of cancer cells [111], these findings raise the intriguing possibility that a portion of a tumor mass may be due to the ability of MVs to transiently transform the normal cells that surround the tumor and stimulate their aberrant growth.

3.2 MVs and the Promotion of Cancer Cell Survival

The inherent or developed resistance of many cancer cells to chemotherapy, targeted types of therapy, and irradiation, is one of the major hurdles confronted by oncologists as they try to devise treatment regiments for their patients [112]. Through basic research efforts aimed at gaining a better understanding of the mechanisms responsible for allowing cancer cells to evade the effects of apoptotic-inducing stimuli (i.e., chemotherapeutic agents), it is hoped that more effective strategies to manage this disease will be developed. That is why recent findings suggesting that MVs can promote cell survival and contribute to drug resistance are possibly of significant value. The uptake of MVs by cells has the potential to protect them from a variety of apoptotic challenges by upregulating the expression and/or activation of proteins that work to counter the actions of cell death machinery. For example, the phosphatidylinositide 3 (PI3)-kinase-AKT/PKB pathway and the extracellular signal-regulated protein kinase (ERK) pathway, two of the most important signaling systems responsible for promoting cell survival and cancer progression [112, 113], are consistently activated in cells after they have been incubated with MVs derived from cancer cells [26–30]. The activation of either of these pathways upregulates the expression of genes that are critical for protecting cells exposed to a host of apoptotic-inducing stimuli or conditions [112, 113]. Alternatively, MVs shed from cancer cells harbor proteins that can directly impact cell survival, and they are presumably transferred to recipient cells where they function to stave off apoptotic challenges. One such protein is survivin, a member of the inhibitor of apoptosis (IAP) family that can directly bind to caspases and prevent them from triggering apoptosis by cleaving cellular proteins essential for cell viability [93, 94]. Survivin is preferentially expressed across a broad range of tumor types but not in their normal tissue counterparts and an increasing number of studies have identified it as a major protein cargo in MVs [95–98]. Several members of the heat shock protein (HSP) family, including HSP90, HSP70, and HSP60, are also similarly enriched in MVs [57, 114–116]. HSPs are well known for their roles in helping maintain protein integrity under conditions of cellular stress and enhancing signaling events that promote cell survival [117]. The targeted inactivation of survivin and HSP family members using small molecule inhibitors are currently being evaluated in clinical trials as a treatment for a variety of cancer types [94, 117, 118], underscoring the importance of these proteins in promoting cancer progression. Because cancer cell-derived MVs contain survivin and HSPs, it is likely that their transfer to recipient cells will assist in protecting them from stressful conditions.

In addition to promoting cell survival by functioning as "satellites of communication" that share information between cells, the shedding of MVs by cancer cells can also serve as a defense mechanism that helps cells cope with various "insults." Cancer cells that are insensitive to a host of apoptotic challenges including exposure to chemotherapeutic agents or radiation, or to the adverse conditions of being cultured in low serum or limiting oxygen (hypoxia), generate more MVs than cells that are susceptible to these types of stress [30, 31, 39, 86–89]. The reason for this appears to be dependent, at least in part, on the ability of these cells to more efficiently scavenge apoptotic-inducing proteins or cytotoxic compounds and release them in MVs or exosomes. The net result of this process is the depletion of harmful substances or proteins from cells, giving them a better chance at tolerating an insult that under normal conditions would likely kill the cell. As a case in point, it was shown that MVs shed by a chemoresistant ovarian cancer cell line exposed to cisplatin contained over 2.5-fold more of the drug than MVs produced by a cisplatin-sensitive ovarian cancer cell line [119]. Moreover, critical cell death machinery, such as the executioner enzyme caspase 3 [112], is actively removed from cells in MVs [120, 121]. Blocking the ability of MVs containing catalytically active caspase 3 to be shed from cells, led to increased rates of cell death, suggesting that MVs provide an efflux mechanism to rid cancer cells of harmful proteins and compounds [122].

3.3 MVs and Tumor Vascularization

A major tipping point in cancer progression is the development of a tumor's vasculature. Tumors can typically grow to a certain size (~1–2 mm^3) using primarily the oxygen and nutrients provided by an organism's inherent blood vessel network [108, 123]. However, for a tumor to exhibit sustained growth it needs to generate its own blood supply. Although a complicated and multifaceted process that involves the coordinated actions of cancer cells, as well as several types of normal cells and the extracellular matrix, tumor vascularization can be divided into two fundamental stages. In the first stage, the tumor must remodel its microenvironment in such a way as to create paths between the tumor and the organism's vasculature system. The second stage involves the ability of cancer cells to send signals that recruit endothelial cells from preexisting blood vessels to the tumor using these newly created access points. The endothelial cells then proliferate and eventually form a new blood vessel network (angiogenesis) that surrounds, and even penetrates, the tumor mass. This new blood supply provides all the necessary nutrients and oxygen that the tumor needs to sustain its growth.

Multiple lines of evidence have implicated MVs as playing a critical role in each step of the tumor vascularization process. This process is initiated by cancer cells in response to the starvation conditions that they encounter because they have expanded into a tumor mass that the inherent vasculature of the organism can no longer adequately supply with the nutrients and oxygen required for cell survival and growth [108, 123]. Exposing cultured cancer cells to these same types of stressful conditions, through serum starvation or growing them under hypoxic conditions, strongly stimulates MV formation and shedding, suggesting that at least a casual correlation, and possibly an important functional relationship, exists between the ability of cancer cells to make MVs and to initiate tumor vascularization [30, 31, 39, 89]. In order for the starving cancer cells in a tumor to promote angiogenesis, they need to either directly secrete proteases that degrade components of the extracellular matrix or they activate other cells in the tumor microenvironment via paracrine signaling. Perhaps the most important and widely studied of these extracellular matrix-degrading enzymes are matrix metalloproteases (MMPs), a large family of zinc-dependent endopeptidases that collectively cleave nearly every major protein constituent of the extracellular matrix [124]. By degrading the matrix that immediately surrounds the tumor and compromising the integrity of the physical barrier or basement membrane that separates the tumor from the organism's vasculature, the tumor generates the space required to accommodate the new blood vessels that will form, as well as creates migration routes for endothelial cells to travel from a preexisting blood vessel to the tumor [108, 123]. Since MVs have been shown to express the MMP family members MMP-1 and MMP-9, their shedding by cancer cells has been proposed as one mechanism by which they help facilitate extracellular matrix degradation [60, 125–127]. In this context, MVs containing MMPs are thought to travel a certain distance from their cell of origin where they rupture, releasing their contents into the local environment to degrade the matrix.

Cancer cell-derived MVs contain several additional factors that promote angiogenesis (both proteins and RNA transcripts encoding pro-angiogenic proteins) and can be transferred to recipient cells that reside in the tumor microenvironment, reprogramming these cells to overproduce MMPs as well as several

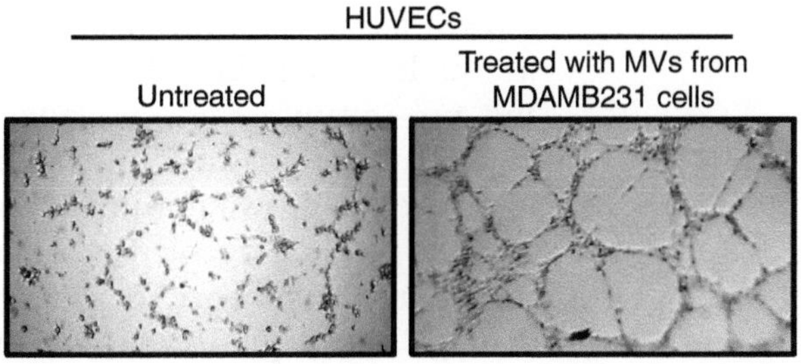

Fig. 5 MVs stimulate angiogenesis. Human umbilical vein endothelial cells (HUVECs) cultured on Matrigel were treated without (untreated control) or with MVs derived from MDAMB231 breast cancer cells. Four hours later, pictures were taken of the cells. Note that the MV-treated HUVECs formed interconnecting tube-like structures (a read-out of angiogenesis), while the control cells did not

other proteolytic enzymes [33]. For example, the uptake of MVs harboring the protein EMMPRIN (for extracellular matrix metalloproteinase inducer) and interleukin-1 (IL-1) by fibroblasts was shown to increase their transcript levels of MMP-1, MMP-2, and MMP-3 [128]. Similarly, the addition of MVs containing EMMPRIN isolated from human ovarian carcinoma cell lines, like OVCAR3 cells, to endothelial cell cultures, stimulated MMP production (i.e., MMP-1, MMP-2, and MT1-MMP) by the recipient cells and enhanced their ability to migrate and invade through an artificial basement membrane matrix (Matrigel), an outcome that is dependent on MMP activity [129]. The net result of MVs stimulating the production of proteolytic enzymes by cells in the tumor microenvironment is to flood the local environment with these proteins, causing the massive degradation and reorganization of the extracellular matrix that is required for the formation of tumor vascular networks.

MVs can also influence angiogenesis by promoting the recruitment of endothelial cells to the tumor, stimulating their growth, and causing them to form into blood vessels [108, 123]. One way to gauge angiogenesis is by using an in vitro assay that monitors the ability of endothelial cells to undergo tubulogenesis, a key step in blood vessel formation [130]. Figure 5 shows how MVs from MDAMB231 breast cancer cells, when collected and then added back to cultures of human umbilical vein endothelial cells (HUVECs), can dramatically impact angiogenesis.

In contrast to untreated (control) cells, which remain as single or small clumps of cells, treatment of HUVECs with MVs for as little as 4 h causes them to move and rearrange themselves into elongated and interconnected tube-like structures (tubulogenesis). In some cases, the ability of cancer cell-derived MVs to induce this phenotype is due to the fact that they contain various combinations of vascular endothelial growth factor (VEGF), basic fibroblast growth factor (bFGF), platelet-derived growth factor (PDGF), as well as IL-1 and IL8 [49, 51, 89, 107, 131, 132], growth factors and pro-inflammatory cytokines that are known to function as important chemoattractants for endothelial cells [123]. Within the setting of a growing tumor, these MVs, along with classically secreted pro-angiogenic factors, establish a gradient that endothelial cells use to enhance their migration capabilities and guide them through the degraded extracellular matrix to the tumor. These same pro-angiogenic factors stimulate endothelial cell growth and their formation into blood vessels. Alternatively, some cancer cells promote endothelial cell migration by shedding MVs that express the EGF-receptor [28]. A549 lung carcinoma cells, DLD-1 colon carcinoma cells, and A431 epidermoid carcinoma cells, all produce MVs that contain activated EGF-receptors. When these MVs are added to cultures of endothelial cells, signaling pathways are activated that increase the production of VEGF. Working in an autocrine signaling fashion, VEGF was then shown to bind to and activate the VEGF-receptor expressed on the surfaces of the endothelial cells, causing them to migrate and undergo tubulogenesis in vitro, as well as promote blood vessel formation in animal tumor models. Since a large percentage of human tumors aberrantly express the EGF-receptor or one of its closely related family members [4], it is interesting to speculate that the EGF-receptor may not only promote cancer progression by transducing signals in cancer cells to stimulate their growth and survival. Specifically, as an outcome of its horizontal transfer to endothelial cells, the EGF-receptor may also participate in tumor angiogenesis by stimulating endothelial cells to migrate and form a vascular network.

3.4 The Long-range Effects of MVs; Promoting Cancer Metastasis

The blood vessel networks that form around tumors as an outcome of angiogenesis are far from normal [108, 123]. They typically are poorly formed and organized, dilated, and most importantly are leaky, providing cancer cells with easy access to an organism's circulation. Indeed, a very small population of cancer cells shed by solid tumors into the circulation of cancer patients can be routinely identified in blood samples taken from these patients [133]. While these circulating tumor cells are being sought after as a source of biomarkers, it is their role in seeding secondary sites of tumor growth that makes circulating tumor cells so detrimental to cancer patients. This is because cancer spreading or metastasis to other parts of the body, rather than the primary tumor itself, is the most common cause of cancer-related deaths [79, 108].

If cancer cells derived from solid tumors can escape into the circulation, then it stands to reason that the MVs generated by cancer cells can do the same. Actually, this have now been shown by several laboratories for a wide variety of cancer types [26, 27, 41, 51, 55, 98, 134, 135]. But why are MVs in the blood of cancer patients? Recent provocative findings using the highly metastatic mouse B16-F10 melanoma cell line as a model suggests that the reason might be to promote metastatic disease by readying the microenvironment at the future site of metastasis (the pre-metastatic niche) to receive circulating cancer cells and stimulate their growth [134]. Some of the strongest support for this notion came from the finding that injecting fluorescently labeled MVs isolated from B16-F10 cells into the tail vein of immunocompromised mice led to their rapid accumulation in bone marrow as well as the same organs that represent the metastatic target sites of B16-F10 cells (i.e., lung). These MVs were shown to trigger extracellular matrix degradation and promote vascular leakiness in the lung, two key events that underlie the formation of the pre-metastatic niche [79]. While MVs can presumably transfer information to recipient cells that reside at the future sites of metastasis and alter their behavior to favor the metastatic process, they also "educate" progenitor cells in the bone marrow [134, 136]. This enables the progenitor cells to translocate to sites where pre-metastatic niches are forming to help further prepare the microenvironment for the arrival of circulating tumor cells by potentiating vascular leakiness and promoting the expression of growth factors that the circulating tumor cells will use to grow and survive. However, the real evidence demonstrating the importance of MVs in promoting metastasis was the demonstration that injecting MVs from B16-F10 cells into the blood stream of mice, prior to injecting the melanoma cells themselves, resulted in the formation of more and larger metastatic nodules on the lungs of the animals compared to when injecting the cancer cells alone [134].

4 Future Perspectives: MVs as Diagnostic Markers and Targets for Therapeutic Intervention

The study of MVs in cancer biology has literally exploded in just the past few years. Although still a young field with many more questions than answers, the ability of MVs to potentially induce or accentuate virtually every major characteristic associated with the oncogenic phenotype has certainly attracted the attention of a variety of research groups. On the one hand, many in the basic cancer research community are trying to gain a deeper appreciation of the mechanisms that underlie MV biogenesis and shedding, how cargo is sorted and trafficked into MVs, as well as identify the specific cargo important for mediating a specific biological outcome. Work currently being done on these fronts will continue to shed more light on the how MVs are regulated and function to drive human cancer progression. Moreover, these lines of investigation could potentially lead to the development of novel approaches to block MV formation/function, and thus limit cancer progression.

Translational researchers are also increasingly pursuing MVs because of their diagnostic potential. In addition to their presence in the blood stream of cancer patients, MVs have also been identified in other biological fluids including cerebral fluid, saliva, and urine [50, 53, 98, 135, 137]. Because they often contain proteins, RNA transcripts, and microRNAs that reflect their cell of origin, MV isolation from biological fluids is now considered an excellent source of biomarkers. Efforts to develop methods to efficiently isolate MV from biological fluids, as well as define and establish markers that can be reliably used for diagnostic purposes, are both considered to be primary goals [22]. Moreover, the levels of cancer cell-derived MVs detected in blood samples taken from cancer patients have been shown to correlate with disease progression [88, 135], raising the possibility that determining MV levels could provide a non-intrusive alternative to biopsies for determining a tumor's grade/aggressiveness.

References

1. Baraniak PR, McDevitt TC (2010) Stem cell paracrine actions and tissue regeneration. Regen Med 5:121–143
2. Perrimon N, Pitsouli C, Shilo BZ (2012) Signaling mechanisms controlling cell fate and embryonic patterning. Cold Spring Harb Perspect Biol 4:a005975
3. Wilson KJ, Mill C, Lambert S et al (2012) EGFR ligands exhibit functional differences in models of paracrine and autocrine signaling. Growth Factors 30:107–116
4. Seshacharyulu P, Ponnusamy MP, Haridas D et al (2012) Targeting the EGFR signaling pathway in cancer therapy. Expert Opin Ther Targets 16:15–31
5. Shilo BZ (2005) Regulating the dynamics of EGF receptor signaling in space and time. Development 132:4017–4027
6. Han W, Lo HW (2012) Landscape of EGFR signaling network in human cancers: biology and therapeutic response in relation to receptor subcellular locations. Cancer Lett 318: 124–134
7. Hopkins S, Linderoth E, Hantschel O et al (2012) Mig6 is a sensor of EGF receptor inactivation that directly activates c-Abl to induce apoptosis during epithelial homeostasis. Dev Cell 23:547–559
8. Moscatello DK, Montgomery RB, Sundareshan P et al (1996) Transformational and altered signal transduction by a naturally occurring mutant EGF receptor. Oncogene 13:85–96
9. Moscatello DK, Ramirez G, Wong AJ (1997) A naturally occurring mutant human epidermal growth factor receptor as a target for peptide vaccine immunotherapy of tumors. Cancer Res 57:1419–1424
10. Miettinen PJ, Berger JE, Meneses J et al (1995) Epithelial immaturity and multiorgan failure in mice lacking epidermal growth factor receptor. Nature 376:337–341
11. Miettinen PJ, Chin JR, Shum L et al (1999) Epidermal growth factor receptor function is necessary for normal craniofacial development and palate closure. Nat Genet 22:69–73
12. Threadgill DW, Dlugosz AA, Hansen LA et al (1995) Targeted disruption of mouse EGF receptor: effect of genetic background on mutant phenotype. Science 269:230–234
13. Kramer C, Klasmeyer K, Bojar H et al (2007) Heparin-binding epidermal growth factor-like growth factor isoforms and epidermal growth factor receptor/ErbB1 expression in bladder cancer and their relation to clinical outcome. Cancer 109:2016–2024
14. Moscatello DK, Holgado-Madruga M, Godwin AK et al (1995) Frequent expression of a mutant epidermal growth factor receptor in multiple human tumors. Cancer Res 55: 5536–5539
15. Paez JG, Janne PA, Lee JC et al (2004) EGFR mutations in lung cancer: correlation with clinical response to gefitinib therapy. Science 304:1497–1500
16. Verhaak RG, Hoadley KA, Purdom E et al (2010) Integrated genomic analysis identifies clinically relevant subtypes of glioblastoma characterized by abnormalities in PDGFRA, IDH1, EGFR, and NF1. Cancer Cell 17:98–110

17. Antonyak MA, Kenyon LC, Godwin AK et al (2002) Elevated JNK activation contributes to the pathogenesis of human brain tumors. Oncogene 21:5038–5046
18. Boroughs LK, Antonyak MA, Johnson JL et al (2011) A unique role for heat shock protein 70 and its binding partner tissue transglutaminase in cancer cell migration. J Biol Chem 286:37094–37107
19. Antonyak MA, Wilson KF, Cerione RA (2012) R(h)oads to microvesicles. Small GTPases 3: 219–224
20. Al-Nedawi K, Meehan B, Rak J (2009) Microvesicles: messengers and mediators of tumor progression. Cell Cycle 8:2014–2018
21. Cocucci E, Meldolesi J (2011) Ectosomes. Curr Biol 21:R940–R941
22. D'Souza-Schorey C, Clancy JW (2012) Tumor-derived microvesicles: shedding light on novel microenvironment modulators and prospective cancer biomarkers. Genes Dev 26:1287–1299
23. Cocucci E, Racchetti G, Meldolesi J (2009) Shedding microvesicles: artefacts no more. Trends Cell Biol 19:43–51
24. Di Vizio D, Kim J, Hager MH et al (2009) Oncosome formation in prostate cancer: association with a region of frequent chromosomal deletion in metastatic disease. Cancer Res 69:5601–5609
25. Varon D, Shai E (2009) Role of platelet-derived microparticles in angiogenesis and tumor progression. Discov Med 8:237–241
26. Skog J, Wurdinger T, van Rijn S et al (2008) Glioblastoma microvesicles transport RNA and proteins that promote tumour growth and provide diagnostic biomarkers. Nat Cell Biol 10:1470–1476
27. Al-Nedawi K, Meehan B, Micallef J et al (2008) Intercellular transfer of the oncogenic receptor EGFRvIII by microvesicles derived from tumour cells. Nat Cell Biol 10: 619–624
28. Al-Nedawi K, Meehan B, Kerbel RS et al (2009) Endothelial expression of autocrine VEGF upon the uptake of tumor-derived microvesicles containing oncogenic EGFR. Proc Natl Acad Sci U S A 106:3794–3799
29. van der Vos KE, Balaj L, Skog J et al (2011) Brain tumor microvesicles: insights into intercellular communication in the nervous system. Cell Mol Neurobiol 31:949–959
30. Wysoczynski M, Ratajczak MZ (2009) Lung cancer secreted microvesicles: underappreciated modulators of microenvironment in expanding tumors. Int J Cancer 125: 1595–1603
31. Antonyak MA, Li B, Boroughs LK et al (2011) Cancer cell-derived microvesicles induce transformation by transferring tissue transglutaminase and fibronectin to recipient cells. Proc Natl Acad Sci U S A 108: 4852–4857
32. Tian T, Wang Y, Wang H et al (2010) Visualizing of the cellular uptake and intracellular trafficking of exosomes by live-cell microscopy. J Cell Biochem 111:488–496
33. Rak J (2010) Microparticles in cancer. Semin Thromb Hemost 36:888–906
34. Ge R, Tan E, Sharghi-Namini S et al (2012) Exosomes in cancer microenvironment and beyond: have we overlooked these extracellular messengers? Cancer Microenviron 5:323–332
35. Di Vizio D, Morello M, Dudley AC et al (2012) Large oncosomes in human prostate cancer tissues and in the circulation of mice with metastatic disease. Am J Pathol 181: 1573–1584
36. Ginestra A, La Placa MD, Saladino F et al (1998) The amount and proteolytic content of vesicles shed by human cancer cell lines correlates with their in vitro invasiveness. Anticancer Res 18:3433–3437
37. Mathivanan S, Ji H, Simpson RJ (2010) Exosomes: extracellular organelles important in intercellular communication. J Proteomics 73:1907–1920
38. Teis D, Saksena S, Emr SD (2009) SnapShot: the ESCRT machinery. Cell 137:182–182 e181
39. Ceruti S, Colombo L, Magni G et al (2011) Oxygen-glucose deprivation increases the enzymatic activity and the microvesicle-mediated release of ectonucleotidases in the cells composing the blood–brain barrier. Neurochem Int 59:259–271
40. Hanson PI, Cashikar A (2012) Multivesicular body morphogenesis. Annu Rev Cell Dev Biol 28:337–362
41. Piccin A, Murphy WG, Smith OP (2007) Circulating microparticles: pathophysiology and clinical implications. Blood Rev 21:157–171
42. Muralidharan-Chari V, Clancy JW, Sedgwick A et al (2010) Microvesicles: mediators of extracellular communication during cancer progression. J Cell Sci 123:1603–1611
43. Muralidharan-Chari V, Clancy J, Plou C et al (2009) ARF6-regulated shedding of tumor cell-derived plasma membrane microvesicles. Curr Biol 19:1875–1885
44. Hosseini-Beheshti E, Pham S, Adomat H et al (2012) Exosomes as biomarker enriched microvesicles: characterization of exosomal proteins derived from a panel of prostate cell

lines with distinct AR phenotypes. Mol Cell Proteomics 11:863–885

45. Li B, Antonyak MA, Zhang J et al (2012) RhoA triggers a specific signaling pathway that generates transforming microvesicles in cancer cells. Oncogene 31:4740–4749
46. Scott G (2012) Demonstration of melanosome transfer by a shedding microvesicle mechanism. J Invest Dermatol 132:1073–1074
47. Shen B, Wu N, Yang JM et al (2011) Protein targeting to exosomes/microvesicles by plasma membrane anchors. J Biol Chem 286: 14383–14395
48. Balaj L, Lessard R, Dai L et al (2011) Tumour microvesicles contain retrotransposon elements and amplified oncogene sequences. Nat Commun 2:180
49. Choi DS, Lee JM, Park GW et al (2007) Proteomic analysis of microvesicles derived from human colorectal cancer cells. J Proteome Res 6:4646–4655
50. Gallo A, Tandon M, Alevizos I et al (2012) The majority of microRNAs detectable in serum and saliva is concentrated in exosomes. PLoS One 7:e30679
51. Graner MW, Alzate O, Dechkovskaia AM et al (2009) Proteomic and immunologic analyses of brain tumor exosomes. FASEB J 23:1541–1557
52. Grange C, Tapparo M, Collino F et al (2011) Microvesicles released from human renal cancer stem cells stimulate angiogenesis and formation of lung premetastatic niche. Cancer Res 71:5346–5356
53. Noerholm M, Balaj L, Limperg T et al (2012) RNA expression patterns in serum microvesicles from patients with glioblastoma multiforme and controls. BMC Cancer 12:22
54. Zhuang G, Wu X, Jiang Z et al (2012) Tumour-secreted miR-9 promotes endothelial cell migration and angiogenesis by activating the JAK-STAT pathway. EMBO J 31:3513–3523
55. Baran J, Baj-Krzyworzeka M, Weglarczyk K et al (2010) Circulating tumour-derived microvesicles in plasma of gastric cancer patients. Cancer Immunol Immunother 59: 841–850
56. Marcucci G, Mrozek K, Radmacher MD et al (2009) MicroRNA expression profiling in acute myeloid and chronic lymphocytic leukaemias. Best Pract Res Clin Haematol 22:239–248
57. Sandvig K, Llorente A (2012) Proteomic analysis of microvesicles released by the human prostate cancer cell line PC-3. Mol Cell Proteomics 11:M111.012914
58. Zhou Q, Souba WW, Croce CM et al (2010) MicroRNA-29a regulates intestinal membrane permeability in patients with irritable bowel syndrome. Gut 59:775–784
59. Del Conde I, Bharwani LD, Dietzen DJ et al (2007) Microvesicle-associated tissue factor and Trousseau's syndrome. J Thromb Haemost 5:70–74
60. Graves LE, Ariztia EV, Navari JR et al (2004) Proinvasive properties of ovarian cancer ascites-derived membrane vesicles. Cancer Res 64:7045–7049
61. Chen C, Skog J, Hsu CH et al (2010) Microfluidic isolation and transcriptome analysis of serum microvesicles. Lab Chip 10: 505–511
62. Owen DM, Magenau A, Williamson D et al (2012) The lipid raft hypothesis revisited: new insights on raft composition and function from super-resolution fluorescence microscopy. Bioessays 34:739–747
63. Simons K, Ikonen E (1997) Functional rafts in cell membranes. Nature 387:569–572
64. Gangalum RK, Atanasov IC, Zhou ZH et al (2011) AlphaB-crystallin is found in detergent-resistant membrane microdomains and is secreted via exosomes from human retinal pigment epithelial cells. J Biol Chem 286: 3261–3269
65. Lopez JA, del Conde I, Shrimpton CN (2005) Receptors, rafts, and microvesicles in thrombosis and inflammation. J Thromb Haemost 3:1737–1744
66. Mairhofer M, Steiner M, Mosgoeller W et al (2002) Stomatin is a major lipid-raft component of platelet alpha granules. Blood 100: 897–904
67. Del Conde I, Shrimpton CN, Thiagarajan P et al (2005) Tissue-factor-bearing microvesicles arise from lipid rafts and fuse with activated platelets to initiate coagulation. Blood 106:1604–1611
68. Liu ML, Scalia R, Mehta JL et al (2012) Cholesterol-induced membrane microvesicles as novel carriers of damage-associated molecular patterns: mechanisms of formation, action, and detoxification. Arterioscler Thromb Vasc Biol 32:2113–2121
69. Pomorski T, Menon AK (2006) Lipid flippases and their biological functions. Cell Mol Life Sci 63:2908–2921
70. Seigneuret M, Devaux PF (1984) ATP-dependent asymmetric distribution of spin-labeled phospholipids in the erythrocyte membrane: relation to shape changes. Proc Natl Acad Sci U S A 81:3751–3755

71. Contreras FX, Sanchez-Magraner L, Alonso A et al (2010) Transbilayer (flip-flop) lipid motion and lipid scrambling in membranes. FEBS Lett 584:1779–1786
72. Dasgupta SK, Abdel-Monem H, Niravath P et al (2009) Lactadherin and clearance of platelet-derived microvesicles. Blood 113: 1332–1339
73. Lima LG, Chammas R, Monteiro RQ et al (2009) Tumor-derived microvesicles modulate the establishment of metastatic melanoma in a phosphatidylserine-dependent manner. Cancer Lett 283:168–175
74. Collino F, Deregibus MC, Bruno S et al (2010) Microvesicles derived from adult human bone marrow and tissue specific mesenchymal stem cells shuttle selected pattern of miRNAs. PLoS One 5:e11803
75. Meng Y, Kang S, Fishman DA (2005) Lysophosphatidic acid stimulates fas ligand microvesicle release from ovarian cancer cells. Cancer Immunol Immunother 54:807–814
76. Charras GT, Hu CK, Coughlin M et al (2006) Reassembly of contractile actin cortex in cell blebs. J Cell Biol 175:477–490
77. Fackler OT, Grosse R (2008) Cell motility through plasma membrane blebbing. J Cell Biol 181:879–884
78. Heasman SJ, Ridley AJ (2008) Mammalian Rho GTPases: new insights into their functions from in vivo studies. Nat Rev Mol Cell Biol 9:690–701
79. Sahai E (2007) Illuminating the metastatic process. Nat Rev Cancer 7:737–749
80. Hahmann C, Schroeter T (2010) Rho-kinase inhibitors as therapeutics: from pan inhibition to isoform selectivity. Cell Mol Life Sci 67: 171–177
81. Mizuno K (2013) Signaling mechanisms and functional roles of cofilin phosphorylation and dephosphorylation. Cell Signal 25: 457–469
82. Narumiya S, Ishizaki T, Uehata M (2000) Use and properties of ROCK-specific inhibitor Y-27632. Methods Enzymol 325:273–284
83. Narumiya S, Tanji M, Ishizaki T (2009) Rho signaling, ROCK and mDia1, in transformation, metastasis and invasion. Cancer Metastasis Rev 28:65–76
84. Sahai E, Ishizaki T, Narumiya S et al (1999) Transformation mediated by RhoA requires activity of ROCK kinases. Curr Biol 9:136–145
85. Del Vecchio CA, Li G, Wong AJ (2012) Targeting EGF receptor variant III: tumor-specific peptide vaccination for malignant gliomas. Expert Rev Vaccines 11:133–144
86. Corcoran C, Rani S, O'Brien K et al (2012) Docetaxel-resistance in prostate cancer: evaluating associated phenotypic changes and potential for resistance transfer via exosomes. PLoS One 7:e50999
87. Lehmann BD, Paine MS, Brooks AM et al (2008) Senescence-associated exosome release from human prostate cancer cells. Cancer Res 68:7864–7871
88. Shedden K, Xie XT, Chandaroy P et al (2003) Expulsion of small molecules in vesicles shed by cancer cells: association with gene expression and chemosensitivity profiles. Cancer Res 63:4331–4337
89. Svensson KJ, Kucharzewska P, Christianson HC et al (2011) Hypoxia triggers a proangiogenic pathway involving cancer cell microvesicles and PAR-2-mediated heparin-binding EGF signaling in endothelial cells. Proc Natl Acad Sci U S A 108:13147–13152
90. Erickson JW, Cerione RA (2010) Glutaminase: a hot spot for regulation of cancer cell metabolism? Oncotarget 1:734–740
91. Koppenol WH, Bounds PL, Dang CV (2011) Otto Warburg's contributions to current concepts of cancer metabolism. Nat Rev Cancer 11:325–337
92. Wilson KF, Erickson JW, Antonyak MA et al (2013) Rho GTPases and their roles in cancer metabolism. Trends Mol Med 19:74–82
93. Altieri DC (2008) New wirings in the survivin networks. Oncogene 27:6276–6284
94. Cheung CH, Cheng L, Chang KY et al (2011) Investigations of survivin: the past, present and future. Front Biosci 16:952–961
95. Honegger A, Leitz J, Bulkescher J et al (2013) Silencing of human papillomavirus (HPV) E6/E7 oncogene expression affects both the contents and amounts of extracellular microvesicles released from HPV-positive cancer cells. Int J Cancer 133:1631–1642. doi:10.1002/ijc.28164
96. Khan S, Jutzy JM, Aspe JR et al (2011) Survivin is released from cancer cells via exosomes. Apoptosis 16:1–12
97. Khan S, Jutzy JM, Valenzuela MM et al (2012) Plasma-derived exosomal survivin, a plausible biomarker for early detection of prostate cancer. PLoS One 7:e46737
98. Taylor DD, Gercel-Taylor C, Parker LP (2009) Patient-derived tumor-reactive antibodies as diagnostic markers for ovarian cancer. Gynecol Oncol 115:112–120
99. Fujita M, Kinoshita T (2012) GPI-anchor remodeling: potential functions of GPI-anchors in intracellular trafficking and membrane

dynamics. Biochim Biophys Acta 1821: 1050–1058

100. Muller G, Schneider M, Biemer-Daub G et al (2011) Microvesicles released from rat adipocytes and harboring glycosylphosphatidylinositol-anchored proteins transfer RNA stimulating lipid synthesis. Cell Signal 23: 1207–1223

101. Hao S, Ye Z, Li F et al (2006) Epigenetic transfer of metastatic activity by uptake of highly metastatic B16 melanoma cell-released exosomes. Exp Oncol 28:126–131

102. Hessvik NP, Phuyal S, Brech A et al (2012) Profiling of microRNAs in exosomes released from PC-3 prostate cancer cells. Biochim Biophys Acta 1819:1154–1163

103. Chiba M, Kimura M, Asari S (2012) Exosomes secreted from human colorectal cancer cell lines contain mRNAs, microRNAs and natural antisense RNAs, that can transfer into the human hepatoma HepG2 and lung cancer A549 cell lines. Oncol Rep 28:1551–1558

104. Bolukbasi MF, Mizrak A, Ozdener GB et al (2012) miR-1289 and "Zipcode"-like sequence enrich mRNAs in microvesicles. Mol Ther Nucleic Acids 1:e10

105. Faini M, Beck R, Wieland FT et al (2013) Vesicle coats: structure, function, and general principles of assembly. Trends Cell Biol 23:279–288. doi:10.1016/j.tcb.2013.01.005

106. Donaldson JG, Jackson CL (2011) ARF family G proteins and their regulators: roles in membrane transport, development and disease. Nat Rev Mol Cell Biol 12:362–375

107. Martins VR, Dias MS, Hainaut P (2013) Tumor-cell-derived microvesicles as carriers of molecular information in cancer. Curr Opin Oncol 25:66–75

108. Egeblad M, Nakasone ES, Werb Z (2010) Tumors as organs: complex tissues that interface with the entire organism. Dev Cell 18: 884–901

109. Bhowmick NA, Neilson EG, Moses HL (2004) Stromal fibroblasts in cancer initiation and progression. Nature 432:332–337

110. Perou CM, Sorlie T, Eisen MB et al (2000) Molecular portraits of human breast tumours. Nature 406:747–752

111. Greaves M, Maley CC (2012) Clonal evolution in cancer. Nature 481:306–313

112. Redmond KM, Wilson TR, Johnston PG et al (2008) Resistance mechanisms to cancer chemotherapy. Front Biosci 13:5138–5154

113. Roberts PJ, Der CJ (2007) Targeting the Raf-MEK-ERK mitogen-activated protein kinase cascade for the treatment of cancer. Oncogene 26:3291–3310

114. Campanella C, Bucchieri F, Merendino AM et al (2012) The odyssey of Hsp60 from tumor cells to other destinations includes plasma membrane-associated stages and Golgi and exosomal protein-trafficking modalities. PLoS One 7:e42008

115. Chalmin F, Ladoire S, Mignot G et al (2010) Membrane-associated Hsp72 from tumor-derived exosomes mediates STAT3-dependent immunosuppressive function of mouse and human myeloid-derived suppressor cells. J Clin Invest 120:457–471

116. McCready J, Sims JD, Chan D et al (2010) Secretion of extracellular hsp90alpha via exosomes increases cancer cell motility: a role for plasminogen activation. BMC Cancer 10:294

117. Morimoto RI (2011) The heat shock response: systems biology of proteotoxic stress in aging and disease. Cold Spring Harb Symp Quant Biol 76:91–99

118. Rappa F, Farina F, Zummo G et al (2012) HSP-molecular chaperones in cancer biogenesis and tumor therapy: an overview. Anticancer Res 32:5139–5150

119. Safaei R, Larson BJ, Cheng TC et al (2005) Abnormal lysosomal trafficking and enhanced exosomal export of cisplatin in drug-resistant human ovarian carcinoma cells. Mol Cancer Ther 4:1595–1604

120. Boing AN, Hau CM, Sturk A et al (2008) Platelet microparticles contain active caspase 3. Platelets 19:96–103

121. Sapet C, Simoncini S, Loriod B et al (2006) Thrombin-induced endothelial microparticle generation: identification of a novel pathway involving ROCK-II activation by caspase-2. Blood 108:1868–1876

122. Abid Hussein MN, Boing AN, Sturk A et al (2007) Inhibition of microparticle release triggers endothelial cell apoptosis and detachment. Thromb Haemost 98:1096–1107

123. Bergers G, Benjamin LE (2003) Tumorigenesis and the angiogenic switch. Nat Rev Cancer 3:401–410

124. Gialeli C, Theocharis AD, Karamanos NK (2011) Roles of matrix metalloproteinases in cancer progression and their pharmacological targeting. FEBS J 278:16–27

125. Dolo V, D'Ascenzo S, Violini S et al (1999) Matrix-degrading proteinases are shed in membrane vesicles by ovarian cancer cells in vivo and in vitro. Clin Exp Metastasis 17:131–140

126. Janowska-Wieczorek A, Marquez-Curtis LA, Wysoczynski M et al (2006) Enhancing effect of platelet-derived microvesicles on the invasive potential of breast cancer cells. Transfusion 46:1199–1209

127. Janowska-Wieczorek A, Wysoczynski M, Kijowski J et al (2005) Microvesicles derived from activated platelets induce metastasis and angiogenesis in lung cancer. Int J Cancer 113:752–760
128. Braundmeier AG, Dayger CA, Mehrotra P et al (2012) EMMPRIN is secreted by human uterine epithelial cells in microvesicles and stimulates metalloproteinase production by human uterine fibroblast cells. Reprod Sci 19:1292–1301
129. Millimaggi D, Mari M, D'Ascenzo S et al (2007) Tumor vesicle-associated CD147 modulates the angiogenic capability of endothelial cells. Neoplasia 9:349–357
130. Abe T, Okamura K, Ono M et al (1993) Induction of vascular endothelial tubular morphogenesis by human glioma cells. A model system for tumor angiogenesis. J Clin Invest 92:54–61
131. Baj-Krzyworzeka M, Szatanek R, Weglarczyk K et al (2006) Tumour-derived microvesicles carry several surface determinants and mRNA of tumour cells and transfer some of these determinants to monocytes. Cancer Immunol Immunother 55:808–818
132. Mause SF, Weber C (2010) Microparticles: protagonists of a novel communication network for intercellular information exchange. Circ Res 107:1047–1057
133. Joosse SA, Pantel K (2013) Biologic challenges in the detection of circulating tumor cells. Cancer Res 73:8–11
134. Peinado H, Aleckovic M, Lavotshkin S et al (2012) Melanoma exosomes educate bone marrow progenitor cells toward a pro-metastatic phenotype through MET. Nat Med 18:883–891
135. Shao H, Chung J, Balaj L et al (2012) Protein typing of circulating microvesicles allows real-time monitoring of glioblastoma therapy. Nat Med 18:1835–1840
136. Aliotta JM, Pereira M, Johnson KW et al (2010) Microvesicle entry into marrow cells mediates tissue-specific changes in mRNA by direct delivery of mRNA and induction of transcription. Exp Hematol 38: 233–245
137. Nilsson J, Skog J, Nordstrand A et al (2009) Prostate cancer-derived urine exosomes: a novel approach to biomarkers for prostate cancer. Br J Cancer 100:1603–1607

Chapter 12

In Vivo Rat Model to Study Horizontal Tumor Progression

Catalina Trejo-Becerril, Enrique Pérez-Cárdenas, and Alfonso Dueñas-González

Abstract

Most cancer deaths are due to metastases. Metastasis is an extraordinarily complex process by which cancer cells complete a sequential series of steps before they transform into a clinically detectable lesion. These steps typically include separation from the primary tumor, invasion through surrounding tissues and basement membranes, entry and survival in the circulation, lymphatic or peritoneal space, and arrest in a distant target organ and the formation of secondary tumors in distant organs.

While proposed or accepted models and mechanisms of metastatic progression, have been demonstrated in experimental systems, none of them sufficiently explain all of the complexities associated with this process. These models can broadly be classified into two types, those occurring by vertical gene transfer (Darwinian) and those involving horizontal or lateral DNA transfer. Here, we describe an experimental system to study the metastatic process involving the horizontal transfer of circulating DNA.

Key words Horizontal tumor progression, Circulating DNA, Horizontal transfer

1 Introduction

The spread of cancer cells from a primary tumor to distant sites in the body, known as metastasis, is responsible for most cancer patient´s morbidity and mortality [1]. It is accepted that cancer cells leave the primary tumor, disseminate, and colonize distant organs to form solid metastasis. This process occurs through a series of sequential steps that include the invasion of adjacent tissues, intravasation, transporting through the circulatory system, arresting at a secondary site, extravasation, and growth in a secondary organ [2]. Many models and mechanisms of metastatic progression have been proposed in an attempt to explain the biological complexities of metastasis which include (1) transient compartment, (2) fusion model, (3) early oncogenesis model, (4) genetic predisposition model, (5) progression model, and (6) horizontal gene transfer model [3].

Martha Robles-Flores (ed.), *Cancer Cell Signaling: Methods and Protocols*, Methods in Molecular Biology, vol. 1165, DOI 10.1007/978-1-4939-0856-1_12,

The most commonly accepted model of metastasis for the past 35 years has been the progression model [4]. This model suggests that a series of mutational events occur either in subpopulations of the primary tumor or disseminated cells, resulting in a small fraction of cells that acquire full metastatic potential. This means that the offspring of an initiating tumor cell inherit the genetic and epigenetic alterations leading to tumor progression via vertical gene transfer.

Horizontal gene transfer has been reported in bacteria and fungi, where it plays an important role in the generation of resistance to antibiotic drugs as well as for adaptation to new environments. Transfer of DNA from bacteria to somatic cells may also occur, as shown by in vitro experiments that demonstrate efficient uptake of a β-gal reporter plasmid from attenuated bacteria into the nucleus of the phagocytic cell [5, 6]. In 1981 Klein [7] proposed a hypothetical step on the route to tumorigenesis, which may be common to many of the tumorigenic events taking place in vivo. According to this hypothesis, portions of DNA from dying tumorigenic cells may escape enzymatic degradation, "*transfect*" other cells and, under appropriate conditions, also induce transformation. This hypothesis is related to the acquisition of metastatic capacity by horizontal gene transfer. That proposal is supported by the results of Holmgrem and colleagues who found that [8] cocultured apoptotic bodies from lymphoid cell lines containing integrated, but not episomal copies of EBV, with either human fibroblasts, macrophages, or bovine aortic endothelial cells, result in expression of the EBV-encoded genes EBER and EBNA1 in these recipient cells at a high frequency. These results confirm that the DNA may be rescued and reused from apoptotic bodies by somatic cells. On the other hand, Goldenberg et al. [9] reported in vivo cell–cell fusion of human lymphoma and rodent host cells, resulting in heterosynkaryon tumor cells that display both human and hamster DNA. According to the authors, it may be a method to disclose genes resulting both organoid and metastasis signatures, suggesting that the horizontal transfer of tumor DNA to adjacent stromal cells may be implicated in tumor heterogeneity and progression.

At present, the horizontal gene transfer model has aroused considerable interest after it was recognized that circulating tumor DNA is present in the bloodstream of animal tumor models [10–12] and cancer patients [13–16]. Also within an organism, the circulating DNA, such as exosomes that contain transcriptionally active mRNA and microRNA, may potentially act as an endocrine or paracrine messenger, able to affect the functionality of recipient cells [17]. According to this, it has been proposed that circulating DNA could participate in the development of metastases via "passive" transfection-like uptake of such nucleic acids by susceptible cells [18]. García-Olmo et al. [19] have shown the biological

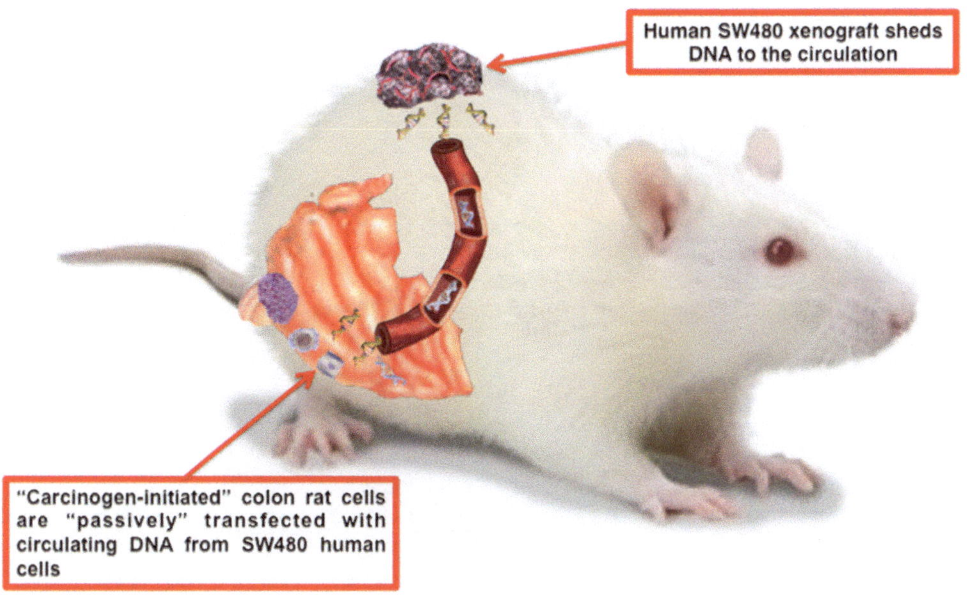

Fig. 1 Experimental models in vitro and in vivo, where it is shown that the circulating DNA, which comes from tumor cells (SW480) transplanted into the flank of rat, is capable of horizontally transfer to the colon cells, which were previously subjected to treatment with a carcinogen (1,2-dimethylhydrazine, DMH)

feasibility of gene transfer and the transformation of cells by cell-free tumor-derived nucleic acids in the plasma of cancer patients.

We implemented an experimental model in immunocompetent rats to show that circulating DNA derived from tumor cells (SW480) transplanted in the flanks of rats, transfers into colonic rat cells previously subjected to treatment with the carcinogen 1,2-dimethylhydrazine (DMH) (Fig. 1).

We also show that these carcinogen-induced rat colonic tumors grow faster if rats are xenografted with human cancer cells as a source of malignant circulating DNA. Interestingly, depletion of circulating DNA by enzymatic treatment ameliorates tumor progression [20] (Fig. 2). Here, we focus on describing the experimental system where circulating DNA participates in horizontal tumor progression in an in vivo model.

2 Materials

2.1 In Vivo Experiments

1. SW480 human colon cancer cell line which has a codon 12 mutation at *K-ras* (*from* American Type Culture Collection).
2. DMEM-F12 medium for cancer cell line.
3. Fetal Bovine Serum.
4. Phosphate-buffered saline (PBS: 137 mM NaCl, 2.7 mM KCl, 10 mM Na_2HPO_4, 2 mM KH_2PO_4, pH 7.4).

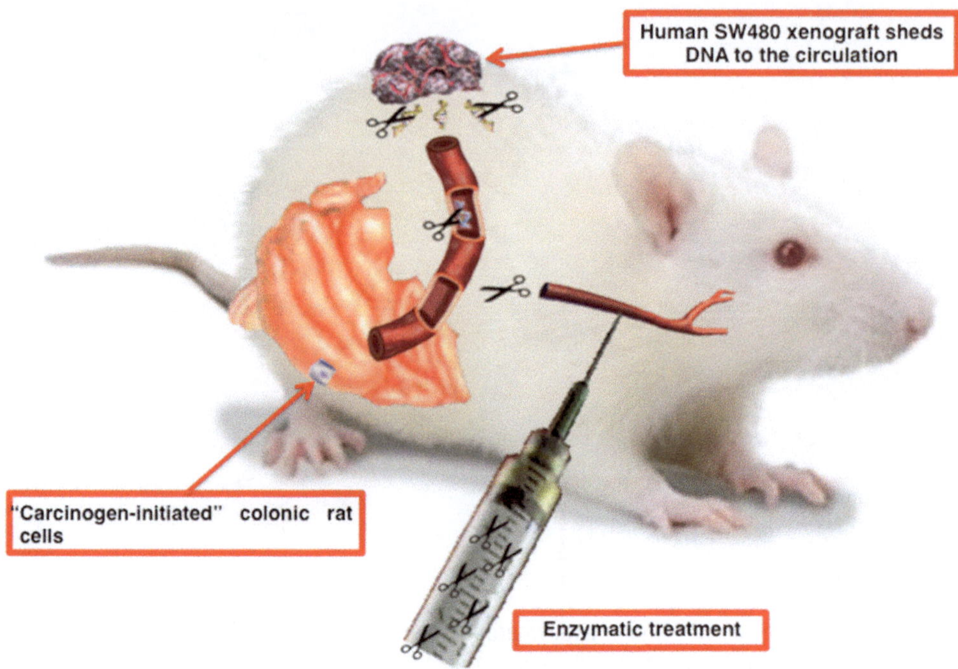

Fig. 2 Enzyme mixture (DNase I, trypsin, papain, and chymotrypsin) depletes circulating DNA and prevents tumor progression

5. Trypsin–EDTA 0.25 % solution.
6. Antibiotics: Penicillin (100 U/ml) and Streptomycin (100 μg/ml).
7. Wistar female rats, 5–6 weeks old.
8. DMH.
9. 70 % (vol/vol) ethanol.
10. 0.9 % (w/v) NaCl solution.
11. DNAse I and protease mix in 0.9 % NaCl solution: DNAseI 2.3 mg/kg body weight, trypsin 0.2 mg/20 g, chymotrypsin 0.2 mg/20 g, and papain 0.5 mg/20 g of body weight, respectively.
12. Isoflurane.
13. Trypan blue.
14. 10 % neutral formalin.

2.2 Tissue Preparation and Microdissection

1. Microtome with a disposable blade.
2. PALM Laser-MicroBeam System (P.A.L.M., Wolfratshausen, Germany) was used for microdissection.
3. For deparaffinization: Xylene, 100 % ethanol, 90 % ethanol, and 70 % ethanol.
4. Hematoxylin and eosin stains.

5. RNase-free H_2O.
6. DNA isolation kit (such as Arcturus®PicoPure®, ARCTURUS Mountain View USA).

2.3 Horizontal Transfer of Circulating DNA

1. dNTP, Taq polymerase (Applied Biosystems).
2. Thermal cycler (Applied Biosystems).
3. 0.2 M HCl.
4. 8 % Sodium thiocyanate.
5. 0.5 % Pepsin.
6. Red-fluorescein-labelled ALU human (FISHBright, Kreatech Biotechnology) and green-fluorescein-labelled LINE 1 rat (FISHBright, Kreatech Biotechnology) probes.
7. Rubber cement.
8. 20× SSC: 3 M NaCl (175.3 g/l), 0.3 M sodium citrate-2 H_2O (88.2 g/l). Adjust pH to 7.0 with 1 M HCl.
9. NP-40.
10. 4′,6′-Diamino-2-phenylindole (DAPI) at a concentration of 0.1-mg/ml in Antifade (Vector Laboratories).
11. Hybridizer (Life Technologies).
12. Fluorescence microscope (Zeiss Axioplan).

2.4 General Equipment

1. Anesthesia Classic System with vaporizer unit.
2. Euthanasia system for small animals.
3. Clinical centrifuge.
4. Laminar flow hood.
5. Analytical balance.
6. Cell counting.
7. 1.5 ml sterile Eppendorf tubes.
8. 50 ml centrifuge tubes.
9. 15 ml centrifuge tubes.
10. 175 cm^2 culture flasks.
11. 75 cm^2 culture flasks.
12. Syringes with 30G or 25G needle.

3 Methods

3.1 In Vivo Experiments

1. Obtain sufficient number of female Wistar rats for the experiments (*see* **Note 1**).
2. After 10 days of acclimation, the rats are randomly assigned to each study group in environmentally controlled room maintained at 22 °C and 50 % relative humidity with a 12-h light–dark cycle.

3. Dissolve the DMH in a 0.9 % (w/v) NaCl solution, immediately before use. This freshly made solution must be filter-sterilized.
4. Every week the animals must be weighted and the dose of DMH adjusted according to the concurred dose of body weight (20 mg/kg) [21]. DMH must be administered intraperitoneally for 20 weeks. The control animals must receive an equivalent number of injections of the saline vehicle.
5. Grow SW480 cells in a monolayer in DMEM-F12 supplemented with 5 % fetal bovine serum.
6. Remove and discard culture medium, wash the cells with 3 ml of cool PBS and incubate the cells in 1 ml of 0.25 % Trypsin–EDTA solution. Add 3 ml of culture medium and centrifuge at $200 \times g$ for 5 min. Then, prepare cell suspension in a concentration of 5×10^6 cells in 200 μl of fetal bovine serum-free DMEM/F12, and leave at room temperature with occasional agitation until they are inoculated (*see* **Note 2**).
7. Before the subcutaneous injection of cells, the animal's flanks must be shaved, to ensure a successful inoculation.
8. Introduce the animal into the anesthesia chamber gas (isoflurane) and turn it to level 2. Before starting the inoculation, make sure that the rats have completely lost consciousness by stimulating the abdominal skin.
9. Clean with ethanol (70 %) the inoculation site (both left and right dorsal flanks) and inject the cells slowly (5×10^6 cells per site).
10. Cell inoculation must be done in 2 phases (Fig. 3): (a) The first is at 28 days of starting treatment with DMH. (b) The second is at 49 days of starting treatment with DMH. At the end, a total amount of 20×10^6 cells must be inoculated per rat.
11. Prepare enzymes in a 0.9 % (w/v) NaCl solution, immediately before use. This freshly made solution must be filter-sterilized (*see* **Note 3**).
12. Inject intramuscularly (hindlimbs) the DNAse I and the protease solutions intraperitoneally. The control animals must receive an equivalent number of injections of the saline vehicle only. Every week the animals must be weighted and the dose of each enzyme adjusted so that it always be 2.3 mg/kg for DNAse I [22] and 0.9 mg/20 g per body weight for the mix of Proteases (trypsin 0.2 mg/20 g, chymotrypsin 0.2 mg/20 g, and papain 0.5 mg/20 g per body weight, *see* **Note 3**).
13. Enzymes must be administered daily except weekends, and the treatment is for 7 weeks (Fig. 3).
14. From week one of treatment with the carcinogen, all rats must be monitored at least twice per week in order to detect the general signs of disease such as ruffled fur, anal bleeding, diarrhea, weight loss, and/or abdominal inflammation.

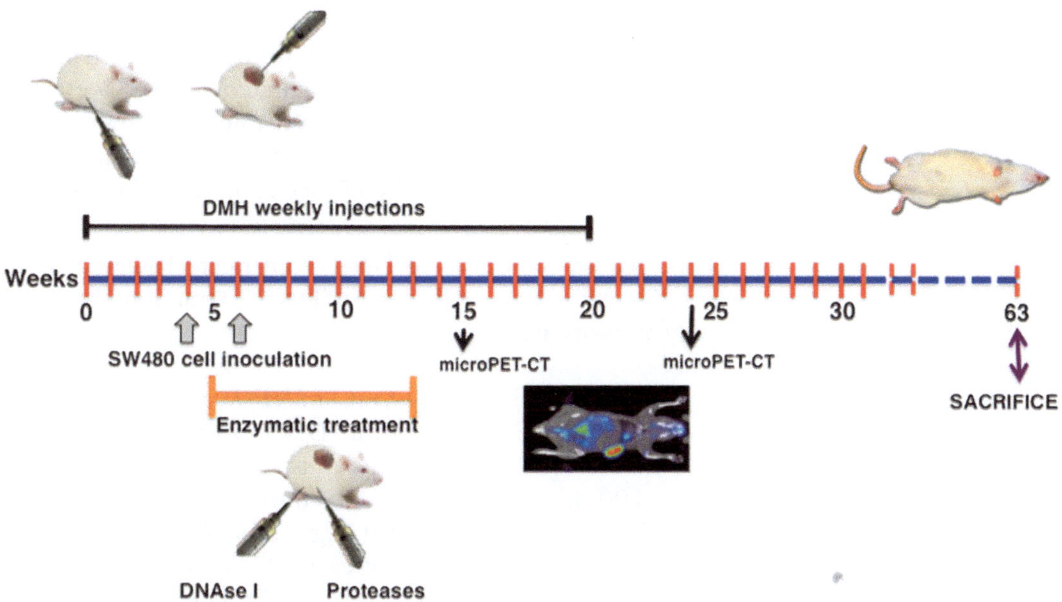

Fig. 3 Experimental design to demonstrate horizontal tumor progression in Wistar rats. Animals were treated with the colon carcinogen DMH, subcutaneously (s.c.) injected with human SW480 colon cancer cells and treated with an enzyme mixture (DNase I, trypsin, papain, and chymotrypsin)

15. If possible, monitor with micro-PET-CT the tumor formation in the colon, from the 15th week of starting treatment with DMH. Repeat the study at week 4 and 10 after completing the carcinogen treatment (Fig. 3) (*see* **Note 4**).

3.2 Tissue Preparation and Microdissection

1. Animals must be killed and autopsies performed 6 months after finishing the 20 weeks of the carcinogen (DMH). The stomach must be removed, opened from the greater curvature. and extended on a cardboard for inspection. The small and large intestinal tracts must be opened and extended as well. The location of individual tumors in the small intestine must be recorded by measuring the distance from pyloric ring. The length of the large intestine must be measured, and the number and size of individual tumors within must be recorded and their locations noted by measuring the distance from the anus. The colorectal tissue must be analyzed from 3 different sites: proximal ascending colon, medium colon and distal descending colon. Whole tissue must be fixed in 10 % neutral formalin.
2. In order to demonstrate the circulating DNA horizontal transfer (that comes from SW480 cells inoculated into the flank of the rats) to the colon cells previously treated with the carcinogen, first, isolate the tumor cells from the colon tumors of each rat group by microdissection technique. To this end, the formalin-fixed, paraffin-embedded colonic tumors must be cut

into 5-μm-thick sections on a microtome with a disposable blade. For microdissection, sections must be deparaffinized in two changes of xylene for 10 min, rehydrated in 100 % ethanol, 90 % ethanol, and 70 % ethanol for 5 min each, stained with hematoxylin and eosin (H and E) for 45 s, rinsed in RNase-free H_2O for 30 s, and finally immersed in 100 % ethanol for 1 min (*see* **Note 5**).

3. Then, extract the genomic DNA from microdissected tumor cells. The kit Arcturus®PicoPure® DNA Extraction Kit can be used (*see* **Note 6**).

3.3 Methodologies Used to Demonstrate the Horizontal Transfer of Circulating DNA

3.3.1 PCR

Detect by PCR specific genes of rat (*LINE-1*) and human (*YD6-Alu, K-ras* and *RAB30*), in genomic DNA extracted from microdissected tumor cells. All reactions must be performed in 20 μl containing 100 ng of template DNA, 10 mM Tris–HCl (pH 8.3), 40 mM KCl, 2 mM $MgCl_2$ (1 mM $MgCl_2$ for *RAB30*, and 5 mM $MgCl_2$ for *ALU*-YD6), 200 μM of each dNTP, 0.25 U Taq polymerase, and 1 μM of each specific primer. Primers used are as follows: human *K-ras* (5′-gactgaatataaacttgtggtagt-3′, and 3′-ggacgaatatgatccaacaatag-5′), 107 bp amplicon; human *RAB30* (5′-gtccattacccagagttactaccg-3′, and 3′-gaccttgttgctggcatattgttc-5′), 130 bp amplicon; *ALU*-YD6 (5′-gagatcgagaccacggtgaaa-3′, and 3′-ttgctctgaggcagagttt-5′), 200 bp amplicon [23]. An initial denaturation at 94 °C for 5 min is followed by 40 cycles of amplification and a final extension step of 5 min at 72 °C (*Alu-YD6* have extension time for 30 s).

The cycles include denaturation at 94 °C for 30 s, 30 s of annealing (60 °C for *K-ras* and *RAB30*; and 61 °C for *ALU*-YD6). For rat specific *LINE 1* sequences the amplification also must be performed in 20 μl reactions containing 100 ng of template DNA, 10 mM Tris–HCl (pH 8.3), 40 mM KCl, 1 mM $MgCl_2$, 200 μM of each dNTP, 0.25 U Taq polymerase, and 1 μM of each primer specific for rat *LINE-1* (5′-aaatcagggactagacaaggctgc-3′, and 3′-cccagccactttgctgaagttgt-5′) [24]. Initial denaturation at 94 °C for 5 min is followed by 40 cycles of amplification and a final extension step of 5 min at 72 °C. Amplification cycles consist of denaturation at 94 °C for 30 s, annealing at 59 °C for 30 s, and extension at 72 °C for 30 s (*see* **Note 7**).

3.3.2 DNA-Sequencing

Detect by sequencing specific bases of rat and human on PCR products of the genes analyzed previously (*K-ras* and *RAB30*). PCR amplicons can be purified using isopropanol precipitation and then sequenced in both forward and reverse directions from at least two independent amplification products (*see* **Note 8**).

3.3.3 FISH Analysis

Evaluate by FISH analysis in paraffin-embedded tumors, the transfer of human DNA to rat colonic tumors (*see* **Note 9**). Sections must be deparaffinized in xylene and rehydrated in graded ethanol series. The slides must be pretreated with 0.2 M HCl, 8 % sodium thiocyanate, and 0.5 % pepsin. Then, red-fluorescein-labelled *ALU*-YD6 human and green-fluorescein-labelled *LINE-1* rat probes must be added simultaneously. Cover the slides with coverslips and seal with rubber cement. Sections must be denatured and hybridized in a Hybridizer at 37 °C overnight. The rubber cement and the coverslips are removed and the sections must be washed stringently using SSC: 2× SSC for 30 min at room temperature, 0.4× SSC/0.3 % NP40 for 5 min at 75 °C, and 2× SSC/0.1 % NP40 for 4 min at room temperature. Nuclei must be counterstained using DAPI at a concentration of 0.1-µg/ml in Antifade. Analyses must be performed using a fluorescence microscope.

4 Notes

1. The selection of appropriate strain of rat (Wistar, Fisher, Sprague-Dawley, etc.) to study horizontal tumor progression, was as reported in the literature, where most authors use the Wistar outbred strain for the model of colon carcinogenesis with DMH [25–27].
2. The cell viability of tumor cell line must be assessed by trypan blue exclusion before inoculation. More than 95 % of the cells for injection should be viable. It is important that the cell line used for implantation must be routinely tested for mycoplasma contamination to prevent skewed experimental results.
3. The dose of each of the enzymes of protease mixture, is the following: trypsin 0.2 mg/20 g; chymotrypsin 0.2 mg/20 g, and papain 0.5 mg/20 g of body weight [28].

   DNase I, trypsin, chymotrypsin, and papain, are soluble in water; but it is important to first dissolve papain, because it must be fully dissolved in a maximal concentration of 10 mg/ml. Once the papain is dissolved, add the trypsin and chymotrypsin to the solution of papain and mix well.
4. Tumor formation in the animals can be monitored using molecular imaging techniques with a micro PET-CT (Albira ARS, Oncovision) and ^{18}F-FDG. Tumor monitoring can be conducted at weeks 15 and 24 after the start of treatment with DMH.

   To quantify tumor activity, the standard uptake value (SUV) can be calculated using Albira system software tools. SUV is a quantitative tool for PET-CT studies that allows the determination of the ^{18}F-FDG concentration in a specific region of interest (i.e., tumor activity). Tumor presence is defined when the SUV (tumor/liver) ratio is ≥1.5.
5. The PALM Laser-MicroBeam System can be used for microdissection (P.A.L.M., Wolfratshausen, Germany). After selecting

the cells of interest, adjacent cells must be photolysed by the microbeam. To retrieve the selected cells from the slide, a computer-controlled micromanipulator and conventional sterile needles must be used to pick up and transfer the cells into a reaction tube.

6. DNA from microdissected tumors from paraffin-embedded tissue can be extracted with the Arcturus®PicoPure® DNA Extraction Kit following the manufacturer's instructions.
7. All the amplification products must be verified by agarose gel electrophoresis, and DNA from SW480 human cells and from rat tail must be used as controls for PCR reactions.
8. Purified DNA can be diluted and cycle-sequenced using the ABI BigDye Terminator kit v3.1 (ABI, Foster City, CA, USA) according to manufacturer's instructions. Sequencing reactions can be electrophoresed in an ABI3100 genetic analyzer. Electropherograms are analyzed in both sense and antisense directions. The sequences obtained are compared with the reference *K-Ras* sequence (GenBank DQ893829) and the *RAB-30* sequence (GenBank NM_014488).
9. The formalin-fixed, paraffin-embedded colonic tumors (one of each group) from each group of DMH-treated rats must be cut into 8-μm-thick sections on a microtome with a disposable blade. Analyses can be performed using a Zeiss Axioplan fluorescence microscope interfaced with the CytoVision system.

References

1. Liotta LA, Kohn E (2004) Anoikis; cancer and the homeless cell. Nature 430:973–974
2. Geiger TR, Peeper DS (2009) Metastasis mechanisms. Biochim Biophys Acta 1796:293–308
3. Hunter KW, Crawford NP, Alsarraj J (2008) Mechanisms of metastasis. Breast Cancer Res 10(Suppl 1):S2
4. Nowell PC (1976) The clonal evolution of tumor cell populations. Science 194:23–28
5. Porter RD (1988) Modes of gene transfer in bacteria. In: Kucherlapati R, Smith GR (eds) Genetic recombination. ASM, Washington, DC, p 1
6. Mishra NC (1985) Gene transfer in fungi. Adv Genet 23:73
7. Klein BY (1981) A suggested mechanism for changing tumor cell phenotype: transfection of host cells with DNA sequences of dead tumor cells. Med Hypotheses 7:645–650
8. Holmgrem L, Szeles A, Rajnavolgyi E et al (1999) Horizontal transfer of DNA by the uptake of apoptotic bodies. Blood 93: 3956–3963
9. Goldenberg DM, Gold DV, Loo M et al (2013) Horizontal transmission of malignancy: in-vivo fusion of human lymphomas with hamster stroma produces tumors retaining human genes and lymphoid pathology. Plos One 8:e55324
10. Bendich A, Wilczok T, Borenfreund E (1965) Circulating DNA as a possible factor in oncogenesis. Science 148:374–376
11. García-Olmo D, García-Olmo DC, Ontañón J et al (1999) Tumor DNA circulating in the plasma might play a role in metastasis. The hypothesis of the genometastasis. Histol Histopathol 14:1159–1164
12. Trejo-Becerril C, Pérez-Cárdenas E, Treviño-Cuevas H et al (2003) Circulating nucleosomes and response to chemotherapy: an in vitro, in vivo and clinical study on cervical cancer patients. Int J Cancer 104:663–668
13. Leon SA, Shapiro B, Sklaroff DM et al (1977) Free DNA in the serum of cancer patients and the effect of therapy. Cancer Res 37:646–650
14. Trejo-Becerril C, Oñate-Ocaña LF, Taja-Chayeb L et al (2005) Serum nucleosomes

during neoadjuvant chemotherapy in patients with cervical cancer. Predictive and prognostic significance. BMC Cancer 5:65

15. Vlassov VV, Laktionov PP, Rykova EY (2010) Circulating nucleic acids as a potential source for cancer biomarkers. Curr Mol Med 10:142–165
16. Gahan PB (2012) Biology of circulating nucleic acids and possible roles in diagnosis and treatment in diabetes and cancer. Infect Disord Drug Targets 12:360–370
17. Record M, Subra C, Silvente-Poirot S et al (2011) Exosomes as intercellular signalosomes and pharmacological effectors. Biochem Pharmacol 81:1171–1182
18. Stroun M (1989) Neoplastic characteristics of the DNA found in the plasma of cancer patients. Oncology 46:318–322
19. García-Olmo DC, Domínguez C, García-Arranz M et al (2010) Cell-free nucleic acids circulating in the plasma of colorectal cancer patients induce the oncogenic transformation of susceptible cultured cells. Cancer Res 70:560–567
20. Trejo-Becerril C, Pérez-Cárdenas E, Taja-Chayeb L et al (2012) Cancer progression mediated by horizontal gene transfer in an in vivo model. Plos One 7:e52754
21. Perse M, Cerar A (2005) The dimethylhydrazine induced colorectal tumors in rat-experimental colorectal carcinogenesis. Radiol and Oncol 39:61–70
22. Patutina O, Mironova N, Ryabchikova E et al (2011) Inhibition of metastasis development by daily administration of ultralow doses of RNase A and DNase I. Biochimie 93: 689–696
23. Walker JA, Kilroy GE, Xing J et al (2003) Human DNA quantitation using Alu element-based polymerase chain reaction. Anal Biochem 315:122–128
24. Soares MB, Schon E, Efstratiadis A (1985) Rat LINE1: the origin and evolution of a family of long interspersed middle repetitive DNA elements. J Mol Evol 22:117–133
25. Madara JL, Harte P, Deasy J et al (1983) Evidence for an adenomacarcinoma sequence in dimethylhydrazine-induced neoplasms of rat intestinal epithelium. Am J Pathol 110:230–235
26. Watanabe H, Uesaka T, Kido S et al (1999) Gastric tumor induction by 1,2-dimethyl hydrazine in Wistar rats with intestinal metaplasia caused by X-irradiation. Jpn J Cancer Res 90:1207–1211
27. Ravnik-Glavac M, Cerar A, Glavac D (2000) Animal model in the study of colorectal carcinogenesis. Eur J Physiol 440:R55–R57
28. Wald M, Olejár T, Poucková P et al (1998) Proteinases reduce metastatic dissemination and increase survival time in $C_{57}B_{16}$ mice with the Lewis lung carcinoma. Life Sciences 63: L237–L243

Chapter 13

An In Vivo Model to Study the Effects of Tumoral Soluble Factors on the Vascular Permeability in Mice

César Alejandro Guzmán-Pérez, Alfredo Ibarra-Sánchez, José Luis Ventura-Gallegos, Claudia González-Espinosa, Jonathan García-Román, and Alejandro Zentella-Dehesa

Abstract

Some cancer cell lines release soluble factors that activate the endothelial cells in vitro; also endothelial activation in vivo includes an increased expression of adhesion molecules on the apical membrane, and an increased permeability, which may contribute to the extravasation process of circulating cells. We have adapted the Miles assay into a protocol that uses IgE/antigen complex and VEGF-1 as controls. The Miles assay comprises the intradermic injection of a pro-inflammatory agent into the skin and the intravenous introduction of a dye; the increase in vascular permeability will allow for the extravasation of the dye and thus the skin will be stained. The dye is then extracted from the dissected skin and quantified by spectrophotometry. The use of localized treatments will allow for testing a larger number of experimental samples in the same animal. With this model, the effects of tumoral soluble factors (TSFs) on endothelial permeability can be studied, as well as the signaling pathways involved. It can also serve to study the interactions between endothelial, immune, and cancer cells during the extravasation process.

Key words Cancer, Endothelium, Vascular permeability, Metastasis, Extravasation, Tumoral soluble factors

1 Introduction

An increase in endothelial permeability is considered to contribute to the metastasis process thus facilitating the transmigration of cancer cells [1–3]. Since endothelial cells regulate both vascular permeability and the transendothelial migration processes, it has been postulated that during the extravasation process the tumor cells must interact with them [4, 5].

It has been demonstrated that some cancer cell lines release soluble factors that activate the endothelium in vitro, with no interaction between cells; this paracrine activation of the endothelium has been defined as pre-metastatic state and it is associated to the extravasation of circulating tumor cells [6–8].

Martha Robles-Flores (ed.), *Cancer Cell Signaling: Methods and Protocols*, Methods in Molecular Biology, vol. 1165, DOI 10.1007/978-1-4939-0856-1_13,

Most of the assays to determine changes on permeability of endothelial cells have been performed in vitro, quantifying the permeability of ions, fluids, and proteins through the endothelial layer, using transwells or Boyden chambers, as well as measuring the transendothelial electrical resistance [9]. Production, solubility, and diffusion of endothelial activation mediators through the artificial matrix that separates compartments in the transwell can produce results that do not reflect the in vivo situation. This becomes more complex when one considers the tissue microenvironment that entails: inflammatory response, matrix remodeling, production of reactive oxygen species and other bioactive molecules, and the presence of additional cell populations. It has been suggested that all these variables can be critical components of the metastasis-specific tissue; in general, these changes are associated with endothelium activation and immune cells recruited during inflammation [10]. For example, macrophage infiltration is associated with a poor prognosis in more than 80 % of cancers, and metastasis has been correlated with macrophage infiltration [11, 12].

An in vivo model allows for studying the interactions between different cell types through the selective release of soluble factors. Some in vivo vascular permeability techniques are based on the ability of agents of interest to cause pulmonary edemas; these techniques quantify the amount of water in the lungs as an indirect measure of endothelial permeability induced by the agent [9]. Also, Evans blue dye is used as a vascular permeability marker because it has high affinity for albumin and its measure is an accurate indicator of how much albumin can pass through the pulmonary endothelium [9]. Systemic treatment with VEGF-1 can be used as a positive control of vascular permeability in acute and chronic treatments [13]. However, with these techniques, specific agents can only be tested at a systemic level. To date, the effect of tumoral soluble factors (TSFs) on vascular permeability in vivo has not yet been fully studied, mainly because the lack of an adequate animal model.

Here, we propose an in vivo assay to quantify vascular permeability based on two variants of the Miles assay [14]: the first is a passive anaphylaxis test with subsequent quantification by spectrophotometry developed by Teshima and collaborators in the 1990s [15]. The second one is a study about the effect of VEGF-1 on skin vascular permeability developed by Mamluk and collaborators [16]. With those studies as a background, we developed a system to measure changes on vascular permeability induced by TSFs. This method takes advantage of the facts that monomeric IgEs can bind the FcεRI receptor located on the surface of mast cells and that FcεRI receptor can be activated by the administration of a specific antigen to which IgE was directed [17]. The stimulation of mast cells loaded with specific IgEs induces a large and sustained extravasation of Evans Blue dye, giving a maximal value that can be taken as a parameter to compare the results with TSFs. On the other hand, VEGF-1 has been reported to be able to induce changes on vascular permeability in a scale comparable with other low-level activators of endothelial cells [16].

The factors that influence the area and intensity of the skin marks were identified, allowing us to test up to six agents per animal, without risk of skin marks overlapping. With this model, the effects of TSFs on endothelial permeability can be studied in vivo, as well as the signaling pathways involved. It can also serve to study the interactions between endothelial, immune, and cancer cells during the metastasis.

2 Materials

All the solutions are prepared with deionized water, and the concentrations of the components are stated in parentheses.

2.1 Cell Culture and the Generation of TSFs

1. RPMI 1640: Dissolve a package of powdered RPMI 1640 medium, without glutamine and sodium bicarbonate in 1 l of water; supplement with 2 g of sodium bicarbonate and 0.3 g of glutamine (2 mM), pH is adjusted to 7.2–7.4 with HCl or NaOH. Pass the solution through a 0.22 μm filter; finally, add antibiotic (100 U/ml of penicillin, 100 μg/ml of streptomycin), antimycotic (2.5 μg/ml of amphotericin B), and 10 % of fetal bovine serum. Store at 4 °C.
2. RPMI 1640 medium without phenol red and without fetal bovine serum: Dissolve a package of powdered RPMI 1640 medium, without glutamine, sodium bicarbonate, and phenol red, in 1 l of water; supplement with 2 g of sodium bicarbonate and 0.3 g of glutamine (2 mM). Adjust the pH to 7.2–7.4 with HCl or NaOH. Pass the solution through a 0.22 μm filter; finally, add antibiotic (100 U/ml of penicillin, 100 μg/ml of streptomycin) and antimycotic (2.5 μg/ml of amphotericin B). Store at 4 °C.
3. Phosphate buffer saline (PBS): Dissolve 8 g of NaCl (137 mM), 0.2 g of KCl (2.7 mM), 0.2 g of KH_2PO_4 (4.3 mM), 2.16 g of Na_2HPO_4 anhydrous (1.4 mM), and bring it up to 1 l with water, adjust pH to 7.2–7.4, sterilize in an autoclave and store at room temperature.
4. PBS/serum free medium without phenol red: A mixture of PBS and serum free medium without phenol red with a ratio 1/1.
5. Dialysis membrane: with an approximate molecular weight cut off of 3.5 kDa (Spectrapore).
6. Incubator: at 37 °C with a relative humidity of 100 and 5 % CO_2.

7. Lyophilizer machine.
8. Additional materials: 50 ml centrifuge tubes, lyophilization flask, 100 mm diameter cell culture dishes, Bradford reagent, clinical centrifuge, spectrophotometer.

2.2 Analysis of Conditioned Medium-Induced Vascular Permeability In Vivo

1. Mice: Swiss Webster mice or from another strain (*see* **Note 1**), 8 weeks old, with a weight of 30 g. All the protocols that require animal use must be approved by a bioethical committee.
2. Evans blue dye: Dissolve 0.5 g Evans blue (Sigma-Aldrich) in 100 ml of water (*see* **Note 2**), sterilize the mixture by filtration and store at 4 °C. Add 100 ng of albumin-DNP (Sigma-Aldrich) per 100 μl of Evans blue (0.5 % w/v) right before use.
3. Saline Solution: Commercial sterile isotonic solution of sodium chloride 0.9 %.
4. VEGF-1: Recombinant mouse or recombinant human vascular endothelial growth factor-165 (PeproTech), reconstituted in sterile saline solution. Store frozen at −20 °C. Dilute in sterile saline solution just before use.
5. Anti-DNP IgE: Commercial monoclonal anti-DNP antibody produced in mouse (Sigma-Aldrich). Dilute to 40 ng/μl in sterile water, right before use.
6. Formamide for molecular biology, ≥99.5 % purity.
7. Anesthesia: a mixture of ketamine (94 mg/kg) and xylazine (5.66 mg/kg).
8. Shaver: use a commercial clipper/trimmer.
9. Spectrophotometer: calibrate to read at 620 nm on visible light.
10. CO_2 chamber: plastic container with lid with an inlet for the CO_2 line with capacity for up to 20 mice.
11. Microcentrifuge.
12. Additional materials: 0.3 ml syringes with 31G needles, 1.5 ml microcentrifuge tubes, tweezers and surgical scissors, expanded polystyrene board, needles or tacks.

3 Methods

3.1 Cell Culture and Generation of TSFs

1. Seed the cells in 100 mm diameter culture dishes. Maintain in RPMI 1640 supplemented medium at 37 °C with a relative humidity of 100 and 5 % CO_2.
2. Maintain cells until they reach 80–90 % confluence (*see* **Note 3**).
3. Wash ten times with 10 ml of PBS/serum free medium without phenol red to remove serum components without removing the cells.
4. Add 10 ml of serum free medium without phenol red.

5. Maintain at 37 °C in the incubator for 48 h (*see* **Note 4**).
6. Collect the medium and centrifuge at 2,100 ×*g* for 15 min.
7. Transfer the supernatant into a new tube or directly into the lyophilization flask (*see* **Note 5**).
8. Freeze the samples at −70 °C and lyophilize for 24–48 h.
9. Resuspend the resulting powder in sterile water, in one tenth of the original volume. Hydrate for 1–2 h at room temperature.
10. Centrifuge at 2,100 ×*g* for 30 min at 4 °C (*see* **Note 6**).
11. Introduce a section of ultrafiltration membrane with pores of 3.5 kDa into a boiling water container for 5 min.
12. Close down one end of the tube of membrane and fill with conditioned media, close down the other end (*see* **Note** 7).
13. Dialyze for 24 h at 4 °C against 1 l of PBS solution.
14. Change the PBS solution and dialyze for further 24 h. So that the osmotic balance of the conditioned medium is recovered.
15. Centrifuge the obtained solution at 2,100 ×*g* for 15 min at 4 °C.
16. Pass the supernatant through a 0.22 μm filter and store at 4 °C until use.
17. Determine the protein content of the TSFs using the Bradford reagent, reading absorbance at 595 nm by spectrophotometry.

3.2 Determination of Vascular Permeability Using the TSFs

The assay was designed with the following main steps: (1) application in the ear of anti-dinitrophenyl IgE (anti-DNP IgE), used as a control of an adequate caudal vein injection, (2) administration of Evans blue dye and the DNP through the caudal vein, (3) intradermic injection on the dorsal area of TSFs and vascular endothelial growth factor (VEGF-1) used as a control of permeability, (4) then, evaluation of permeability by the extraction of the extravasated dye from the dissected areas with treatment, and (5) spectrophotometric quantification at 620 nm.

1. Mice are anesthetized intraperitoneally with a mixture of ketamine–xylazine.
2. Shave the mice's back (dorsal area) (*see* **Note 8**).
3. Inject 280 ng of anti-DNP IgE in the ear (*see* **Note 9**), this will be the control of adequate injection in tail vein.
4. Let the anti-DNP IgE interact with the mast cells in the skin for a period of 18–24 h.
5. Inject 100 μl of Evans blue (0.5 % in water) with 100 ng of DNP via the tail vein (*see* **Note 10**).
6. Wait 2 min and *continue the experiment only if the ear turns blue* (Fig. 1) (*see* **Note 11**).

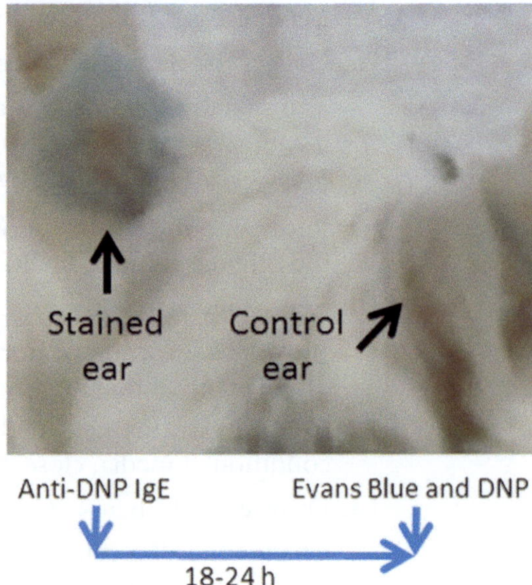

Fig. 1 Control of the caudal vein injection. Ear with anti-DNP IgE turns blue only if the caudal vein injection of Evans blue dye and DNP was done right. Anti-DNP IgE was injected only in the left ear. The *arrows below* describe the experimental protocol

7. Anesthetize the mice again with a mixture of ketamine–xylazine.
8. Apply the pharmacological inhibitors of signaling pathways (*see* **Note 12**).
9. Let the Evans blue and DNP circulate through the vasculature for 15 min following Evans blue injection (*see* **Note 13**).
10. Inject the TSFs (40 μl) and the VEGF-1 (positive control) (25 ng) into the dorsum skin (intradermally) (*see* **Note 14**).
11. Wait for the agents to act for a period of 15–45 min (Fig. 2).
12. Sacrifice the animals in a CO_2 chamber.
13. Take skin samples of 5 mm in diameter from the treated areas and weigh them (approximately 0.020 g) (*see* **Note 15**).
14. Place the skin samples in microcentrifuge tubes with 150 μl of formamide. Keep the ratio between the sample's weight and the amount of formamide (0.02 g skin/150 μl formamide).
15. Incubate the samples in 150 μl of formamide at 56 °C for a period of 18–24 h (*see* **Note 16**).
16. Centrifuge at 2,400 × *g* for 5 min.
17. Quantify by spectrophotometry at 620 nm.
18. The expected values range from 0.05 to 0.12 OD (*see* **Note 17**), for negative and positive controls, respectively (Fig. 3).

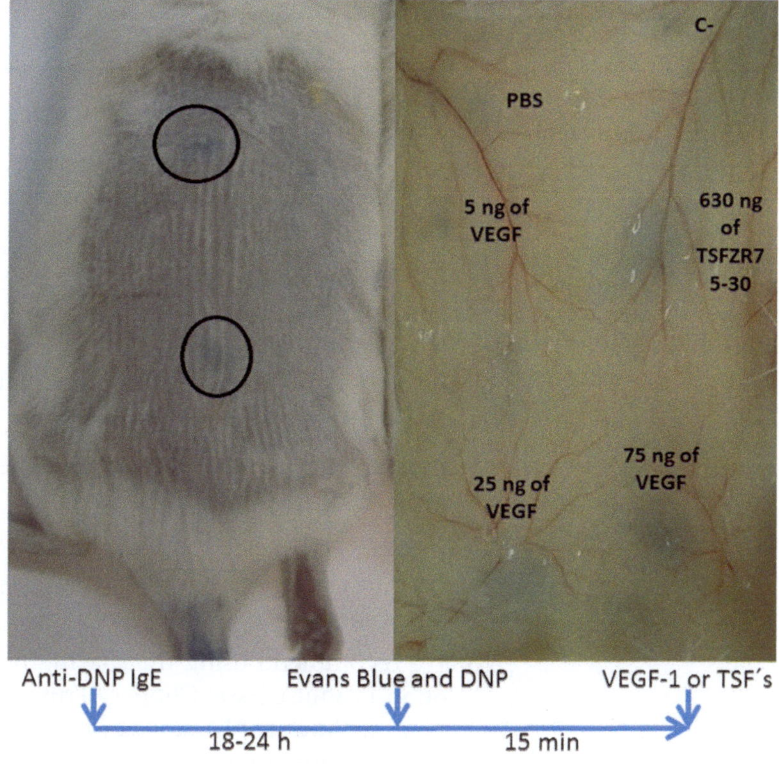

Fig. 2 Expected staining skin. *Left panel*: the *upper circle* shows the staining of the skin generated by 25 ng of VEGF-1, the *lower circle* shows the staining generated by 630 ng of tumoral soluble factors (TSFs) derived from ZR75-30 cells (TSFZR75-30). *Right panel*: staining generated on the inner side of the skin by administering medium that was not in contact with cells (C-), PBS, 5, 25, and 75 ng of VEGF and 630 ng of TSFZR75-30. The *arrows below* describe the experimental protocol

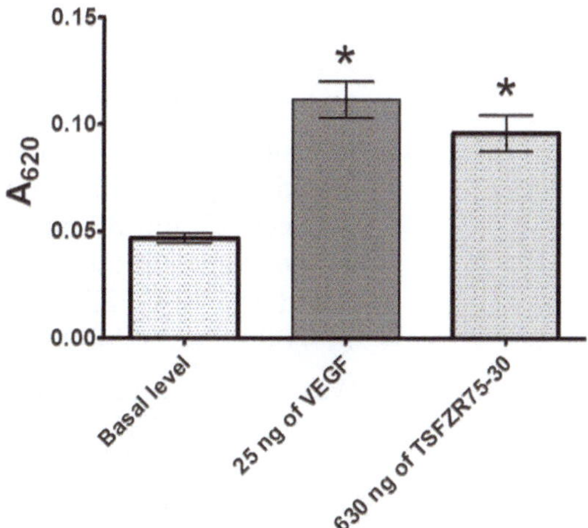

Fig. 3 Expected results of TSFs induced vascular permeability. Vascular permeability in the back of Swiss Webster mice using 25 ng of VEGF-1 or 630 ng of TSFs derived from ZR75-30 cells (TSFZR75-30) as stimuli; $^{*}P<0.05$, the values denote mean ± S.E.M. For bar level controls $n=20$, for 25 ng of VEGF n = 12, and for 630 ng of TSFZR75-30 $n=8$

4 Notes

1. The mice strain is important, specifically the color of the mice. The caudal vein injection and visualization of the skin stain are easier to perform in white mice.
2. Evans blue dye can be dissolved in water, PBS or saline solution.
3. For production of TSFs the cell confluence may vary. TSFs can be collected during the phase of exponential growth or once the cells have reached 100 % confluence.
4. Verify 100 % of cell viability at the end of the 48 h incubation.
5. Incline the lyophilization vials while the content freezes to increase the exposed area and reduce lyophilization time.
6. The centrifuge must be at 4 °C at least 10 min before use.
7. Be sure that the content is not leaking from the dialysis membrane.
8. We recommend using the shaving machine Wahl Professional 8685 Peanut Classic Clipper/Trimmer. The shaving should be done without applying excessive pressure. The excess in pressure can cause injuries to the skin, which will interfere with the results of permeability. The use of depilatory creams is not recommended because they can cause skin irritation.
9. The ear injection can only be intradermal, try to steer clear from the ear veins.
10. We suggest to filter the Evans blue dye solution right before use.
11. Ear with anti-DNP IgE turns blue only if the caudal vein injection was performed in the right way; if the ear does not turned blue, the experiment should not be continued. Without the dye in the bloodstream it is not possible to detect the increase on vascular permeability.
12. If pharmacological inhibitors of signaling pathways or other blocking agents are going to be used, they should be added in between 15 and 30 min before the TSFs; they can be applied in the skin or intravenously. Examples of agents that can be tested are: VEGFR2 inhibitor SU1498, NF-κB inhibitor BAY 11-7085, Jak3 inhibitor WHI-P131.
13. We applied the TSFs 15 min after the Evans blue dye injection. The time of TSFs administration (for all samples) should be amended if an inhibitor with more time of incubation is tested.
14. We recommend to perform the dorsal injections in a controlled environment, because mice experience vasoconstriction at low temperatures, which would affect the dye distribution and the agents tested. Injections in the back must be intradermal, if during the insertion of the needle resistance is lost it means that the injection is at a subdermal level. For better results all samples should be applied at the same volume (40 μl).

15. On the inner side of skin the marks are more visible. We suggest to dissect all dorsal and to place it on an expanded polystyrene board, and to fix it with tacks or syringe needles.
16. You must ensure that no dye remains in the tissue after the extraction. The tissue should be white. If the dye is still present, incubate for a few additional hours.
17. Vascular permeability values can vary depending on the strain of mice used. We use the Swiss Webster strain.

Acknowledgements

César Alejandro Guzmán Pérez received a master's scholarship from the Mexican National Council for Science and Technology (CONACyT, Grant 245198). The authors thank M. Sc. Layla Ortiz and Dr. Melissa Boyd for their assistance in translating the manuscript.

References

1. Criscuoli ML, Nguyen M, Eliceiri BP (2005) Tumor metastasis but not tumor growth is dependent on Src-mediated vascular permeability. Blood 105:1508–1514
2. Eum SY, Lee YW, Hennig B et al (2004) VEGF regulates PCB 104-mediated stimulation of permeability and transmigration of breast cancer cells in human microvascular endothelial cells. Exp Cell Res 296:231–244
3. García-Román J, Zentella-Dehesa A (2013) Vascular permeability changes involved in tumor metastasis. Cancer Lett 335(2):259–269. doi:10.1016/j.canlet.2013.03.005
4. Aghajanian A, Wittchen ES, Allingham MJ et al (2008) Endothelial cell junctions and the regulation of vascular permeability and leukocyte transmigration. J Thromb Haemost 6:1453–1460
5. Nguyen DX, Bos PD, Massagué J (2009) Metastasis: from dissemination to organ-specific colonization. Nat Rev Cancer 9:274–284
6. Perelmuter VM, Manskikh VN (2012) Preniche as missing link of the metastatic niche concept explaining organ-preferential metastasis of malignant tumors and the type of metastatic disease. Biochemistry (Mosc) 77: 111–118
7. Estrada-Bernal A, Alcántara-Meléndez MA, Mendoza-Milla C et al (2003) NF-kappaB dependent activation of human endothelial cells treated with soluble products derived from human lymphomas. Cancer Lett 191:239–248
8. Gupta GP, Massagué J (2006) Cancer metastasis: building a framework. Cell 127:679–695
9. Garcia AN, Vogel SM, Komarova YA et al (2011) Permeability of endothelial barrier: cell culture and in vivo models. Methods Mol Biol 763:333–354
10. Coussens LM, Werb Z (2002) Inflammation and cancer. Nature 420:860–867
11. Joyce JA, Pollard JW (2009) Microenvironmental regulation of metastasis. Nat Rev Cancer 9:239–252
12. Calorini L, Bianchini F (2010) Environmental control of invasiveness and metastatic dissemination of tumor cells: the role of tumor cell-host cell interactions. Cell Commun Signal 2010, 8:24. doi:10.1186/1478-811X-8-24
13. Weis SM (2011) Evaluation of VEGF-induced vascular permeability in mice. Methods Mol Biol 763:403–415
14. Miles AA, Miles EM (1952) Vascular reactions to histamine, histamine-liberator and leukotaxine in the skin of guinea-pigs. J Physiol 118: 228–257
15. Teshima R, Akiyama H, Akasaka R et al (1998) Simple spectrophotometric analysis of passive and active ear cutaneous anaphylaxis in the mouse. Toxicol Lett 95:109–115
16. Mamluk R, Klagsbrun M, Detmar M et al (2005) Soluble neuropilin targeted to the skin inhibits vascular permeability. Angiogenesis 8:217–227
17. Galli SJ, Kalesnikoff J, Grimbaldeston MA et al (2010) Mast cells as "tunable" effector and immunoregulatory cells: recent advances. Annu Rev Immunol 23:749–786

Chapter 14

Use of the Tumor Repressor DEDD as a Prognostic Marker of Cancer Metastasis

Qi Lv, Fang Hua, and Zhuo-Wei Hu

Abstract

DEDD, a member of a family of death effector domain-containing proteins, plays crucial roles in mediating apoptosis, regulating cell cycle, and inhibiting cell mitosis. Our recent work demonstrates that DEDD is a novel tumor repressor, which impedes metastasis by reversing the epithelial–mesenchymal transition (EMT) process in breast and colon cancers. DEDD expression therefore may represent a prognostic marker and potential therapeutic target for the prevention and treatment of cancer metastasis. To reveal the anti-metastatic roles of DEDD in these cancer cells, a number of experiments, including immunohistochemistry, the establishment of stably overexpressing or silencing cancer cells, chemoinvasion assay, soft agar assay, protein degradation, and protein–protein interaction were used in our in vitro and in vivo studies. This chapter focuses on the details of these experiments to provide references for the researchers to investigate the function of a gene in the regulation of tumor metastasis.

Key words Anchorage independent growth, DEDD, Immunohistochemistry, Invasion assay, Metastasis, Protein degradation, Protein interaction

1 Introduction

Death effector domains (DEDs) were originally described as protein–protein interaction domains involved in death receptor mediated extrinsic programmed cell death. The DEDs have no enzymatic function. Rather, they transduce the cellular signals by forming homotypic interactions [1]. The death effect domains containing proteins are a group of proteins including FADD, Caspase-8, Caspase-10, c-FLIP, PEA-15, DEDD1, and DEDD2 [2]. As a member of DEDs, the death-effector domain-containing DNA-binding protein (DEDD), hereafter referred to DEDD1, was originally recognized as an apoptotic scaffold protein located in both nucleus and cytoplasm [1]. Indeed, Arai et al. reported that DEDD can regulate cell cycle and inhibit cell mitosis by suppressing the activity of Cdk1/cyclin B1 complexes via direct binding to cyclin B1 [3].

Martha Robles-Flores (ed.), *Cancer Cell Signaling: Methods and Protocols*, Methods in Molecular Biology, vol. 1165,
DOI 10.1007/978-1-4939-0856-1_14,

Using a yeast 2-hybrid screening system, we have recently found that DEDD can negatively regulate the cancer cell invasion by physically interacting with the TGF-β1 signal molecule SMAD3 [4]. In the following studies, we demonstrated that the expression level of DEDD in breast and colon cancers correlates conversely with the poor prognosis of these cancers. Indeed, the expression of DEDD is negatively correlated with the aggressive/invasive phenotype of breast cancer cell lines. We next generated stable cells including stable over-expressing DEDD-MDA-MB-231 (DEDD expression lost) cells and stable silencing DEDD-MCF-7 cells (DEDD high expression). We found that the overexpression of DEDD resulted in a loss of metastatic phenotype in metastatic MDA-MB-231 cells, whereas silencing of DEDD promoted the cells to acquire the invasion phenotype in the non-metastatic MCF-7 cells [5].

Epithelial–mesenchymal transition (EMT) firstly recognized as a feature of embryogenesis, is characterized by loss of cell adhesion, repression of CDH1/E-cadherin expression, and increased cell motility. It is well known that the EMT process plays a critical role in the regulation of tumor progression and metastasis [6–8]. Snail and Twist are two key regulators that promote the EMT process by mainly repressing the transcription of CDH1 [9–12]. By using protein degradation assay, protein interaction assays such as yeast two-hybrid, GST pull-down, and co-immunoprecipitation, we found that DEDD can stabilize and activate PIK3C3 through a physical interaction with this protein, leading to the autophagy-lysosome degradation of Snail/Twist and suppressing the EMT process [5].

Our findings indicate that DEDD is a prognostic marker and a potential therapeutic target for the prevention and treatment of cancer metastasis. In this chapter, we present several methods that have been used in our research, including immunohistochemistry, the establishment of stable overexpression or knockdown cells, chemoinvasion assay, soft agar assay, protein degradation assay, yeast two-hybrid, GST pull-down, and co-immunoprecipitation. These methods can be extensively applied in the study of tumor metastasis and progress.

2 Materials

2.1 Immunohistochemistry Staining

1. Tissue array: Human breast cancer tissue microarray slides (US Biomax).
2. Phosphate-buffered saline (PBS): Dissolve 8.01 g NaCl, 0.20 g KCl, 1.78 g $NaH_2PO4 \cdot 2H_2O$, 0.27 g KH_2PO_4 into 900 mL of ddH_2O (pH to 7.2) and adjust final volume to 1 L with ddH_2O.

3. Citrate Buffer (10 mM Citric Acid, 0.05 % Tween 20, pH 6.0): Dissolve citric acid 1.92 g into 900 mL of ddH_2O, adjust pH to 6.0 with 1 N NaOH and then add 0.5 mL of Tween 20 and mix well. Store this solution at room temperature for 3 months or at 4 °C for longer storage.
4. 70–100 % alcohols: Mix dehydrated alcohol with ddH_2O to the concentration of alcohol at 70, 85, 95, 100 %.
5. Xylene.
6. 1 % (v/v) of HCl–alcohol solution: Mix 5 mL hydrochloric acid (HCl 37 %) with 495 mL alcohol to total volume of 500 mL.
7. DEDD antibody: anti-DEDD for immunohistochemistry (Protein Tech.) (*see* **Note 1**).
8. PBST (PBS + 0.1 % Ttriton X-100): Mix 1 mL Triton X-100 in 1 L 1× PBS (*see* **Note 2**).
9. Immunohistochemistry kit (SP method).
10. DAB detection kit (*see* **Note 3**).
11. Neutral balsam.

2.2 Establishment of Stable Cell Lines

1. MDA-MB-231 breast cancer cell line (American Type Culture Collection, ATCC).
2. MCF-7 breast cancer cell line (American Type Culture Collection, ATCC).
3. Dulbecco's Modified Eagle Medium and Fetal Bovin Serum (FBS).
4. Trypsin–EDTA solution (0.25 % trypsin + 0.02 % EDTA in PBS): Dissolve 1 g trypsin, 0.08 g EDTA into 400 mL PBS (pH 7.2), then follow aseptic filtration using 0.22 μm filter.
5. pcDNA 3.1-Flag-DEDD (gifted by Dr. Marcus E. Peter, University of Chicago, USA).
6. FuGENE®HD Transfection Reagent (Roche).
7. G418 (100 mg/mL): Dissolve G418 powder into 1 M HEPES, adjusts final volume to 10 mL with ultrapure water, then follow aseptic filtration by 0.22 μm filter (*see* **Note 4**).
8. pSliencer 3.1-H1 hygro DEDD shRNAs: three shRNAs were designed on the basis of the DEDD sequence (NM_032998) identified with siRNA Target Finder (Ambion).
9. Lipofectamine 2000 (Invitrogen).
10. Hygromycin B (50 mg/mL) (*see* **Note 5**).

2.3 Chemoinvasion Assay

1. 8.0 μm pore, 6.5-mm diameter Transwells (*see* **Note 6**).
2. Breast cancer cell lines (MDA-MB-231, MCF-7).
3. DMEM medium, FBS, Trypsin–EDTA (*see* Subheading 2.2).

4. Fibronectin solution: dilute 10 μL 0.1 % fibronectin liquid into 990 μL DMEM to a final concentration of 10 ng/mL.
5. Matrigel (BD Biosciences).
6. Toluidine blue stain (0.5 % toluidine blue + 2 % Na_2CO_3): Dissolve 0.05 g toluidine blue powder, 0.2 g Na_2CO_3 into 10 mL ultrapure water (*see* **Note 7**).
7. 4 % paraformaldehyde: Place 4 g paraformaldehyde into 80 mL PBS (pH 7.4), heat and stir for a while to complete dissolving. Once paraformaldehyde is in solution, allow it to cool to room temperature, and then adjust the volume up to 100 mL with more PBS.
8. 24-Well culture plates.
9. Cotton swab.

2.4 Soft Agar Assay for Colony Formation

1. 1 % (w/v) Agar (DNA grade): Place 1 g agar into 100 mL ddH_2O. Heat by microwave to complete dissolving, then sterilize in autoclave and store at 4 °C.
2. 0.7 % (w/v) Agarose (DNA grade): Place 0.7 g agar into 100 mL ddH_2O, heat by microwave to complete dissolving, then sterilize in autoclave and store at 4 °C.
3. 2× DMEM medium: Dissolve DMEM powder, 1.85 g $NaHCO_3$ into 450 mL ddH_2O, adjust pH to 7.2 by 1 N HCl, then fill ddH_2O to final volume of 500 mL. Sterilize by 0.22 μm filter and store at 4 °C.
4. Breast cancer cell lines (MDA-MB-231, MCF-7).
5. DMEM medium, FBS, Trypsin–EDTA (*see* Subheading 2.2).
6. Toluidine blue stain (0.5 % toluidine blue + 2 % Na_2CO_3): Dissolve 0.05 g toluidine blue powder, 0.2 g Na_2CO_3 into 10 mL ultrapure water (*see* **Note 7**).
7. 6-Well culture plates.

2.5 Degradation of Exogenous Protein Assay

1. Mammalian expressing plasmids (pCMV2-Flag-Snail, pEGFP vector).
2. Breast cancer cell lines (MCF-7 control-shRNA, MCF-7 DEDD-shRNA stable cells).
3. DMEM medium, FBS, Trypsin–EDTA (*see* Subheading 2.2).
4. FuGENE®HD Transfection Reagent (Roche).
5. CHX (cycloheximide, 20 mM): Dissolve 0.281 g CHX powder into ddH_2O to a final volume of 50 mL, and then follow aseptic filtration by 0.22 μm filter.
6. 35 mm culture dishes.
7. MG132 solution (10 mM): MG-132 ready-made solution is supplied as a 10 mM (1 mg/210 μL) 0.2 μm-filter sterilized solution in DMSO.

8. 3-MA (3-methyladenine, 1 M): Dissolve 0.149 g 3-MA powder into ddH_2O to a final volume of 1 mL, and then follow aseptic filtration by 0.22 μm filter.
9. Primary antibodies: mouse anti-Flag, mouse anti-GFP, HRP conjugated anti-β-actin antibody.
10. Goat anti-mouse HRP-conjugated antibody.
11. RIPA cell lysis buffer: 150 mM NaCl, 1 % NP-40, 0.5 % DOC (deoxycholate), 0.1 % SDS (sodium dodecylsulfate; electrophoresis grade), 50 mM Tris (pH 8.0).
12. 2× SDS loading buffer: 100 mM Tris–Cl (pH 6.8), 4 % (w/v) SDS, 0.2 % (w/v) bromophenol blue, 20 % (v/v) glycerol, 200 mM DTT (dithiothreitol).
13. SDS Polyacrylamide Gel Components: 30 % acrylamide–Bis-acrylamide solution, 1.5 M Tris–HCl (pH 8.8), 0.5 M Tris–HCl (pH 6.8), 10 % SDS, 10 % APS (ammonium persulfate), TEMED.
14. 10× SDS-PAGE running buffer: Dissolve 30.0 g of Tris base, 144.0 g of glycine, and 10.0 g of SDS in 1,000 mL of ddH_2O. The pH of the buffer should be 8.3 and no pH adjustment is required. Store the running buffer at room temperature and dilute to 1× before use.
15. PVDF transfer membrane (0.2 μm).
16. Transfer buffer (25 mM Tris, 192 mM glycine, 20 % methanol): Dissolve 28.8 g glycine, 6.04 g Tris base, and 200 mL methanol into 700 mL ddH_2O, then add ddH_2O to a final volume of 1 L.
17. Tris buffered saline (TBS; 10×): 1.5 M NaCl, 0.1 M Tris–HCl, pH 7.4.
18. TBST (1× TBS + 0.1 % Tween-20): Dilute 100 mL 10× TBS into 900 mL ddH_2O, add 1 mL Tween-20, stir constantly until homogenous.
19. 5 % nonfat milk in TBST: Weigh 5 g nonfat milk and dissolve in 100 mL 1× TBST (0.1 %) to a final mixture of 5 % nonfat milk/TBST (0.1 %).
20. ECL detection kit (Thermo Scientific).

2.6 Yeast 2-Hybrid Screening for DEDD Interacting Proteins

1. *E. coli* strain: DH5α.
2. Pfu DNA Polymerase, restriction enzymes (EcoR I and BamH I) and Rapid DNA ligation Kit (Fermentas).
3. Yeast strain AH109: *Matα*, *trp1-901*, *leu2-3*, *112*, *ura3-52*, *his3-200*, *gal4Δ*, *gal80Δ*, *LYS2::GAL1*$_{UAS}$-*GAL1*$_{TATA}$-*HIS3*, *GAL2*$_{UAS}$-*GAL2*$_{TATA}$-*ADE2*, *URA3::MEL1*$_{UAS}$-*MEL1*$_{TATA}$-*lacZ*.
4. DNA BD vector (pGBKT7-DEDD): GAL4 DNA-binding domain fusion vector with *TRP1* marker.

5. Human Fetal Kidney MATCHMAKER cDNA Library (CLONTECH): GAL4 activation domain fusion vector (pACT2) with LEU2 marker.
6. YPDA Medium: 20 g/L Difco Bacto Peptone, 10 g/L Bacto Yeast extract, 20 g/L Bacto Agar. After autoclaving, allow medium to cool to ~55 °C and then add dextrose (glucose) to 2 % (50 mL of a sterile 40 % stock solution). For adenine-supplemented YPD (YPDA), add 15 mL of a 0.2 % adenine hemisulfate solution per liter of medium (final concentration is 0.003 %).
7. SD/–Ade/–His/–Leu/–Trp/X-α-Gal medium

   46.7 g/L Minimal SD Agar Base, 0.6 g/L –Ade/–His/–Leu/–Trp DO Supplement. After autoclaving, allow medium to cool to ~55 °C and then add X-α-Gal (1 mL of a 20 mg/mL stock solution to 1.0 L). Pour plates and allow medium to harden at room temperature.
8. Stock solutions:
   - 50 % PEG 3350 (Polyethylene glycol, avg. mol. wt. = 3,350), prepare with deionized H_2O, and sterilize by filtration.
   - 100 % DMSO (Dimethyl sulfoxide).
   - 10× TE buffer: 0.1 M Tris–HCl, 10 mM EDTA, pH 7.5. Autoclave.
   - 10× LiAc: 1 M lithium acetate, adjust to pH 7.5 with dilute acetic acid and autoclave.
9. TE–LiAc solution, prepare fresh just prior to use.
   Final concentration to prepare 10 mL of solution:

ddH_2O	8 mL
TE buffer 1×	1 mL of 10× TE
LiAc 1×	1 mL of 10× LiAc

10. PEG–LiAc solution, prepare fresh just prior to use.
    Final concentration to prepare 10 mL of solution:

PEG 3350 40 %	8 mL of 50 % PEG
TE buffer 1×	1 mL of 10× TE
LiAc 1×	1 mL of 10× LiAc

11. Carrier DNA from salmon testes (Deoxyribonucleic acid, single stranded from salmon testes).

2.7 Co-immunoprecipitation Assay

1. HEK293T cell line (American Type Culture Collection, ATCC).
2. PBS buffer (*see* Subheading 2.1, **item 2**).

3. CO-IP lysis buffer: 25 mM Tris (pH 7.6), 150 mM NaCl, 2.5 mM $MgCl_2$, 0.5 mM EDTA, 0.5 % NP-40, 5 mM β-glycerophosphate, 1 mM DTT, 5 % glycerol, and proteinase inhibitors (1 μg/mL aprotinin, 1 μg/mL leupeptin, 1 μg/mL pepstatin, and 1 mM phenylmethylsulfonyl fluoride (PMSF)).
4. BCA Protein assay kit.
5. Protein A/G PLUS-Agarose.
6. Antibodies: mouse anti-FLAG, rabbit anti-Beclin 1, rabbit anti-PI3KC3.
7. CO-IP wash buffer: 25 mM Tris (pH 7.6), 150 mM NaCl, 2.5 mM $MgCl_2$, 0.5 mM EDTA, 0.05 % NP-40, 5 mM β-glycerophosphate, 1 mM DTT, 5 % glycerol, and proteinase inhibitors.

2.8 GST Pull-Down Assay

1. *E. coli* strain: BL21 (DH3).
2. pGEX-4T-1-DEDD prokaryotic expression plasmid.
3. LB medium: 10 g Bacto-Tryptone, 5 g Bacto-yeast extract, 10 g NaCl. Add ddH_2O to 1 L, adjust pH to 7.0 and autoclave to sterilize.
4. Lysis Buffer A: 50 mM Tris pH 7.5, 150 mM NaCl, 0.05 % NP-40. Just before use, add protease inhibitors (*see* Subheading 2.7, **item 3**).
5. Lysis buffer B: 25 mM Tris (pH 7.6), 150 mM NaCl, 2.5 mM $MgCl_2$, 0.5 mM EDTA, 0.5 % NP-40, 5 mM β-glycerophosphate, 1 mM DTT, 5 % glycerol, and proteinase inhibitors (*see* Subheading 2.7, **item 3**).
6. 100 mM IPTG: Dissolve 500 mg of isopropyl-β-D-thiogalactoside (IPTG) in 20 mL of distilled H_2O. Filter-sterilize and store in small aliquots at −20 °C.
7. PBS buffer (*see* Subheading 2.1, **item 2**).
8. Glutathione Sepharose 4 Fast Flow.
9. Coomassie Blue Staining:
   - Fixing solution (50 % methanol and 10 % glacial acetic acid).
   - Staining solution (0.1 % Coomassie Brilliant Blue R-250, 50 % methanol and 10 % glacial acetic acid).
   - Destaining solution (40 % methanol and 10 % glacial acetic acid).

3 Methods

3.1 Immuno-histochemisty Staining

1. Deparaffinization and rehydrating. Place the slides in a rack, and perform the following washes before preheat in incubator at 60 °C for 2 h:

Xylene: 2×3 min; 100 % alcohol: 2×3 min; 95 % alcohol: 3 min; 85 % alcohol: 3 min; 70 % alcohol: 3 min; Running tap water to rinse (*see* **Note 8**).

2. Antigen retrieval (*see* **Note 9**): Add the antigen retrieval buffer (Citrate Buffer) to the pressure cooker. Place the pressure cooker on the hotplate and turn it on full power. Do not secure the lid of the pressure cooker at this point. Simply rest it on top. While waiting for the pressure cooker to come to a boil, deparaffinize and rehydrate the sections as above. Once boiling, transfer the slides from the tap water to the pressure cooker. Secure the pressure cooker lid as in the manufacturer's instructions.

   As soon as the cooker has reached full pressure, leave it for 2 min. When 2 min has elapsed, turn off the hotplate and place the pressure cooker in an empty sink.

   Activate the pressure release valve and run cold water over the cooker. Once de-pressurized, open the lid and run cold water into the cooker for 10 min. USE CARE WITH HOT SOLUTION!

3. Immunohistochemical staining (*see* **Note 10**): Rinse array slide twice with PBS for 5 min each. Blocked the endogenous peroxidase activity at room temperature by a 15-min incubation in the final developmental 3 % H_2O_2 in distilled water or PBS (pH 7.4). Rinse the array slide twice with PBS for 5 min.

4. Permeabilization: incubate the slide in 0.25 % Triton X-100 in PBS for 10 min at room temperature (*see* **Note 11**). Rinse the array slide twice with PBS for 5 min. Apply the blocking antibody (normal goat serum), incubate for 20 min at room temperature, and throw off residual fluid (do not wash) (*see* **Note 12**).

5. Apply primary antibody diluted in normal goat serum. We recommend antibody dilution at 1:25. Incubate overnight at 4 °C (*see* **Note 13**). Rinse the array slide three times by PBST for 5 min with gentle agitation.

6. Incubate the array slide with a biotin-conjugated secondary antibody at 37 °C for 1 h. Rinse the array slide three times by PBST for 5 min with gentle agitation.

7. Incubate the array slide with streptavidin–HRP at 37 °C for 30 min. Rinse the array slide three times by PBST for 5 min with gentle agitation.

8. Add DAB solution and incubate for 3 min at room temperature (*see* **Note 14**). Rinse in running tap water for 5 min.

9. Counterstaining: counterstain the array slide in hematoxylin for 3 min. Rinse in running tap water for 5 min. Differentiate the array slide by immersing it in 1 % (v/v) of HCl–alcohol for 20 s. Immediately, rinse in running tap water for 10 min.

10. Dehydrate, clear, and mount. Dehydrate sequentially in the following solutions:

    70 % alcohol (20 s); 85 % alcohol (30 s); 95 % alcohol (1 min); 100 % alcohol (3 min); 100 % alcohol II (3 min).
11. Clear in Xylene I and II (3 min).
12. Mount the array slides with neutral balsam.
13. Observe the expression of DEDD under a microscope.

3.2 Establishment of MDA-MB-231 Cells with Stable Expression of DEDD

3.2.1 Determine the G418 Titration (Kill Curve) in MDA-MB-231 Cells

Plate 20,000 MDA-MB-231 cells into each well of a 24-well dish containing 1 mL of culture medium. After 24 h, add 500 μL culture medium containing 100–1,500 μg/mL G418 (100, 200, 300, 400, 500, 600, 700, 800, 900, 1,000, 1,100, 1,200, 1,300, 1,400, 1,500 μg/mL). Culture the cells for 10–14 days, replacing the G418-containing medium every 3 days. Examine the dishes for viable cells every 2 days. Identify the lowest G418 concentration that begins to give massive cell death in approximately 5–7 days, and kills all cells within 2 weeks. Use this G418 concentration to select cells containing the pcDNA 3.1 neo$^+$ plasmid after transfection.

3.2.2 Transfecting pCDNA 3.1-DEDD into MDA-MB-231 Cells

One day before transfection, plate 100,000 MDA-MB-231 cells into each well of a 6-well plate containing 2 mL of culture medium (*see* **Note 15**). Dilute 2 μg DNA (pcDNA 3.1-Flag DEDD, or -control vector) in 100 μL DMEM medium without serum. Pipet the FuGENE® HD Transfection Reagent 5 μL directly into the medium containing DNA without allowing contact with the walls of the plastic tubes (*see* **Note 16**). Vigorously tap the tube or vortex for 1–2 s to mix the transfection complex, and then incubate for 15 min at room temperature. Add the transfection complex to the cells in a drop-wise manner. Swirl the wells to ensure distribution over the entire plate surface (*see* **Note 17**).

3.2.3 Selecting G418-Resistant Transfected Cells

After transfection for 24 h, add culture medium containing the concentration of G418 identified in Subheading 3.2.1 (*see* **Note 18**). Replace the G418-containing medium every 3 days. Culture the cells in medium containing G418 until all of the cells in the non-transfected control culture are killed. At this point, the selection is complete and the cells can be grown without hygromycin until they repopulate the culture plate. Analyze expression of the target gene by immunoblotting to evaluate the elevating levels of DEDD expression in DEDD overexpressing cells.

3.3 Establishment of MCF-7 Cells Stably Silencing DEDD

3.3.1 Hygromycin Titration (Kill Curve)

Plate 25,000 MCF-7 cells into each well of a 24-well dish containing 1 mL of culture medium. After 24 h, add 500 μL culture medium containing 50–1,000 μg/mL hygromycin (50, 100, 200, 300, 400, 500, 600, 700, 800, 900, 1,000 μg/mL). Culture the cells for 10–14 days, replacing the hygromycin-containing medium every 3 days. Examine the dishes for viable cells every 2 days.

Identify the lowest hygromycin concentration that begins to give massive cell death in approximately 5–7 days, and kills all cells within 2 weeks. Use this hygromycin concentration to select cells containing the pSilencer hygro plasmid after transfection.

3.3.2 Transfecting pSilencer into Mammalian Cells

One day before transfection, plate 100,000 MCF-7 cells into each well of a 6-well plate containing 2 mL of culture medium (*see* **Note 19**). Mix Lipofectamine 2000 gently before use; then dilute the appropriate amount in 100 μL of DMEM medium without serum. Incubate for 5 min at room temperature. Dilute 4 μg of plasmid DNA (pSliencer 3.1-H1 hygro DEDD shRNAs or control shRNA) in 100 μL of DMEM medium without serum. After 5 min incubation, combine the diluted DNA with diluted Lipofectamine 2000 (total volume = 200 μL). Mix gently and incubate for 20 min at room temperature. Add the 200 μL of complexes to each well containing the cells and medium. Mix gently by rocking the plate back-forth and left-right. Incubate the cells at 37 °C in a CO_2 incubator. The medium may be changed after 4–6 h (*see* **Note 20**).

3.3.3 Selecting Hygromycin-Resistant Transfected Cells

After transfection for 24 h, add culture medium containing the concentration of hygromycin identified in Subheading 3.3.1 (*see* **Note 21**). Replace the hygromycin-containing medium every 3 days. Culture the cells in medium containing hygromycin until all of the cells in the non-transfected control culture are killed. At this point, the selection is complete and the cells can be grown without hygromycin until they repopulate the culture plate. Analyze expression of the target gene by immunoblotting to evaluate the silencing efficiency of each shRNA.

3.4 Chemoinvasion Assay

The ability of tumor cells to invade is one of the hallmarks of the metastatic phenotype. Invasion is a multistep process that involves adhesion, directed migration, and proteolytic activity to degrade extracellular matrix (ECM) barriers [13]. To elucidate the mechanisms by which tumor cells acquire an invasive phenotype, in vitro assays have been developed that mimic the in vivo process. The most commonly used in vitro invasion assay is a modified Boyden chamber assay using a basement membrane matrix preparation as the matrix barrier.

1. Dilute 10 μL fibronectin (1 mg/mL) into 990 μL DMEM medium to the final concentration of fibronectin at 10 ng/mL.
2. Add 30 μL fibronectin (10 ng/mL) on the back-side of the transwell chambers. Air-dry in super clean bench, then place the chambers in 24-well plates.
3. Thaw Matrigel at 4 °C and keep on ice.
4. Dilute Matrigel into cold DMEM medium to a final concentration of 0.125 μg/μL (the volume of matrigel: DMEM = 1:7) using precooled pipet tips (*see* **Note 22**).

5. Add 40 μL of diluted Matrigel to the upper chamber of each transwell. Tap the sides of the plate gently to ensure the liquid covers the bottom of each well completely.
6. Allow the wells to sit without a cover until the Matrigel has completely dried onto the porous membranes. This usually takes 4–5 h.
7. 1 h prior to the invasion assay, reconstitute the Matrigel by adding 200 μL of DMEM to the upper chamber of each transwell; then, incubate the wells at 37 °C.
8. Seed 2×10^5 cells into 6-well plates or 35 mm dishes. Culture 1–2 days to the confluence of cells at 70–80 %.
9. Starving cells prior to the invasion assay (optional step): replace the completed culture medium (DMEM + 10 % FBS) with DMEM serum-free medium. Incubate at 37 °C, 5 % CO_2 for 12 h.
10. Trypsinize the experimental cells and add them to a 5 mL centrifuge tube containing 2 mL of DMEM containing 0.5 % FBS.
11. Collect the cells by centrifugation (125 × *g*, 5 min). Wash the cells once resuspending the cell pellet in fresh DMEM with serum. Take an aliquot of cells to count.
12. Regulate the concentration of cells to 2.5×10^5/mL.
13. Remove the coated Transwells from the 37 °C incubator. Add 100 μL of the cell suspensions to the upper chamber of each coated Transwell insert (*see* **Note 23**).
14. Incubate the plates containing the Transwells at 37 °C for a suitable time.
15. Add 700 μL per well cold PBS into a new 24-well plate.
16. Discard the medium of upper chamber, and then remove the transwell inserts into cold PBS.
17. Remove the transwell inserts into new wells containing 700 μL 4 % paraformaldehyde to fix the cells. Incubate for 20 min at room temperature.
18. Remove the transwell inserts into new wells containing 700 μL ddH_2O for 5 min.
19. Repeat **step 18** for twice.
20. Add 700 μL toluidine blue stain (0.5 % toluidine blue + 2 % Na_2CO_3) into new wells. Then remove the transwell inserts into the stain and incubate for 1 h at 37 °C.
21. Repeat **steps 4** and **5** to wash the stained Transwell inserts.
22. Place the cotton swab into the well and swab the filter using a rotating motion while applying moderate pressure (*see* **Note 24**).
23. Place the swabbed Transwell inserts into the wells containing ddH_2O and add 50 μL ddH_2O into the upper chamber.

Table 1
The suggested amounts of soft agar assay for colony formation

Culture dish	24-Well plate	6-Well plate	35-mm Dish	60-mm Dish	100-mm Dish
Bottom and top agar volume (mL/well)	0.5	1.0	1.5	3.0	5.0
Cells/well	1,250	2,500	5,000	7,500	12,500
Medium volume (mL/well) for feeding	0.25	0.5	0.75	1.5	2.5

24. Quantify invasion by counting the cells that have invaded onto the lower surface of the porous membrane. Visualize the cells with an inverted microscope using bright-field optics (*see* **Note 25**).

3.5 Soft Agar Assay for Colony Formation

Anchorage-independent proliferation is a hallmark of oncogenic transformation and is thought to be conductive to proliferation of cancer cells away from their site of origin. The process of anchorage independent growth of cancer cells in vitro has been characterized as a key aspect of the tumor phenotype, particularly with respect to metastatic potential [14]. The soft agar assay for colony formation is an anchorage independent growth assay in soft agar, which is considered the most stringent assay for detecting malignant transformation of cells. For this assay, cells are cultured with appropriate controls in soft agar medium for 21–28 days. Following this incubation period, formed colonies can be analyzed morphologically using cell stain, and the number of colonies formed per well be quantified.

3.5.1 Preparation of Bottom Agar (See Note 26)

Mix equal volumes of the 1.0 % agar and 2× DMEM containing 20 % FBS to give 0.5 % agar + 1× DMEM + 10 % FBS. Add 1.0 mL of mixture from **step 1** to each well of 6-well culture plate, and set aside for 10 min to allow agar to solidify (*see* **Note 27**).

3.5.2 Preparation of Top Agarose

Trypsinize cultured cells, centrifuge, and resuspend. Count the number of cells per mL (*see* Subheading 3.3.2). Adjust the cell count to 100,000 cells/mL (*see* Table 1). Add 100 μL of cell suspension to 5 mL tubes. Add 2 mL 2× DMEM + 20 % FBS and 2 mL 0.7 % agarose to the tube containing cells from **step 3**, mix gently by pipette, and add 1 mL to each of the three or four replicate wells. Incubate plates at 37 °C for 10–30 days. Feed the cells 1–2 times per week with 500 μL DMEM containing 10 % FBS.

3.5.3 Staining and Observing

Stain plates with 0.5 mL toluidine blue stain (0.5 % toluidine blue + 2 % Na_2CO_3) for 1 h at 37 °C. Take picture or count colonies using a dissecting microscope.

3.6 Degradation of Exogenous Protein Assay

Proteolysis in eukaryotic cells occurs through two major pathways: one mediated by the proteasome and the other by the lysosome. The proteasome, in collaboration with the sophisticated ubiquitin system used for the selection of target proteins, plays crucial roles in rapid and aggressive degradation of not only numerous short-lived regulatory proteins, but also abnormal proteins with aberrant structures that should be eliminated from the cell to maintain cellular homeostasis [15, 16].

Previous work indicated that the degradation of Snail and Twist depends on the ubiquitin-proteasome degradation pathway [17, 18]. In our current study, we have found that knockdown of DEDD can attenuate the degradation of Snail/Twist. However, blocking the ubiquitin-proteasome degradation by a proteasome inhibitor such as MG132, did not change the difference of Snail/Twist degradation between MCF-7 control shRNA and MCF-7 DEDD shRNA cells [5]. Autophagy-lysosome is another intracellular degradation system for the majority of proteins and some organelles [19]. Inhibiting autophagy by 3-MA (a PI3K class III inhibitor), can significantly attenuate the degradation of Snail/Twist because a similar degradation rate was observed in MCF-7 control shRNA and MCF-7 DEDD shRNA cells [5].

3.6.1 Cotransfection of pCMV2-Flag-Snail and pEGFP (See **Note 28**)

One day before transfection, plate 200,000 MCF-7 cells (control-shRNA and DEDD-shRNA) into a 35 mm culture dish containing 2 mL of culture medium. Each kind of cells is seeded for six dishes. Dilute 24 μg DNA (12 μg pCMV2-Flag-Snail + 12 μg pEGFP) in 1,200 μL DMEM-serum free medium. Pipet 60 μL FuGENE® HD transfection reagent directly into the medium containing DNA without allowing contact with the walls of the plastic tubes. Vigorously tap the tube or vortex for 1–2 s to mix the transfection complex, and then incubate for 15 min at room temperature. Add 100 μL/dish of transfection complex to the cells in a drop-wise manner. Swirl the wells to ensure distribution over the entire plate surface. Incubate at 37 °C for 36 h (*see* **Note 29**).

3.6.2 Blocking Protein Biosynthesis by Treating with CHX (See **Note 30**)

Replace the medium containing transfection complex with 2 mL fresh complete DMEM medium. Add 2 μL CHX (20 mM) to the transfected cells to a final concentration of 20 μM. Incubate at 37 °C. After the cells are treated with CHX for the indicated time, remove the cells from incubator, wash once with ice-cold PBS and homogenize in ice-cold RIPA lysis buffer (containing protein inhibitors and PMSF). After centrifugation, the supernatants are collected. Dilute cell lysate with 2× SDS loading buffer in the ratio 1:1. Proteins are denatured at 100 °C for 3 min.

3.6.3 SDS-PAGE Gel Electrophoresis

Prepare SDS Polyacrylamide Gel: 12 % resolving gel, 5 % stacking gel (*see* **Note 31**).

Load samples and run gel (80 mA in stacking gel, 120 mA in resolving gel).

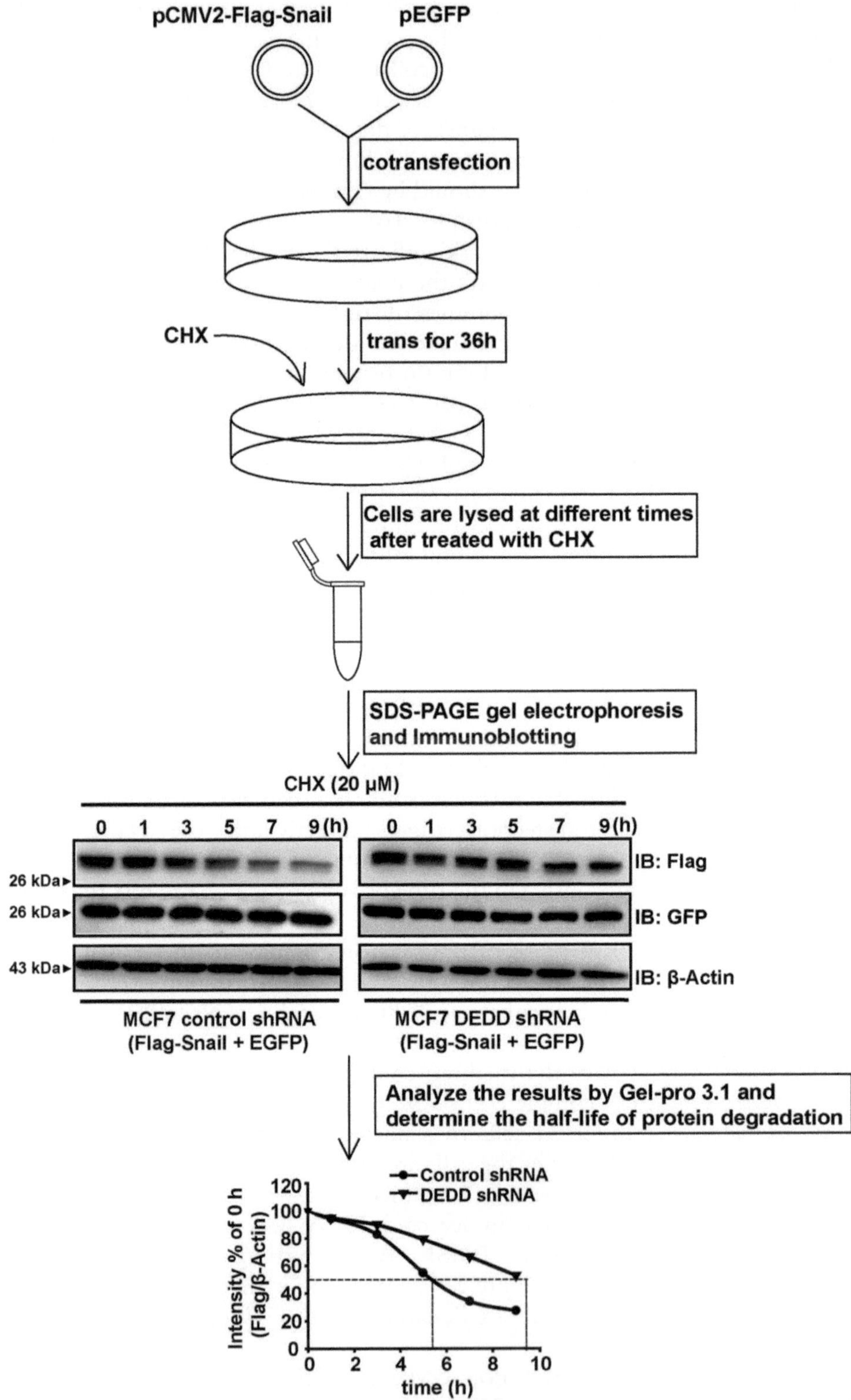

Fig. 1 Determination of the half-life of Flag-Snail in MCF-7 cells expressing control shRNA or DEDD shRNA. MCF-7 cells expressing control shRNA or DEDD shRNA are co-transfected with pCMV2-Flag-Snail or pEGFP for 36 h.

3.6.4 Immunoblotting and Detection

Transfer gel to PVDF membrane by standard procedures. Add adequate amount of 5 % nonfat milk/TBST blocking buffer and leave the sample at room temperature for 1 h. Dilute the primary antibody with fresh blocking buffer to the designated concentration. Remove the membrane from the previous blocking buffer and add the diluted primary antibodies to the membrane. Incubate at 4 °C overnight. Wash the membrane with TBST (0.1 %) for 10 min. Repeat three times. Add adequate amount of Goat anti-mouse-HRP-conjugated secondary antibody. Incubate at room temperature for 1 h. Wash the membrane with TBST (0.1 %) for 10 min. Repeat three times. After washing the sample, add freshly prepared chemiluminescent reagents on the PVDF membrane to coat the entire membrane. Take photographs immediately with CCD camera. Analyze the results by Gel-pro 3.1 and determine the half-life of protein degradation (Fig. 1).

3.6.5 Determining the Degradation Pathway (Ubiqutin-Proteasome or Autophagy-Lysosome)

Cotransfect pCMV2-Flag-snail and pEGFP (*see* Subheading 3.6.1). After 36 h of transfection, add CHX (20 μM) together with MG132 (10 μM) to block ubiqutin-proteasome pathway, or add 3-MA (10 mM) to block ubiqutin-proteasome pathway. Cells are lysed at the corresponding time to Subheading 3.6.2. Proceed then with SDS-PAGE followed by immunoblotting to evaluate the half-life of Snail/Twist in MCF-7 control shRNA or MCF-7 DEDD shRNA cells.

3.7 Yeast 2-Hybrid Screening for DEDD Interacting Proteins

The protein–protein interactions are of central importance for virtually every process in a living cell. Information regarding these interactions improves our understanding of protein biological functions. The yeast 2-hybrid (Y2H) assay is a well-established technique that has been developed as a powerful tool for investigating the protein interactions. Two proteins called bait and prey are separately fused to the independent DNA-binding (DBD) and transcriptional activation domains (AD) of the Gal4 transcription factor. If the proteins interact, they reconstitute a functional Gal4 to activate the expression of reporter genes, which enable growth on specific media or a color reaction. In this way, two individual proteins may be tested for their ability to interact. Furthermore, using such a system, novel interacting partners can be identified by screening a single protein against a library [20].

Fig. 1 (continued) The transfected cells are treated with cycloheximide (CHX) for the indicated time to block protein synthesis. The harvested cells are lysed for immune blotting analysis. Flag-Snail or EGFP are detected with an anti-Flag or an anti-GFP antibody, respectively. The expression of β-Actin is detected as a loading control. The immune blotting results are analyzed by Gel-pro 3.1. The half-life of Flag-Snail can be determined and presented as a plot in the MCF-7 cells expressing control shRNA or DEDD shRNA

Death effector domains (DEDs) are protein–protein interaction structures that are found in proteins that regulate a variety of signal transduction pathways [1]. DEDD is a novel death effector domain (DED) containing protein. In the current study, the full length DEDD was ligated into the pGBKT7 vector to create the pBD-DEDD bait plasmid. The construct was used to screen a human embryonic kidney library to find the interacting proteins of DEDD. In the positive clones, we found that PI3KC3 is a candidate binding partner of DEDD.

3.7.1 Construction of DEDD-pGBKT7 Bait Plasmid

Human DEDD full-length gene was amplified by PCR using forward primer: 5'-GGAA TTC ATG GCG GGC CTA AAG C-3', and reverse primer: 5'-CGG GAT CCG TCA GGG CAA TGC TTG CAG-3'. The PCR product (gel-purified) and the pGBKT7 plasmid were digested with double restriction enzyme (EcoR I and BamH I) and recovered. Set up the ligation mixture by adding the following into a microcentrifuge tube: 100 ng linearized vector, DNA of DEDD (threefold molar excess over vector), 4 μL of 5× Rapid ligation Buffer, 1 μL of T4 DNA Ligase, and add ddH_2O to give a final volume to 20 μL. Vortex and spin briefly to collect drops. Incubate the mixture at 22 °C for 5 min. Use 2–5 μL of the ligation mixture for transformation. Identify insert-containing plasmids by restriction analysis or PCR. Sequence to check the orientation and reading frame of the junctions.

3.7.2 Autonomous Activation Verification

Transform DEDD-pGBKT7 fusion construct into strain AH109. Assay the transformant for MEL1 activation by selecting for transformants on SD/–Trp/X-α-Gal. The pCL1, pGADT7-T+pGBKT7-53, pGADT7-T+pGBKT7-lam were performed as positive and negative controls in parallel. The DEDD-pGBKT7 transforming colonies are white, so the library transformation can be prepared.

3.7.3 Yeast Competent Cell Preparation

Inoculate 1 mL of YPDA with several 2–3 mm colonies. Vortex vigorously to disperse any clumps (*see* **Note 32**). Transfer cells to a flask containing 150 mL of YPDA. Incubate at 30 °C for 16–18 h with shaking (250 rpm) to stationary phase (OD 600 > 1.5). Transfer overnight culture into 1 L of YPDA. Incubate at 30 °C for 2–3 h with shaking (230–270 rpm) until OD 600 reach to 0.5 ± 0.1. Place cells in 50-mL tubes and centrifuge at 1,000 × *g* for 5 min at room temperature. Discard the supernatant and wash cell pellets by vortexing in 500 mL of sterile TE. Centrifuge cells at 1,000 × *g* for 5 min at room temperature. Decant the supernatant. Resuspend the cell pellet in 8 mL of freshly prepared, sterile 1× TE/LiAc to obtain the competent cells.

3.7.4 Yeast Transformation

In a 500-mL bottle, add and mix the following: DEDD-pGBKT7 (0.2–1.0 mg), human fetal kidney cDNA Library (0.1–0.5 mg), and

Herring testes carrier DNA (20 mg). Add 8 mL of yeast competent cells and mix well by vortexing (*see* **Note 33**). Add 60 mL of sterile PEG/LiAc solution and vortex at high speed to mix. Incubate at 30 °C for 30 min with shaking (200 rpm). Add 7.0 mL of DMSO, mix well by gentle inversion. Do not vortex.

Heat shock cells for 15 min in a 42 °C water bath, then swirl occasionally to mix. Chill cells on ice for 1–2 min. Centrifuge cells for 5 min at room temperature at 1,000 × *g*. Remove the supernatant and resuspend cells in 10 mL of YPDA. Plate transformations on 150-mm plates with SD/–Ade/–His/–Leu/–Trp/X-α-Gal medium (high-stringency) to screen for ADE2, HIS3, and MEL1 expression. Incubate plates upside down at 30 °C for 5–10 days until colonies appear.

Choose $Ade^+/His^+/Mel1^+$ colonies for further analysis (*see* **Note 34**).

3.7.5 Retest Phenotype

Test positive colonies on SD/–Leu/–Trp/X-α-Gal plates to allow loss of some of the AD/library plasmids while maintaining selective pressure on both the DEDD-pGBKT7 and AD vectors. Incubate plates at 30 °C for 4–6 days. A mixture of white and blue colonies indicates segregation. Transfer well-isolated colonies to SD/–Ade/–His/–Leu/–Trp/X-α-Gal plates to verify that they maintain the correct phenotype. Collect the colonies and retest $Ade^+/His^+/Mel1^+$ on SD/–Ade/–His/–Leu/–Trp plates in a grid fashion. Incubate plates at 30 °C for 4–6 days.

3.7.6 Rescue AD Plasmids

Isolate plasmid DNA from yeast by using the Yeast Plasmid Isolation Kit (TianGen) (*see* **Note 35**). Transform the yeast-purified plasmid DNA into *E. coli* by electroporation. The transformants must be plated on LB medium containing ampicillin to select for only the AD/library plasmid (*see* **Note 36**). Isolate from *E. coli* the AD/library plasmid for sequencing test.

3.7.7 Retest DEDD-PI3K3C Interaction in Yeast

Cotransform DEDD-pGBKT7 and PI3K3C-AD plasmids into AH109. Plate on SD/–Ade/–His/–Leu/–Trp/X-α-gal and incubate plates at 30 °C until colonies appear.

3.8 Co-immunoprecipitation

Co-immunoprecipitation is most commonly used to test whether two proteins of interest are associated in vivo. In the form presented here, immunoprecipitation uses proteins expressed in mammalian cells, which allows them to be post-translationally modified in ways not possible to obtain when using prokaryotic sources. The proteins expressed in the mammalian cells can be endogenous proteins, or proteins ectopically expressed by transfection. The cells are harvested and lysed under conditions that preserve protein–protein interactions. The target protein is specifically immunoprecipitated from the cell extracts, and the immunoprecipitates are fractionated by SDS-PAGE and detected by western blotting with an antibody directed against the target protein. In the current study, co-immunoprecipitation and GST pull-down assays (described in Subheading 3.9) are used to confirm the interaction between DEDD and PI3KC3 identified through Y-2H screening.

3.8.1 Preparation of Cell Lysates

Transfect transiently two 100 mm dishes of HEK 293T cells with FLAG-DEDD expression plasmid. Transfect transiently another two dishes with pcDNA 3.1 empty vector as control. Grow the cells 24 h to allow time for protein expression. At the time of harvest, cells should be 80–100 % confluent, if possible. Wash cells three times with phosphate-buffered saline and scrap each plate of cells into 1 mL of ice-cold PBS buffer. Centrifuge cells for 5 min at 4 °C at 1,000 × *g*. Remove the supernatant. Lyse each plate of cells in 300 μL CO-IP lysis buffer on ice for 30 min. Remove unlysed cells, cell nuclei, and debris from the lysates by centrifugation (14,000 × *g*, 20 min, 4 °C). Transfer the supernatant to a clean tube. Determine the protein concentration by BCA protein assay kit. Adjust protein concentration to the same value with lysis buffer.

3.8.2 Immunoprecipitation

Add 2 μg of mouse anti-FLAG antibody to the cleared lysate (5 mg protein). Gently rock the samples overnight at 4 °C to allow the capture antibody to bind the bait–protein complexes (*see* **Note 37**). Add 20 μL 50 % Protein A/G Plus-Agarose slurry to the lysate/antibody mixture. Gently rock the samples at 4 °C for 1.5 h so that the beads remain in suspension (*see* **Note 38**).

Pellet the Protein A/G beads by brief centrifugation (5–10 sec in a microcentrifuge). Remove the supernatant by pipetting or gentle aspiration, and discard. Because the Protein A/G beads pellet is easily disturbed, a small amount (approx. 10–20 μL) of supernatant should be left over the beads. Add 500 μL wash buffer (at 4 °C) to the beads pellet. Gently resuspend the beads in the wash buffer by tapping or inverting the tube. Pellet the beads and discard the wash buffer. Repeat twice for a total of three washes, working quickly to maintain the temperature near 4 °C throughout the procedure. At the last wash, remove the supernatant as completely as possible (*see* **Note 39**).

3.8.3 Analysis of Precipitated Proteins

Remove the liquid portion of the mixture by aspiration. Add 60 μL of 1× SDS gel-loading buffer to the beads, and boil them for 5 min to dissociate the immune-complexes from the beads. Collect beads by centrifugation and perform SDS-PAGE with the supernatant fraction (*see* Subheadings 3.6.3 and 3.6.4).

3.9 GST Pull-Down Assay

GST pull-down assay is a technique to test the interaction between a GST-fusion protein (bait) and another protein (prey). The bait protein, purified from an appropriate expression system (e.g., *E. coli* or mammalian cells), is immobilized on a glutathione affinity gel.

The prey protein can be obtained from multiple sources including recombinant purified proteins, cell lysate or in vitro transcription/translation reactions. The protein–protein interactions can be visualized by SDS-PAGE and detected by Coomassie or silver staining, by western blotting and by [^{35}S] radioisotopic detection depending on the sensitivity requirements of the interacting proteins. Here we only describe the methods being used in the current study.

3.9.1 GST-DEDD Fusion Protein Expression

Transform pGEX4T-1-DEDD expression vector and pGEX4T-1 empty vector plasmids, respectively, into BL21 (DH3) competent cells and plate on an LB/Amp agar plate overnight at 37 °C. Pick an individual colony after 12–18 h of growth and inoculate in 5 mL of LB/Amp (100 μg/mL) liquid medium. Shake the medium overnight in a 37 °C shaker. The next morning pour the 2 mL culture of *E. coli* into 200 mL of LB/Amp medium and grow for 2–5 h with vigorous agitation until an OD 600 of 0.5–0.6 has been reached (*see* **Note 40**). Add IPTG to 0.1 mM and shake cells for 2–6 h at 37 °C (*see* **Note 41**).

3.9.2 Protein Extraction

Centrifuge cells at 7,000×*g* for 15 min at 4 °C. Discard the supernatants.

Resuspend each pellet cells in 10 mL ice-cold lysis buffer A (50 μL of lysis buffer for each mL of culture). Lyse the cells using a sonicator equipped with an appropriate probe. Sonicate 10–20 s/each time and totally 10–15 times. Lysis is complete when the cloudy cell suspension becomes translucent. Keep sample as cold as possible while sonicating. Add 20 % Triton X-100 to a final concentration of 1 %. Mix gently for 30 min for solubilization of the fusion protein. Centrifuge lysate at 12,000×*g* for 20 min at 4 °C. Transfer the supernatant to a fresh container.

3.9.3 Preparation of Glutathione Sepharose 4B

Gently shake the bottle of sepharose 4B to resuspend the matrix. Use a pipet to remove sufficient slurry for use, and transfer to an appropriate tube (1.33 mL of original sepharose slurry per mL of bed volume is required to dispense).

Sediment the matrix by centrifugation at 500×*g* for 5 min. Carefully decant the supernatant. Wash the sepharose 4B by adding 10 mL of cold 1× PBS per 1.33 mL of the original slurry of glutathione sepharose dispensed. Invert to mix (*see* **Note 42**). Sediment again the matrix by centrifugation at 500×*g* for 5 min. Decant the supernatant. For each 1.33 mL of the original slurry, add 1 mL of 1× PBS. This produces a 50 % slurry.

3.9.4 Purification of Fusion Proteins

Add 200 μL of a 50 % slurry of glutathione sepharose 4B equilibrated with 1× PBS to each 10 mL of sonicates. Incubate with gentle agitation at room temperature for 1 h. Centrifuge at 500×*g* for 5 min. Remove the supernatant and wash the matrix with 10 bed volumes of 1× PBS (1 mL 1× PBS for 200 μL 50 % slurry).

Centrifuge at 500×*g* for 5 min. Remove the supernatant. Repeat twice more for a total of three washes. In the last wash or after three washes, the lysis buffer B in which the pull-down experiment will be performed, can be used instead. After last washing extensively, add the lysis buffer to the GST-fusion protein/sepharose 4B beads, resulting a 50 % slurry. Keep the beads on ice. Take 20 μL out for SDS-PAGE gel and the Comassie staining to measure the amount of bound protein. Dilute GST protein in different folds to get similar amounts as for GST-DEDD protein.

3.9.5 Preparation of DEDD Conditional Medium

Transiently transfect four 100 mm dishes of HEK 293T cells with FLAG-DEDD expression plasmid. Obtain HEK 293T cell lysates as described in Subheading 3.7.1 and divide into two equal parts for GST pull-down. Add lysis buffer B to adjust lysates volume to 1 mL. Pre-clear the lysates by adding about 15 μL of undiluted GST beads to 1 mL lysates and incubate at 4 °C for 2 h.

3.9.6 Pull-Down Experiments

Add 20 μL of GST-DEDD beads to 1 mL precleared lysates. Incubate at 4 °C overnight (totally 2–5 μg of fusion protein, *see* **Note 43**). Dilute GST-beads with naked beads appropriately to control the same amount of input for GST and GST fusion protein. Add the same amount of beads to the lysates.

Centrifuge at 500×*g* for 5 min and remove the supernatants. Wash the beads using lysis buffer B for 3–4 times (50 μL of beads need 1 mL lysis buffer per wash). Add 2× SDS loading buffer to the beads and then boil for 10 min (50 μL of beads need 30–40 μL loading buffer). Run samples on an SDS-PAGE gel and subject to a western blot with an anti-FLAG antibody to detect the binding of FLAG-PI3KC3 with DEDD-GST. The equal amount of GST and GST-DEDD used in the binding assay are run on an SDS-PAGE gel and stained with Coomassie blue.

4 Notes

1. There are many kinds of DEDD antibodies currently. In our experience, DEDD antibody from Protein Technology is suitable for the detection of immunohistochemistry but not for immunoblotting. The antibody from Abnova is suitable for the detection by immunoblotting.
2. The use of 0.1 % Triton X-100 in the PBS helps to reduce surface tension, allowing reagents to cover the whole tissue section with ease. It is also believed to dissolve Fc receptors, therefore reducing nonspecific binding. We recommend PBST to give a cleaner background than PBS.
3. DAB is a suspected carcinogen. Wear the appropriate protective clothing. Deactivate it with sodium hypochlorite (bleach)

solution in a sealed container overnight (it produces noxious fumes when chloride is added) and dispose it according to laboratory guidelines.

4. G418 is an aminoglycoside antibiotic similar in structure to gentamicin B1. It is produced by *Micromonospora rhodorangea*. G418 blocks polypeptide synthesis by inhibiting the elongation step in both prokaryotic and eukaryotic cells. Resistance to G418 is conferred by the neo gene from Tn5 encoding an aminoglycoside 3'-phosphotransferase, APT 3' II. G418 is commonly used in laboratory research to select genetically engineered cells.
5. Hygromycin B is an antibiotic produced by the bacterium *Streptomyces hygroscopicus*. It is an aminoglycoside that kills bacteria, fungi and higher eukaryotic cells by inhibiting protein synthesis. In the laboratory it is used for the selection and maintenance of prokaryotic and eukaryotic cells that contain the hygromycin resistance gene. The resistance gene is a kinase that inactivates hygromycin B through phosphorylation.
6. Cell invasion, chemotaxis and motility studies are usually done in Transwell membranes with 3.0 μm or larger pores. The ability of cells to migrate through pores of a membrane is dependent on the cell line used and the culture conditions, as well as the pore size. Cell migration will not occur with pores smaller than 3.0 μm.
7. We find that it is best to prepare this fresh each time.
8. Keep the slides in the tap water until ready to perform antigen retrieval. At no time from this point ahead should the slides be allowed to dry. Drying out may cause nonspecific antibody binding and therefore high background staining.
9. Most formalin-fixed tissues require an antigen retrieval step before immunohistochemical staining can be proceed. This is due to the formation of methylene bridges during fixation, which causes cross-links of proteins and therefore masks the antigenic sites.
10. All incubations should be carried out in a humidified chamber to avoid drying of the tissue. Drying at any stage will lead to nonspecific binding and ultimately high background staining. A shallow, plastic box with a sealed lid and wet tissue paper in the bottom is an adequate chamber, just as long as the slides are kept off the paper and can lay flat so that the reagents do not drain off!
11. Permeabilization should only be required for intracellular epitopes when the antibody required access to the inside of the cell to detect the protein. The expression of DEDD is mainly locating in cytoplasm and nuclear, so permeabilization is needed.

12. The secondary antibody may cross-react with endogenous immunoglobulins in the tissue. This is minimized by pretreating the tissue with normal serum obtained from the species in which the secondary antibody was raised. The use of normal serum before the application of the primary also eliminates Fc receptor binding of both the primary and secondary antibodies.
13. Overnight incubation allows antibodies of lower titer or affinity to be used by simply allowing more time for the antibodies to bind. Also, regardless of the antibody's titer or affinity for its target, once the tissue has reached saturation point, no more binding can take place. Overnight incubation ensures that this occurs.
14. The optimal incubation time, incubation temperature, and antibody dilution, should be determined by the individual laboratory.
15. The cell density should not be too high. If the cells are too crowded, they may not be killed very effectively. Split cultures that are too close to confluency for good antibiotic selection. On the other hand, low cell density cultures typically grow slowly, and may be more sensitive to antibiotics than higher cell density cultures of the same cell line.
16. Do not allow undiluted FuGENE® HD Transfection Reagent to come into contact with plastic surfaces, because chemical residues in plastic vials can significantly decrease the biological activity of the reagent.
17. Because the transfection reagent demonstrates minimal cytotoxicity or changes in morphology when the cells are transfected, there is no need to remove it and replace the medium after transfection.
18. It is important to include two non-transfected control cultures. One is subjected to G418 selection to control for the fraction of cells that survive selection; it will help to determine the effectiveness of the transfection and selection processes. The second control is grown without G418 selection as a positive control for cell viability.
19. According to the guidelines of Lipofectamine 2000, do not add antibiotics to culture medium during transfection as this cause cell death.
20. Because the cytotoxicity of liposome, we recommend replacing the culture medium containing transfect complex after transfection.
21. It is important to include two non-transfected control cultures. One is subjected to hygromycin selection to control for the fraction of cells that survive selection; it will help determine the effectiveness of the transfection and selection processes.

The second control is grown without hygromycin selection as a positive control for cell viability.

22. To increase reproducibility between experiments, multiple wells for several experiments can be coated at the same time and stored at 4 °C after the Matrigel has dried completely.
23. Be careful to do not disrupt the Matrigel when adding the cell suspension by pipetting the cells gently down the inner wall of the upper Transwell chamber.
24. Caution should be taken in applying too much pressure because the swab can detach the membrane from the Transwell insert if it is pushed too hard.
25. To accurately count the cells that have invaded, it is necessary to distinguish individual cells from one another. Cells that are too dense will appear as large aggregates and the exact cell number will be difficult to determine. The conditions must be adjusted (i.e., cell number, length of assay and Matrigel concentration) to obtain cell densities for the quantitative assay.
26. Before the experiment, redissolve the 1.0 % agar and 0.7 % agarose by heating in microwave. Then cool to 40 °C in a waterbath. Warm 2× DMEM with 20 % FBS to 40 °C in a waterbath.
27. These plates can be stored at 4 °C for up to 1 week. Then remove to sit at room temperature for 30 min before using.
28. pEGFP vector is co-transfected to normalize for experimental variation between different time points.
29. The protein expression usually occurs at 24–72 h post-transfection. The optimal time interval depends on the cell type, research goals and specific expression characteristics of the transferred gene.
30. Cycloheximide (CHX) is an inhibitor of protein biosynthesis in eukaryotic organisms. It is produced by the bacterium *Streptomyces griseus*. Cycloheximide exerts its effect by interfering with the translocation step in protein synthesis (movement of two tRNA molecules and mRNA in relation to the ribosome) thus blocking translational elongation. Cycloheximide is widely used in biomedical research to inhibit protein synthesis in eukaryotic cells studied in vitro.
31. Refer to Table 2 for the effective range of separation of SDS-PAGE. Since the molecular weights of Flag-Snail and EGFP are approximately 28 kDa, we chose 12 % resolving gel to separate the proteins.
32. Use a 1–3 week-old colony (2–3 mm) to inoculate each liquid culture. If colonies on the stock plate are <2 mm, use several colonies.

Table 2
Effective range of separation of SDS-PAGE

Acrylamide concentration (%)	Linear range of separation (kDa)
15	10–43
12	12–60
10	20–80
8	36–94
6	57–212

33. Remove two 100-μL aliquots of competent cells to perform control transformations with pCL1, and pGBKT7-53+ pGADT7-T.
34. Ignore small, pale colonies that appear after 2 days but never grow to >2 mm in diameter. True His+ colonies are robust and can grow to >2 mm. Ade+ colonies will remain white to pale pink.
35. The plasmid DNA isolated from each positive yeast colony will be a mixture of the DNA-BD/bait plasmid and at least one type of AD/library plasmid.
36. The transformation efficiency of competent *E. coli* cells must be >10^9 cfu/μg.
37. The IgG antibodies used for immunoprecipitation and western blot are heterotetramers composed of two heavy chains (approx. 50 kDa each) and two light chains (approx. 25 kDa each). When boiling the beads to collect the protein complexes, most of the antibody will co-elute, resulting in a large amount of antibody contamination in the final sample. The secondary antibody for western blot is usually a peroxidase-conjugated anti-IgG. If the immunoprecipitation antibody and the primary western antibody are from the same species, the signal intensity of the precipitation antibody can be extremely high and proteins with molecular weights similar to the antibody components can be masked. To avoid this phenomenon, capture antibody and primary blotting antibody from different species should be used as far as possible. Another potential solution is using a capture antibody that is covalently linked to a bead matrix.
38. It is recommended to cut the tip off of the pipette tip when manipulating agarose beads to avoid disruption of the beads.
39. To avoid unspecific protein binding in the sample, the concentration of NaCl can be adjusted up to 0.5–1 M. If a washing buffer with high salt concentration was used, it is best to wash the beads with 500 μL PBS or 50 mM HEPES, pH 7.5.

The low ionic strength of such buffers helps to prevent gel artifacts during SDS-PAGE.

40. To ensure adequate aeration, fill flasks to only 20–25 % capacity (e.g., 100 mL in a 500 mL flask).

41. The optimal final concentration of IPTG can only be determined empirically. It may range from 0.1 to 1.0 mM. Lower culture temperatures (even as 20 °C) may be used to reduce the formation of inclusion bodies.

42. Sepharose 4B must be thoroughly washed with PBS to remove the 20 % ethanol storage solution. Residual ethanol may interfere with subsequent procedures.

43. Determine the amount of protein bound to the beads per volume. Take 20 μL of beads out (from Subheading 3.8.3) and elute with loading buffer, boil and then run SDS-PAGE gel. As a maker, run BSA with different amounts alongside. Stain the gel with G-blue and then determine the amount of protein attached to the beads per 20 μL volume.

References

1. Valmiki MG, Ramos JW (2009) Death effector domain-containing proteins. Cell Mol Life Sci 66:814–830
2. Tibbetts MD, Zheng L, Lenardo MJ (2003) The death effector domain protein family: regulators of cellular homeostasis. Nat Immunol 4:404–409
3. Arai S, Miyake K, Voit R, Nemoto S, Wakeland EK, Grummt I et al (2007) Death-effector domain-containing protein DEDD is an inhibitor of mitotic Cdk1/cyclin B1. Proc Natl Acad Sci U S A 104:2289–2294
4. Xue JF, Hua F, Lv Q, Lin H, Wang ZY, Yan J et al (2010) DEDD negatively regulates transforming growth factor-beta1 signaling by interacting with Smad3. FEBS Lett 584: 3028–3034
5. Lv Q, Wang W, Xue J, Hua F, Mu R, Lin H et al (2012) DEDD interacts with PI3KC3 to activate autophagy and attenuate epithelial-mesenchymal transition in human breast cancer. Cancer Res 72:3238–3250
6. Ledford H (2011) Cancer theory faces doubts. Nature 472:273
7. Thiery JP, Acloque H, Huang RY, Nieto MA (2009) Epithelial-mesenchymal transitions in development and disease. Cell 139:871–890
8. Reiman JM, Knutson KL, Radisky DC (2010) Immune promotion of epithelial-mesenchymal transition and generation of breast cancer stem cells. Cancer Res 70:3005–3008
9. Foubert E, De Craene B, Berx G (2010) Key signalling nodes in mammary gland development and cancer. The Snail1-Twist1 conspiracy in malignant breast cancer progression. Breast Cancer Res 12:206
10. Tran DD, Corsa CA, Biswas H, Aft R, Longmore GD (2011) Temporal and spatial cooperation of Snail1 and Twist1 during epithelial-mesenchymal transition predicts for human breast cancer recurrence. Mol Cancer Res 9:1644–1657
11. Peinado H, Olmeda D, Cano A (2007) Snail, Zeb and bHLH factors in tumour progression: an alliance against the epithelial phenotype? Nat Rev Cancer 7:415–428
12. Yang J, Mani SA, Donaher JL, Ramaswamy S, Itzykson RA, Come C et al (2004) Twist, a master regulator of morphogenesis, plays an essential role in tumor metastasis. Cell 117:927–939
13. Albini A, Benelli R (2007) The chemoinvasion assay: a method to assess tumor and endothelial cell invasion and its modulation. Nat Protoc 2:504–511
14. Mori S, Chang JT, Andrechek ER, Matsumura N, Baba T, Yao G et al (2009) Anchorage-independent cell growth signature identifies tumors with metastatic potential. Oncogene 28:2796–2805
15. Goldberg AL (2003) Protein degradation and protection against misfolded or damaged proteins. Nature 426:895–899

16. Nath D, Shadan S (2009) The ubiquitin system. Nature 458:421
17. Zhou BP, Deng J, Xia W, Xu J, Li YM, Gunduz M et al (2004) Dual regulation of snail by GSK-3[beta]-mediated phosphorylation in control of epithelial-mesenchymal transition. Nat Cell Biol 6:931–940
18. Demontis S, Rigo C, Piccinin S, Mizzau M, Sonego M, Fabris M et al (2005) Twist is substrate for caspase cleavage and proteasome-mediated degradation. Cell Death Differ 13: 335–345
19. Yang Z, Klionsky DJ (2009) An overview of the molecular mechanism of autophagy. Curr Top Microbiol Immunol 335:1–32
20. Miller J, Stagljar I (2004) Using the yeast two-hybrid system to identify interacting proteins. Methods Mol Biol 261:247–262

Chapter 15

Zebrafish-Based Systems Pharmacology of Cancer Metastasis

Yasuhito Shimada, Yuhei Nishimura, and Toshio Tanaka

Abstract

Because of their small size, high fecundity, and commonality to human genetics and genomics, phenotype-based animal testing using zebrafish (*Danio rerio*) has emerged as a powerful tool for identifying disease mechanisms, drug target molecules and small bioactive compounds over the last decade. Importantly, the immaturity of the zebrafish larvae immune system compared with that of mammals facilitates the implantation of human tumors representing aggressive cancer progression with metastasis. In the current chapter, we describe the methods for human cancer cell xenotransplantation into zebrafish, phenotypic image analysis, and transcriptome analysis using deep sequencing.

Key words Cancer transplantation, Next-generation sequencer, RNA-seq, Transcriptome analysis, Bioinformatics, Pathway analysis

1 Introduction

Cancer metastasis occurs through a complex series of steps in which cancer cells leave the original tumor site and migrate to other sites in the body via the bloodstream, the lymphatic system, or by direct cell migration. These multistep cascades require that animal models are indispensable in order to find new mechanisms, therapeutic genes and for the examination of candidate compounds against cancer metastasis. Since the zebrafish shares pathophysiological pathways common to human cancer progression [1], their application in cancer research is now expanding, primarily with tumor-initiating/promoting gene introduction [2] and with mammalian cancer cell/cancer stem cell xenotransplantation [3, 4]. Their small size is suitable for the 96-well plate format, and their transparent body wall enables in vivo imaging in combination with fluorescent labeling to permit genetic and chemical testing in a high-throughput manner. In fact, the high similarity of amino acid sequences between human and zebrafish proteins alludes to the functional similarity in cancer progression [5], but the nucleotide sequences are apparently different.

Martha Robles-Flores (ed.), *Cancer Cell Signaling: Methods and Protocols*, Methods in Molecular Biology, vol. 1165, DOI 10.1007/978-1-4939-0856-1_15,

Thus, transcriptome analysis using deep sequencing enables distinguishing of human and zebrafish transcripts in cancer-xenotransplanted zebrafish. In addition, combination with 96-well plate-based phenotypic analysis makes it possible to perform genetic and chemical screening in animal-based drug discovery.

2 Materials

Prepare all solutions using ultrapure water (e.g., Milli-Q water) and analytical grade reagents.

2.1 Zebrafish Breeding Resources

Materials for fish breeding have been described previously [6] with some modifications.

1. Fish lines: pigmentless and transparent fish line, *nacre* mutant [7] (Zebrafish International Resource Center, Eugene, OR, USA). Transgenic zebrafish with EGFP expression in endothelial cells, *Tg (fli1:egfp)* [8] (Zebrafish International Resource Center) (*see* **Note 1**).
2. Fish breeding system: MH-F1600M (MeitoSystem, Aichi, Japan), ZEBTEC (Tecniplast, Varese, Italy) or any that can circulate clean water and maintain the water temperature at 28 °C with periodic light switching (14 h light–10 h dark cycle). Water quality measurement instruments and/or reagents (*see* **Note 2**).
3. Incubator (28–34 °C) with periodic light system (14 h light–10 h dark cycle) for embryo and larvae breeding.
4. E3 medium: 5 mM NaCl, 0.17 mM KCl, 0.4 mM $CaCl_2$, and 0.16 mM $MgSO_4$. Store at room temperature or 28–34 °C to pre-warm for breeding larvae (*see* **Note 3**).
5. Bleach water: a solution for bleaching eggs is prepared by placing 0.1 mL of 5.25 % sodium hypochlorite (bleach) and 170 mL of distilled water into a 250 mL beaker.
6. Collection and washing of embryos: tea strainers, petri dishes and 100 mL beakers. Pipettors and wide-bore pipette tips for embryo transfer.

2.2 Cell Culture and Stable Expression of Fluorescent Protein

1. Equipment for cell culture: CO_2 incubator (37 °C, 5 % CO_2) for cell culture, centrifuges, microbiological safety cabinet, and Countess cell counter (Invitrogen, Carlsbad, CA, USA). Consumable supplies include culture flasks (T75), petri dishes, 6-well plate, pipettes, and pipettors.
2. Human cancer cells: KLM-1, human pancreatic cancer cells (Riken Cell Bank, Tokyo, Japan) or any cells to be implanted into zebrafish (*see* **Note 4**).

3. Culture medium: RPMI1640 (Gibco-BRL, Grand Island, NY, USA). Fetal bovine serum (FBS; Gibco-BRL) that has been heat-inactivated at 56 °C for 30 min. Store culture medium at 4 °C.
4. Antibiotics: penicillin–streptomycin liquid (Gibco-BRL) containing 10,000 U/mL penicillin and 10,000 μg/mL streptomycin for routine culture. Geneticin (G418; Roche Diagnostic, Mannheim, Germany) for transformants selection (*see* **Note 5**).
5. Plasmid for fluorescent protein expression: Kusabira-Orange (KOr) fluorescent protein expression vector, phKO1-MC1 plasmid (Amalgaam, Tokyo, Japan) (*see* **Note 6**).
6. Transfection reagent: Lipofectamine 2000 reagent (Invitrogen) and Opti-MEM I medium (Gibco-BRL) (*see* **Note 7**).
7. Cell sorter: FACSAria (BD Biosciences, San Jose, CA, USA) for KOr-positive cell selection (*see* **Note 8**).

2.3 Preparation of Zebrafish for Xenotransplantation

All reagents described below are freshly prepared prior to the experiments.

1. Pronase solution for removing chorions. Twenty milligrams of pronase powder (Roche Diagnostic) is dissolved in 40 mL E3 medium (final 2 mg/mL). This is enough for 500 embryos.
2. Anesthetics: 100 ppm of ethylene glycol monophenyl ether (TokyoKasei, Tokyo, Japan) dissolved in E3 medium (*see* **Note 9**).
3. Embryo holding sheet that can array zebrafish embryos during xenotransplantation.

2.4 Cancer Cell Xenotransplantation

1. For cell collection: trypsin–EDTA (0.05 % trypsin, 0.02 % EDTA; Gibco-BRL) and 2 % EDTA solution in phosphate-buffered saline (PBS) without calcium and magnesium (2 % EDTA–PBS; MP Biomedicals, Irvine, CA, USA).
2. Matrigel (BD Biosciences) and Hank's balanced salt solution (HBSS; Nakarai Chemicals, Kyoto, Japan). Matrigel should be preserved at 4 °C. Tips for pipetting Matrigel should also be stored at 4 °C.
3. Glass needles for cell implantation: GD-1 glass capillary (Narishige, Tokyo, Japan), PP-830 gravity puller (Narishige), and EG-44 microforge (Narishige).
4. SMZ800 stereoscopic microscope (Nikon, Tokyo, Japan), also recommended to be set up at 4 °C.
5. FemtoJet (Eppendorf, Hamburg, Germany) for cell injection. GELoader Tips (Eppendorf,) for loading cell mixture into glass needles. They should also be preserved at 4 °C.

2.5 High-Content Imaging

1. 96-well imaging plate with lid (tissue culture treated black/clear; BD Falcon, Franklin Lakes, NJ, USA).
2. MZ16F stereoscopic microscope (Leica Microsystems, Wetzlar, Germany) equipped with a DP71 digital camera (Olympus, Tokyo, Japan). For fluorescent imaging, the GFP (for fli1:egfp) and DsRed (for KOr) filters are recommended (*see* **Note 10**).
3. ImageXpress Micro system (Molecular Devices, Sunnyvale, CA, USA) for high-content imaging (*see* **Note 11**).
4. MetaXpress software (Molecular Devices) for image analyses (*see* **Note 12**).

2.6 Deep Sequencing of Human and Zebrafish Transcriptome

1. RNA purification: RNase-free tips and microtubes, and RNase AWAY (Molecular Bioproducts, San Diego, CA, USA) for preparation of an RNase-free environment. Isogen reagent (Nippon Gene, Tokyo, Japan) and RNeasy Mini RNA isolation kit (Qiagen, Valencia, CA, USA) (*see* **Note 13**). β-Mercaptoethanol (Sigma-Aldrich, St Louis, MO, USA), 70 % RNase-free ethanol, chloroform (Wako Pure Chemicals, Osaka, Japan), and nuclease-free water (not DEPC-treated). MM300 Mixer Mill (Retsch, Haan, Germany, *see* **Note 14**) with zirconia balls (3 mm diameter; YTZ-3, AsOne, Osaka, Japan).
2. Ribosomal RNA (rRNA) depletion: Ribo-Zero rRNA Removal kit for human, mouse, and rat (Epicentre Biotechnologies, Madison, WI, USA, *see* **Note 15**). Block incubator.
3. ND-1000 Spectrophotometer (Nano-drop Technologies, Wilmington, DE, USA) and Agilent 2100 Bioanalyzer (Agilent Technologies, Palo Alto, CA, USA) to quantify and qualify RNA and DNA. For Agilent 2100 Bioanalyzer, Agilent RNA 6000 Pico Chip Kit and DNA 1000 Kit are necessary.
4. SOLiD Whole Transcriptome Analysis Kit (Applied Biosystems, Foster City, CA, USA). Required equipment includes a thermal cycler with a heated lid, fluorometer, XCell SureLock Mini-Cell (Invitrogen), Agilent 2100 Bioanalyzer, ND-1000 Spectrophotometer, centrifugal vacuum concentrator, microcentrifuge, and transilluminator. Novex 6 % TBE-Urea Gels (1.0 mm, 10 well; Invitrogen), TBE-Urea Sample Buffer (2×; Invitrogen) and TBE-Running Buffer (5×; Invitrogen). PureLink PCR Micro Kit (Invitrogen), PureLink RNA Micro Kit (Invitrogen), Quant-IT RNA Assay Kit (Invitrogen), and SYBR Gold nuclei acid gel stain (10,000×; Invitrogen). Agilent DNA 1000 Kit, MinElute PCR Purification Kit (Qiagen).
5. Deep sequencer: SOLiD4 System (Applied Biosystems, *see* **Note 16**).

2.7 Software for Bioinformatics Analyses

1. Bioscope software (Applied Biosystems) for mapping read-tags.
2. Partek Genomics Suites (Partek, St Louis, MO, USA) for quantification and normalization.

3. GenomeJack viewer software (Mitsubishi Space Software, Tokyo, Japan) (*see* **Note 17**).
4. MultiExperiment Viewer (MeV) software (Dana-Faber Cancer Institute, Boston, MA, USA) for clustering analysis (*see* **Note 18**).
5. Life Science Knowledge Bank (World Fusion, Tokyo, Japan) for conversion of zebrafish genes to human orthologs.
6. Pathway Studio software (Elsevier, Amsterdam, Netherlands) for pathway analysis (*see* **Note 19**).

3 Methods

All animal experiments should be conducted according to the animal laws of each country and research facility, and should comply with international guidelines. After the experiments, the fish are killed by anesthetic overdose.

3.1 Fluorescent Labeling of KLM-1, Human Pancreatic Cancer Cells

Ensure all cell culture procedures are performed under standard sterilized conditions to prevent contamination from bacteria, fungi, and mycoplasma and cross-contamination with other cell lines.

1. KLM-1 cells (5×10^5 cells) are seeded in a 6-well plate in 2 mL of RPMI1640 supplemented with 10 % heat-inactivated FBS without any antibiotics at 37 °C in 5 % CO_2. One or 2 days after cell seeding, confirm that cells have reached more than 70 % confluency before conducting the transfection procedure.
2. Dilute 10 μL Lipofectamine 2000 reagent (LPX) in 100 μL Opti-MEM and incubate for 5 min at room temperature. Within the waiting time, dissolve phKO1-MC1 plasmid (2.5 μg) in 100 μL Opti-MEM I. Add dilute plasmid to each tube of diluted LPX (1:1 ratio) and incubate for 20 min at room temperature. Then add 200 μL DNA–LPX complex mixture to cultured cells.
3. Twenty-four hours after transfection, change medium to complete medium (RPMI1640 + 10 % FBS with 100 U/mL penicillin and 100 μg/mL streptomycin). Add geneticin (final concentration 800 μg/mL) to select the stable transformants (*see* **Note 20**). Culture medium containing geneticin should be changed every day.
4. A week after geneticin treatment, purify the KOr-expressing cells by FACSAria flow cytometry. We denoted KOr-expressing KLM-1 cells as KLM1-KOr. After that, continue culture of KLM1-KOr in T75 flasks for xenotransplantation (*see* **Note 21**).

3.2 Preparation for Xenotransplantation

Care and breeding of zebrafish are conducted as described previously [6] with some modifications according to the environmental status of each research facility.

1. Construction of the transparent and vascular-monitoring zebrafish: because of their greater transparency enabling in vivo imaging for tumor angiogenesis, we crossbred the *nacre* mutants and *fli1:egfp* transgenic zebrafish (*see* Fig. 1).
2. On the evening of 3 days before the xenotransplantation, individual female zebrafish are placed in mating tanks with males. The next morning, the mating is initiated by light stimuli and their fertilized eggs are collected in a petri dish using a tea strainer to remove debris. After collection, the eggs are immersed in bleach water for 5 min, and washed with distilled water five times (*see* **Note 22**). After that, these eggs are incubated in E3 medium at 28 °C. In total, 100–200 eggs per petri dish are appropriate for normal development. E3 medium should be changed every day and dead embryos be removed.
3. Approximately 2–4 h before xenotransplantation (usually in the morning), the 48 h post-fertilization (hpf) embryos are dechorionated using 2 mg/mL Pronase solution. Immerse 500 embryos to 40 mL Pronase solution in a petri dish with stirring occasionally at 28 °C for 5 min (*see* **Note 23**). Confirm 50 % embryos are dechorionated, and wash the embryos with 50 mL E3 medium five times to remove remaining chorions completely.
4. Anesthetize the embryo by immersion in 1× anesthetic solution in E3 medium. Transfer the embryos into appropriate anesthetic-containing E3 medium to an embryo-holding sheet using wide-bore pipette tips. Remove the excess medium from the corner of the sheet that has arrayed embryos in each well.

3.3 Cancer Cell Xenotransplantation

1. Make glass needles from glass capillaries using a gravity puller. Grind the tip of the needle to a diameter of approximately 30 μm. FemtoJet nano-injector, glass needles, pipette, and GELoader tips should be cooled at 4 °C to prevent solidification of the Matrigel (*see* **Note 24**). It is recommended that the xenotransplantation procedure be conducted in a cold room.
2. Preparation for cell–Matrigel mixture: conduct trypsinization procedure to dissociate adherent culture cells. Remove culture medium and add 10 mL of 0.02 % EDTA–PBS solution to KLM1-KOr cells in T75 flask, and shake gently for 15 s. Remove 0.02 % EDTA–PBS, add 10 mL Trypsin–EDTA, and incubate for 2 min at 37 °C in CO_2 incubator. Confirm 50–70 % cell detachment by microscopy, and add 10 mL complete medium (containing 10 % FBS) to neutralize trypsin activity (*see* **Note 25**). Transfer cells to a centrifuge tube and

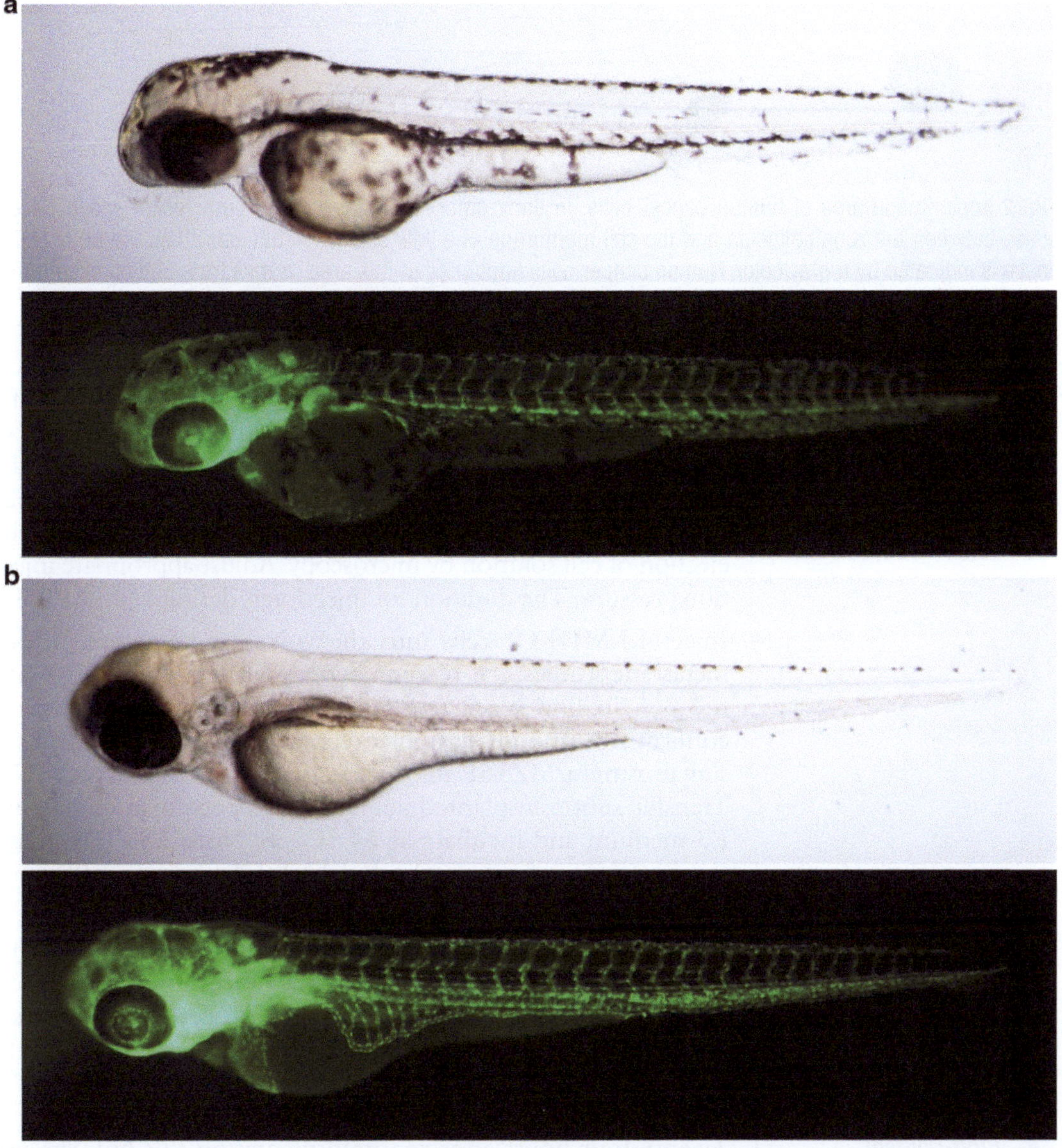

Fig. 1 Transparent zebrafish for cancer xenotransplantation. (**a**) *Tg* (*fli1:egfp*): wild type (AB) with *fli1:egfp* expression. (**b**) *nacre/fli1:egfp*: transparent type (*nacre*) with *fli1:egfp* expression. The pigmentation of the wild type (**a**), which disturbs the fluorescent signals, is completely diminished in the transparent line (**b**). This improves the signal-to-noise ratio during in vivo imaging of cancer xenotransplants

centrifuge at 200×*g* for 5 min at room temperature. Remove supernatant and dissolve cells in 5 mL medium. Count cell numbers using Countess cell counter and remove culture medium again by centrifugation (200×*g*). Adjust cell density to 8×10^7 cells/mL in 20–50 μL HBSS (*see* **Note 26**). Put the cells on ice. Add an equal volume of ice-cold Matrigel and mix gently by pipetting with ice-cold pipette tips.

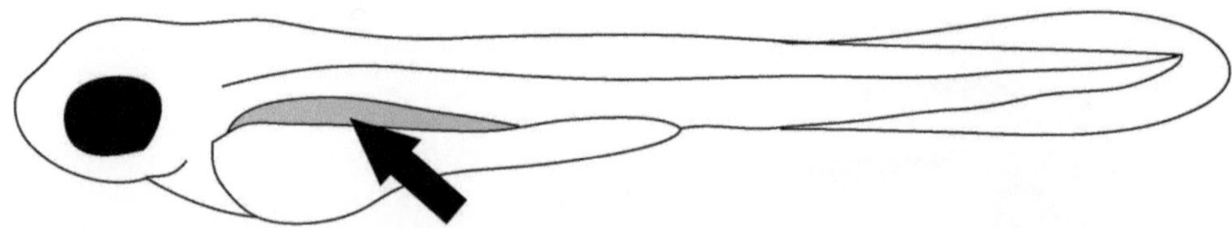

Fig. 2 Implantation area of human cancer cells. In early embryonic stages, the "perivitelline space" is the space between the zona pellucida and the cell membrane of a yolk sac. In 48 hpf zebrafish, the perivitelline space is indicated by a *gray* color. Human cancer cells implanted in this area (*arrow*) frequently induce tumor angiogenesis and metastasis

3. Implantation of KLM1-KOr cells into zebrafish embryos: fill glass needle with 5–10 μL cell–Matrigel mixture using an ice-cold GELoader tip. Next, connect the glass needle to the FemtoJet and inject cells at 300 hPa for 0.1 s. Immerse needle tip in a blank well of the embryo holding sheet, and confirm the ejection of cell solution by microscopy. Adjust appropriate injection pressure. The duration for injection is desirable under 0.3 s.
4. Inject KLM1-KOr cells into the yolk sacs of zebrafish. For metastatic studies, it is recommended that cells be injected into the perivitelline space (*see* Fig. 2). During the experiment, confirm the amount of implanted cells by observing the KOr signal using a MZ16F fluorescent microscopy (*see* **Note 27**). Transfer xenotransplanted zebrafish to a petri dish containing E3 medium and incubate at 34 °C (*see* **Note 28**). This temperature is acceptable for both zebrafish breeding and human cell proliferation.
5. Approximately 1–2 h after injection, transfer living xenotransplants to a new petri dish filled with E3 medium using a pipettor with wide-bore pipette tips. This procedure completely removes dead cells and debris to decrease bacterial pollution to xenotransplanted zebrafish. Continue incubation at 34 °C.

3.4 High-Content Imaging

1. Twenty-four hours after xenotransplantation (72 hpf, larvae), anesthetize the embryos and select xenotransplanted zebrafish under a fluorescent microscope. Then transfer the larvae to a 96-well plate with 50 μL of anesthetic medium.
2. Spin down the plate at 300 × *g* for 15 s to sink the larvae to the plate bottom. Turn the larvae onto their lateral side using the top of a GELoader tip under microscopic observation (*see* **Note 29**).
3. Place the 96-well plate in an ImageXpress MICRO XL and capture images with a montage mode and z-stacking (*see* Fig. 3 and **Note 30**). In detail, first the image of overall well is prescanned by taking four-view photographs using a FITC filter and a 2-power lens. The zebrafish body is recognized by GFP

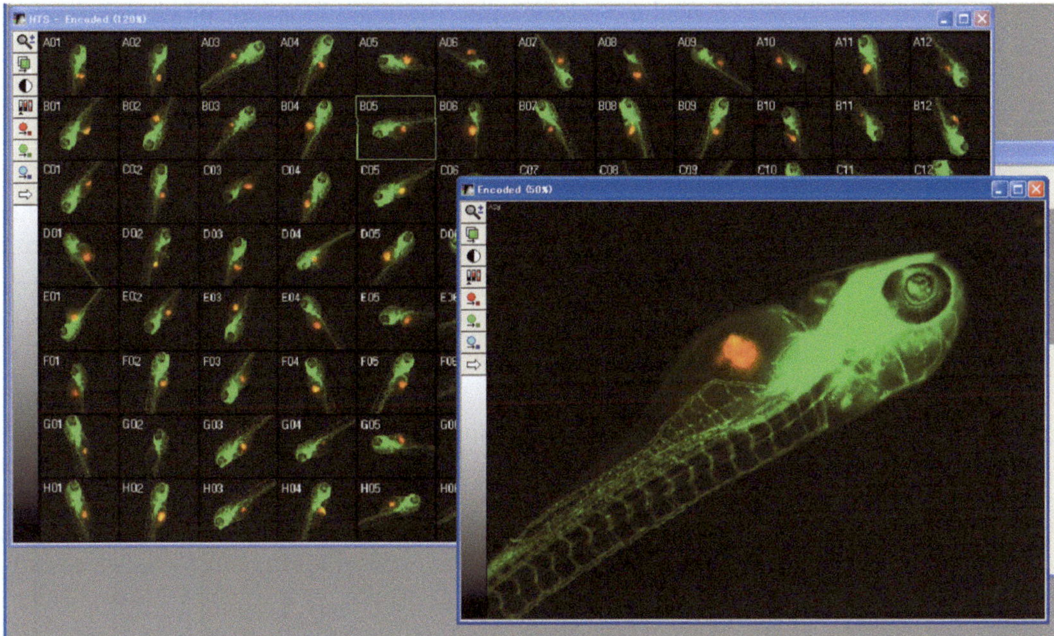

Fig. 3 High-content imaging of cancer-xenotransplanted zebrafish. Zebrafish are transferred into 96-well plates, and then anesthetized during image capture. After image capture, the breeding medium is changed to enable recovery from the anesthesia and to continue breeding

intensity contained in the obtained image, and the stage is moved so that the center of brightness comes into the center of the view. Then, five images are taken continuously with movement in the *Z* direction by 20 μm each time with a FITC filter and a 4-power lens, and the best focused composite image is created. In the case of dual wavelengths (EGFP and KOr), serial radiography is similarly performed to TRITC filter for KOr and the best focused composite image is created. Camera binning 1 and camera gain 1 are recommended.

4. After imaging, add 150 μL E3 medium and aspirate 150 μL. Repeat this step at least three times and finally add 150 μL E3 medium. Dilute anesthetic to a final concentration of 1.6 ppm (in addition this anesthetic, ethylene glycol monophenyl ether, is vaporizable). This procedure is repeatable until 6 dpf without unnecessary damage to zebrafish breeding. Continue incubation or conduct chemical treatment if needed.

5. Image analysis: MetaXpress analysis software is used for image analysis. The transplanted cancer clusters are recognized as nuclei using the Multiwavelength Cell Scoring application module based on KOr (TRITC filter) images, and the area, the total fluorescence value, and the average radius are calculated. Metastatic cancer tumors are recognized using the

Transfluor application module. Concentric circles of 150, 300, and 450 μm radius are drawn from the center of brightness, and the number and area of metastatic tumors in each four domains (0–150, 150–300, 300–450, and over 450 μm) are calculated.

3.5 Deep Sequencing for Transcriptome Analysis

Ribonuclease (RNase) contamination is a significant concern when working with RNA. RNase A is a highly stable and active ubiquitous ribonuclease that can contaminate any laboratory environment and is present on human skin. Create an RNase-free work environment and maintain RNase-free solutions for total RNA purification and rRNA removal reactions (*see* **Note 31**).

1. Sample homogenization: transfer the larvae to a 2 mL tube with E3 medium (*see* **Note 32**). Put the tube on ice to promote sinking of the larvae, and completely remove the medium using a 200 μL pipettor. Add 1 mL ISOGEN reagent and two zirconia balls (3 mm diameter). Homogenize the larvae using an MM300 Mixer Mill at 30 Hz for 2 min (*see* **Note 33**).
2. Total RNA purification: add 200 μL chloroform and mix by tipping tubes upside down (do not vortex). Store for 3 min at room temperature and centrifuge at 12,000 × *g* for 15 min at 4 °C. Transfer the aqueous phase to a new 1.5-mL tube, add 0.5 mL isopropanol and mix by pipetting. Store for 5 min at room temperature. Centrifuge at 12,000 × *g* for 10 min at 4 °C and discard supernatant. Add 70 % ethanol (RNase-free), vortex briefly, and centrifuge at 7,500 × *g* for 5 min at 4 °C. Discard the supernatant and dissolved the pellet in 100 μL RNase-free water. Add 350 μL RLT buffer (β-mercaptoethanol added) and 250 μL ethanol, and mix well. Transfer the samples to an RNeasy Mini column and centrifuge at 10,000 × *g* for 15 s at room temperature. Replace the collection tube with a new tube, add 500 μL RPE buffer to the column, and centrifuge at 10,000 × *g* for 15 s. Repeat this step. Centrifuge (10,000 × *g* for 2 min) again without adding any buffer to completely remove the excess liquid from the column. Change the new collection tube, add 30 μL of RNase-free water, and incubate for 5 min at room temperature. Centrifuge (10,000 × *g* for 1 min) and collect the purified total RNA. Measure the total RNA amount using an ND-1000 spectrophotometer.
3. Conduct rRNA depletion using a Ribo-Zero rRNA removal kit according to manufacturer's protocol (*see* **Note 34**). Ribosomal RNA depletion involves the following procedures: (1) preparation of the Ribo-Zero Microsphere; (2) mixing total RNA sample with Ribo-Zero rRNA Removal Solution; (3) microsphere reaction and rRNA removal; (4) purification of rRNA-depleted RNA; and (5) quantification

and qualification of the rRNA-depleted RNA samples using an Agilent 2100 Bioanalyzer with an RNA6000 Pico Chip kit (*see* **Note 35**). For further details, please see the manufacturer's instructions.

4. Amplified library construction using a SOLiD Whole Transcriptome Analysis Kit involves the following procedures: (1) fragmentation of the RNA; (2) hybridization and ligation of RNA; (3) reverse transcription; (4) cDNA purification; (5) selection of the correctly sized cDNA; (6) cDNA amplification; (7) amplified DNA purification; and (8) assessment of the yield and size distribution of the amplified DNA. These steps are generally conducted by the technical staff of the deep sequencer at each research facility. For further details, please see the manufacturer's instructions.
5. Deep sequencing: when less than 20 % of the amplified DNA is in the 25–150 bp range, proceed with the SOLiD4 System templated bead preparation stage, in which each library template is clonally amplified on SOLiD P1 DNA Beads by emulsion PCR. Refer to the Applied Biosystems SOLiD4 System Templated Bead Preparation Guide (Applied Biosystems). Bar-coded libraries are pooled prior to templated bead preparation.

3.6 Bioinformatics Analyses

1. Map the sequence reads (read-tags) to human (GRCh37) and zebrafish genomes (Zv9) using BioScope software (*see* **Note 36**). We recommend that a zebrafish- and human-fused genome is desirable for the separation of human and zebrafish transcripts.
2. Mapping and numerical conversion to exon and gene expression: import BioScope data to the Partek Genomics Suite software and re-map the read-tags again. After conducting RPKM normalization [9], retrieve numerical expression data (ratio) of each exon and gene (including alternative splicing) and export as tab-eliminated text or Excel files. Statistical significance tests can be also conducted using the Partek Genomics Suite (*see* **Note 37**).
3. Clustering analysis: import expression data to MeV4 and conduct hierarchical clustering with K-means methods and/or self-organizing mapping to retrieve the gene clusters of interest (*see* **Note 38**).
4. Pathway analysis: convert zebrafish genes to human orthologs using the Life Science Knowledge Bank. The gene symbols of human orthologs from zebrafish genes can be applicable for further functional analyses. Import the converted expression data to Pathway Studio and conduct pathway analysis with Gene Set Enrichment Analysis [10] and Sub-Network Enriched Analysis [11] to create pathways (*see* Fig. 4 and **Note 39**).

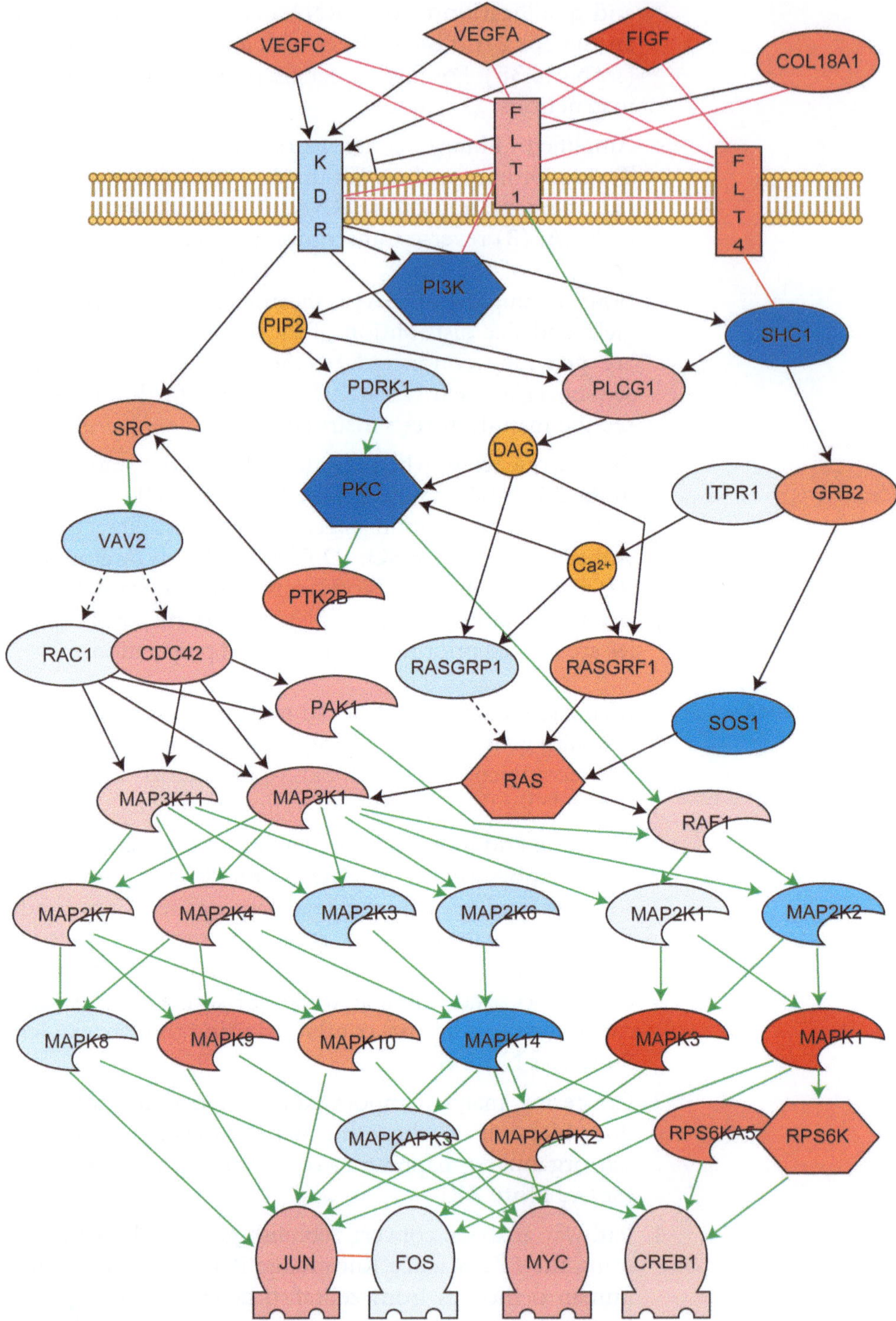

Fig. 4 Metastatic pathway of cancer-xenotransplanted zebrafish. In xenotransplants with a high-metastatic phenotype, the upregulation of the VEGF family and their downstream signaling are illustrated by pathway Analysis. *Red* color indicates increases and *green* indicates decreases in their gene expressions

4 Notes

1. For the tissue-specific imaging during tumor progression in zebrafish xenografts, other transgenic zebrafish are also available, for example *gata1a:egfp* for blood cell imaging.
2. Keep the water quality as listed (underlining indicates the important parameters to keep fish healthy). Alkalinity, 50–150 mg/L; water hardness, 80–300 mg/L; pH, 6.5–8.0; salinity, 0.5–1 g/L; conductivity, 300–1,500 μS; unionized ammonia (NH_3), <0.02 mg/L; nitrite (NO_2–), <1 mg/L; nitrate (NO_3–), <50 mg/L; chlorine, 0 mg/L; DO_2, >6 mg/L; CO_2, <5 mg/L.
3. Other fish-breeding medium can be also used, e.g., egg water, embryo medium, and E2 medium [6]. Of these, egg water is the easiest breeding medium for preparation (dissolve Instant Ocean (Aquarium Systems, Mentor, OH, USA) to a final concentration of 60 mg/L).
4. Cancer cells of your interest are also practicable. We have confirmed the translatability of several human cancer cell lines.
5. Change the antibiotics according to the inserted drug-resistant genes in transfected vectors.
6. Other plasmid, lentiviral, or transposon vectors are available to construct stable cell lines with fluorescent protein expression.
7. Other transfection reagents are also available, e.g., FuGene HD Transfection Reagent (Roche Diagnosis).
8. Other flow cytometry or cell pickers are also available.
9. 160 mg/L of tricaine (3-aminobenzoic acid ethyl ester, MS-222; Sigma-Aldrich, St Louis, MO, USA) solution is also suitable for anesthesia. Dissolve tricaine in E3 medium with 5 mM HEPES and adjust to pH 7.0. Do not use Tris–HCl instead of HEPES, since Tris–HCl has slight toxicity to zebrafish larvae.
10. Change fluorescent filter according to the selection of fluorescent proteins or dyes introduced to the cells.
11. Other imaging devices are available, for example IN Cell Analyzer 3000 (GE Healthcare, Waukesha, WI, USA).
12. Other image analysis software are also available, e.g., MetaMorph Software (Molecular Devices) and Image J (US National Institute of Health, Bethesda, MD, USA).
13. Other RNA isolation protocols based on acid guanidinium thiocyanate–phenol–chloroform extraction are also available. The procedure for RNA cleanup is recommended for deep sequencing.
14. Other homogenizers or sonicators are also suitable.

15. Ribo-Zero rRNA Removal kit has been changed to Ribo-Zero Magnetic Gold Kit (Human/Mouse/Rat). Both kits can remove almost 95 % ribosomal RNA from zebrafish total transcripts.
16. Other deep sequencers are also available for RNA sequencing, for example Ion Proton (Life Technologies).
17. Other viewer software are suitable, for example Integrative Genomics Viewer (IGV; Broad Institute, Cambridge, MA, USA).
18. Other clustering software are available, for example R Software with Bioconductor package [12].
19. Other pathway analysis software can be used, for example Ingenuity Pathway Analysis (IPA; Ingenuity Systems, Redwood City, CA, USA).
20. Select antibiotic according to the drug-resistant gene introduced by each plasmid. Determine the concentration of the antibiotic using a control cell experiment.
21. Cells can be cryopreserved in deep freezer (−80 °C) or in liquid nitrogen until xenotransplantation.
22. Aspirate the debris using a pipettor with a wide-bore pipette tip, if necessary.
23. Do not use pipetting to mix embryos with pronase solution. Pipetting frequently destroys the embryo bodies.
24. Matrigel will solidify rapidly at 22–35 °C. Thaw overnight at 4 °C.
25. Modify the trypsinization procedures according to cell types and the volume of culture medium.
26. If cell numbers are not enough providing a density of 8×10^7 cells/mL, the cells can be directly mixed with Matrigel without HBSS.
27. When the numbers of implanted cells are too small or large, modify the ejection pressure of FemtoJet for the appropriate implantation.
28. The incubation temperature depends on the cell type. However, incubation at more than 34 °C increases anomalies in zebrafish without xenotransplantation.
29. Until 6 dpf larvae (before air bladder development), spinning alone can turn almost 80 % zebrafish onto their lateral side.
30. Ask the technical staff of the manufacturer to create the image capturing program.
31. We strongly recommend the following guidelines: (1) Use RNase-free tubes and pipette tips. (2) Always wear gloves when handling samples containing RNA. Change gloves frequently, especially

after touching potential sources of RNase contamination such as doorknobs, pens, pencils, and human skin. (3) Always wear gloves when handling kit components. Do not pick up any kit component with an ungloved hand. (4) Keep all kit components tightly sealed when not in use. Keep all tubes containing RNA tightly sealed during the incubation steps. It is easy to use RNase AWAY to prepare and maintain an RNase-free environment.

32. This tube should be tolerable to bead homogenization. Less than 100 fish in one tube is recommended for 1 mL ISOGEN reagent.

33. If the larvae are not homogenized completely, increase the time for homogenization.

34. Other ribosomal RNA depletion kits are also available, however, for zebrafish we recommend the Ribo-Zero rRNA Removal Kit.

35. Apply the samples (total RNA and rRNA-depleted RNA) with two patterns of dilution to decrease the measurement errors.

36. Download the most recent genome sequences from Ensembl (http://asia.ensembl.org/index.htmL) or UCSC Genome Bioinformatics (http://hgdownload.soe.ucsc.edu/downloads.htmL).

37. The optional setting of the Partek Genomics Suite enables clustering and IPA pathway analysis.

38. Clustering analyses decrease the gene numbers for the next pathway analysis, thus this step is omissible depending on the experimental conditions. Other clustering methods are also available.

39. Gene Set Enrichment Analysis (GSEA) can identify gene sets defined by prior biological knowledge, such as the Gene Ontology category. GSEA can determine whether members of a gene set belonging to the same biological pathway tend to occur toward the top of a given gene list, such as the gene list established from clustering analysis. Sub-Network Enriched Analysis (SNEA) can identify key molecules regulating the expression of the genes of interest.

References

1. Lam SH, Wu YL, Vega VB et al (2006) Conservation of gene expression signatures between zebrafish and human liver tumors and tumor progression. Nat Biotechnol 24:73–75
2. Langenau DM, Traver D, Ferrando AA et al (2003) Myc-induced T cell leukemia in transgenic zebrafish. Science 299:887–890
3. Haldi M, Ton C, Seng WL et al (2006) Human melanoma cells transplanted into zebrafish proliferate, migrate, produce melanin, form masses and stimulate angiogenesis in zebrafish. Angiogenesis 9:139–151
4. Zhang B, Shimada Y, Kuroyanagi J et al (2014) Quantitative phenotyping-based in vivo chemical screening in a zebrafish model of leukemia stem cell xenotransplantation. PLoS One 9:e85439
5. Feitsma H, Cuppen E (2008) Zebrafish as a cancer model. Mol Cancer Res 6:685–694

6. Westerfield M (2007) The Zebrafish book, 5th ed.; a guide for the laboratory use of Zebrafish (Danio rerio). University of Oregon Press, Eugene
7. Lister JA, Robertson CP, Lepage T et al (1999) Nacre encodes a zebrafish microphthalmia-related protein that regulates neural-crest-derived pigment cell fate. Development 126:3757–3767
8. Lawson ND, Weinstein BM (2002) In vivo imaging of embryonic vascular development using transgenic zebrafish. Dev Biol 248:307–318
9. Bullard JH, Purdom E, Hansen KD et al (2010) Evaluation of statistical methods for normalization and differential expression in mRNA-Seq experiments. BMC Bioinformatics 11:94
10. Subramanian A, Tamayo P, Mootha VK et al (2005) Gene set enrichment analysis: a knowledge-based approach for interpreting genome-wide expression profiles. Proc Natl Acad Sci U S A 102:15545–15550
11. Kotelnikova E, Yuryev A, Mazo I et al (2010) Computational approaches for drug repositioning and combination therapy design. J Bioinform Comput Biol 8:593–606
12. Gentleman RC, Carey VJ, Bates DM et al (2004) Bioconductor: open software development for computational biology and bioinformatics. Genome Biol 5:R80

Part IV

New Technologies in the Study of Cancer

Chapter 16

A Flow Cytometry-Based Assay for the Evaluation of Antibody-Dependent Cell-Mediated Cytotoxicity (ADCC) in Cancer Cells

Nohemí Salinas-Jazmín, Emiliano Hisaki-Itaya, and Marco A. Velasco-Velázquez

Abstract

The development of new therapeutic monoclonal antibodies (mAbs) for cancer therapy will rise in the following years. The evaluation of biological activity of mAbs is required during drug development and during drug production as quality control test. Antibody-dependent cell-mediated cytotoxicity (ADCC) is a desirable activity of anticancer mAbs. Here, we describe a flow cytometry-based method to quantify ADCC that combines the staining of cancer cells, effector cells, and dead cells, with specific dyes. This method is inexpensive, has low background, and avoids the use of radioisotopes.

Key words ADCC, Flow cytometry, Trastuzumab, 7-AAD, Biosimilar

1 Introduction

The use of monoclonal antibodies (mAbs) for cancer therapy is one of the most successful and important strategies for treating patients with cancer. mAbs can kill tumor cells by (1) blocking the function of the target molecule, (2) mediating the delivery of cytotoxic drugs, (3) affecting tumor vasculature or stroma, and/or (4) triggering immune-mediated cell killing mechanisms [1]. The clinical success of mAbs is translated in high volumes of sales. For example, the mAbs bevacizumab (anti-VEGF), rituximab (anti-CD20), and trastuzumab (anti-HER2) had sales of over five billion dollars each during 2010 [2]. However, there are only 12 mAbs (targeting diverse surface molecules) currently approved in the market for cancer treatment in the European Union and the USA. Thus, the development of new therapeutic mAbs as well as that of biosimilars (called follow-up drugs in the USA) will rise in the following years.

Martha Robles-Flores (ed.), *Cancer Cell Signaling: Methods and Protocols*, Methods in Molecular Biology, vol. 1165, DOI 10.1007/978-1-4939-0856-1_16,

The successful development of therapeutic mAbs requires both a deep understanding of cancer biology and the establishment of methods that allow the evaluation of their physicochemical and biological attributes [1]. The in vitro analysis of the activity of the mAb provides key information of its effectiveness and, therefore, can be used as screening test during drug development and/or as quality control test for batch release during drug production.

Antibody-dependent cellular cytotoxicity (ADCC) is a humoral immune response mediated by cells of the innate immune system, including granulocytes, macrophages, and natural killer (NK) cells. In ADCC, the Fc portion of an antigen-specific antibody binds to the Fc receptor CD16A (FcγRIII) on lymphocytes and triggers apoptosis of target cells [3]. Cancerous cells are susceptible to ADCC mediated mainly by NK cells [4].

The chromium-release assay has long been the standard assay to measure cell-mediated cytotoxicity. Though this method has the benefits of being reproducible and relatively easy to perform, it has several drawbacks: (1) it provides only semiquantitative data unless limiting dilution assays are performed; (2) it has a relatively low level of sensitivity; (3) there is poor labeling of some target cell lines; (4) a high spontaneous release from some target cell lines occurs; and (5) there are biohazard and disposal problems associated with radioisotope usage [5]. The use of fluorescent probes offers as an alternative to chromium-release assays [5]. Here, we describe a flow cytometry method to quantify ADCC that combines the staining of target (cancer) cells, effector cells, and dead cells with specific dyes. This method avoids the use of radioisotopes, has lower background than *lactate dehydrogenase* (LDH) activity assays, and is cheaper than commercially available kits.

2 Materials (*See* Note 1)

1. mAb(s) of interest: For example, trastuzumab (Herceptin).
2. Actively growing cancer cells with high expression of the target antigen: For example, the cell line SK-OV-3, which overexpress HER2 (*see* **Note 2** and Fig. 1).
3. Actively growing cancer cells with low/absent expression of the target antigen: For example, the cell line MDA-MB-231, which displays discrete expression of HER2 (*see* **Note 2** and Fig. 1).
4. Complete media appropriate for the cell lines employed, such as McCoy's 5a (SK-OV-3) and Leibovitz's L-15 (MDA-MB-231).
5. Heat-inactivated fetal bovine serum (FBS).
6. Trypsin 0.25 %, 0.53 mM EDTA solution.

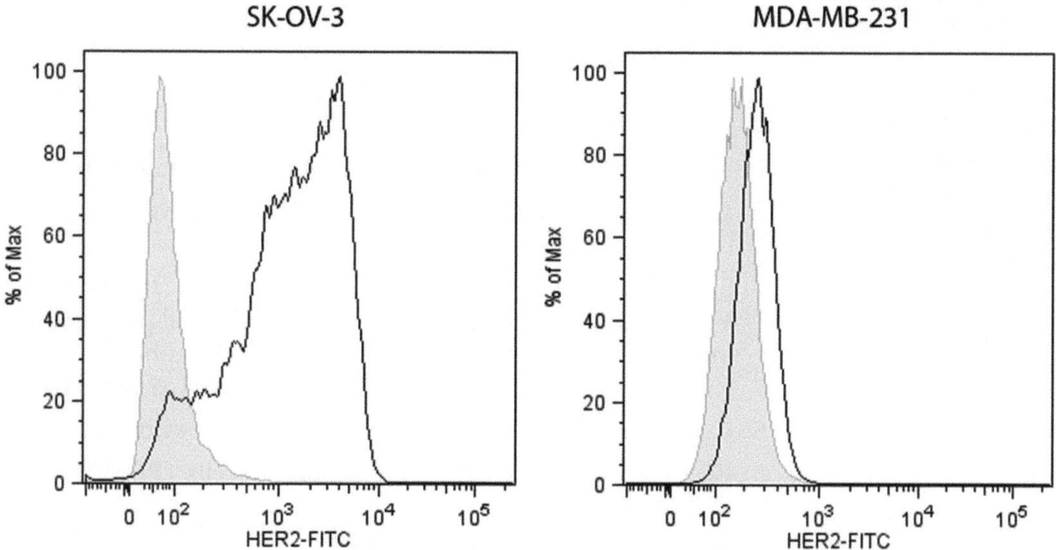

Fig. 1 Expression of HER2 in the two selected tumoral cell lines (SK-OV-3 and MDA-MB-231). Cells were stained with an isotype control (*gray*) or an FITC-labeled anti-HER2 antibody (*black line*)

7. Density gradient medium (Lymphoprep™ 1.077 g/L; *see* **Note 4**).
8. Phosphate-buffered saline (PBS) without Ca^{2+}/Mg^{2+} (137 mM NaCl, 2.7 mM KCl, 8 mM Na_2HPO_4, 1.46 mM KH_2PO_4) and endotoxin free.
9. Peripheral blood collected with anticoagulant.
10. Vacutainer sodium heparin 95 USP units.
11. Anti-CD56-PE.
12. 7-Amino-actinomycin D (7-AAD).
13. 5,6-Carboxyfluorescein diacetate succinimidyl ester (CFSE).
14. 25 cm^2 flasks.
15. 1.5 mL microtubes.
16. 10 mL sterile serological pipets.
17. Conical centrifuge tubes (15 and 50 mL).
18. Biosafety cabinet.
19. Inverted microscope, hemocytometer, and trypan blue, or reagents and equipment for counting cells.
20. Temperature-controlled centrifuge with horizontal rotor.
21. Flow cytometer.
22. Water bath with precise temperature control.

3 Methods

3.1 Isolation of Peripheral Blood Mononuclear Cells by Density Gradient Medium

1. Collect peripheral blood collected in tubes with heparin (*see* **Note 5**).
2. Transfer fresh blood into 50 mL conical centrifuge tubes.
3. Using a sterile serological pipet, add an equal volume of PBS (at room temperature) and mix well.
4. Slowly layer the Lymphoprep™ solution underneath the blood/PBS mixture by placing the tip of the serological pipet containing the Lymphoprep™ at the bottom of the tube. Use 1 mL Lymphoprep™ per 3 mL blood/PBS mixture (*see* **Note 6**).
5. Centrifuge at 800 × *g* for 30 min (18–20 °C) with no brake.
6. Recover peripheral blood mononuclear cells (PBMCs) from the interface.
7. Wash cells with 10 mL of cool PBS.
8. Centrifuge at low velocity (200 × *g*) for 5 min (18–20 °C; *see* **Note 7**).
9. Remove supernatant and resuspend in complete medium used for target cells (L-15 or McCoy's).
10. Quantify the concentration of viable PBMC in the hemocytometer with trypan blue. Only use cultures with cell viability ≥90 %.
11. Adjust cell concentration (using target cell culture medium) to 8×10^6 cells/mL (*see* **Note 8**).
12. Maintain the PBMC at 4 °C until use.

3.2 Label of Target Cells with 5, 6-Carboxyfluorescein-Diacetate Succinimidyl Ester

1. Determine the number of target cells needed for the assay. Consider 10^4 target cells per tube (*see* **Note 9**). Table 1 exemplifies the conditions needed for a typical assay.
2. Remove and discard culture medium from 25 cm^2 flask (s) containing target cells (SK-OV-3 or MDA-MB-231).
3. Wash the cells with 3 mL of cool PBS.
4. Add 1 mL of trypsin/EDTA solution, and incubate to 37 °C for around 3–5 min.
5. Add 3 mL of culture medium, and centrifuge at 200 × *g* for 5 min.
6. Decant supernatant, and resuspend the pellet in complete medium.
7. Quantify the concentration of viable cells by adding trypan blue and counting in hemocytometer. Use cultures with cell viability ≥95 %.
8. Transfer 5×10^6 cells to conical centrifuge tube of 15 mL (*see* **Note 12**).

Table 1
Controls and conditions needed in an ADCC assay (*see* Note 10)

Sample	Target cells	CFSE	7-AAD	Trastuzumab (μg/mL)	PBMC (effector cells)	Anti-CD56-PE
Unstained target cells	+	–	–	–	–	–
Unstained effector cells	–	–	–	–	+	–
CFSE control (target cells)	+	+	–	–	–	–
CD56-PE control (effector cells)	–	–	–	–	+	+
Basal death control (*see* **Note 11**)	+	+	+	–	+	–
Positive control	+	+	+	45 °C, 15 min		
mAb concentration 1	+	+	+	0.001	+	+
mAb concentration 2	+	+	+	0.01	+	+
mAb concentration 3	+	+	+	0.1	+	+
mAb concentration 4	+	+	+	1.0	+	+
mAb concentration 5	+	+	+	10	+	+

9. Wash cells with 10 mL of cool PBS and centrifuge at 400×*g* for 5 min.
10. Aspirate the supernatant, and resuspend the pellet of target cells (SK-OV-3 or MDA-MB-231 cells) in 250 μL of PBS. Now the cell concentration should be approximately 2×10^7 cells/mL (*see* **Note 13**).
11. Stain the target cells with 2.5 μM CFSE in PBS by adding 250 μL of CFSE 5 μM (2×; diluted in PBS) to the target cell suspension. Incubate for 5 min in the dark at room temperature. Manually stir the tube three times during incubation.
12. Stop staining reaction by adding five volumes of cool PBS with 10 % FBS.
13. Wash the cells. Centrifuge them at 400×*g* for 5 min, decant supernatant, and resuspend the pellet in 1 mL of complete medium appropriate for the cell line.
14. Determine again cell concentration by counting with trypan blue. Adjust concentration to 10^5 viable cells/mL (*see* **Note 14**).
15. Add 100 μL of the target cell suspension to a 1.5 mL microtube (10^4 cells/tube).

3.3 Opsonization of Target Cells with mAb

1. Reconstitute the mAb according to the manufacturer's instructions. For trastuzumab concentration after reconstitution is 21 mg/mL.
2. Prepare mAb solutions in complete target cell culture medium. Prepare solutions three times (3×) more concentrated than needed. For the trastuzumab samples listed in Table 1, concentrations would be 30, 3, 0.3, 0.03, and 0.003 μg/mL. The volume prepared for each concentration should be enough to use 100 μL/tube.
3. Add the mAb, medium, or excipient (*see* Table 1 and **Note 11**) to target cells, and incubate for 15 min at room temperature.
4. Add the 100 μL of the 8×10^6 PBMC/mL suspension to the mix of target cells with mAb. At this point final volume per tube is 300 μL.
5. Incubate the co-culture at 37 °C for 4 h with occasional (approximately every hour) manual stirring.
6. Wash cells with 1 mL cool PBS/tube and centrifuge at 400 × *g* for 5 min.

3.4 Staining of Effector Cells

1. Eliminate supernatant and resuspend in 100 μL of anti-CD56-PE diluted in PBS plus 1 % FBS (*see* **Note 15**). Some samples will not require CD56 staining (see Table 1), and in this case, cells should be resuspended in 50 μL of PBS and kept on ice until 7-AAD staining.
2. Incubate for 30 min at 4 °C, in the dark, with occasional (approximately every 10 min) manual stirring.
3. Wash the tube with 1.2 mL of PBS plus 10 % FBS and centrifuge at 400 × *g* for 5 min.
4. Resuspend the pellets in 50 μL/tube of PBS.

3.5 Staining of Dead Cells with 7-AAD

1. Dilute 7-AAD to 2.5 μg/mL in PBS. Volume should be enough to use 50 μL/tube (*see* Table 1).
2. Add 50 μL/tube of 7-AAD solution (0.125 μg/tube).
3. Incubate at 4 °C for 10 min in the dark.
4. Wash with 1 mL of cool PBS plus 10 % FBS, and centrifuge at 400 × *g* for 5 min.
5. Resuspend cell pellet in 200 μL of PBS, and transfer the suspension to flow cytometry tubes. Samples are ready to be analyzed.

3.6 Analysis Strategy for Flow Cytometry

1. Open 3 dot plots on the worksheet. Set FSC-A vs. FSC-H in the first, FSC-A vs. SSA-A in the second, and CFSE detection parameter vs. 7-AAD detection parameter in the third one.
2. Acquire "Unstained target cells" tube at the lowest flow speed.
3. In the FSC-A vs. FSC-H, make a gate (R1) selecting singlet events as shown in Fig. 2a.

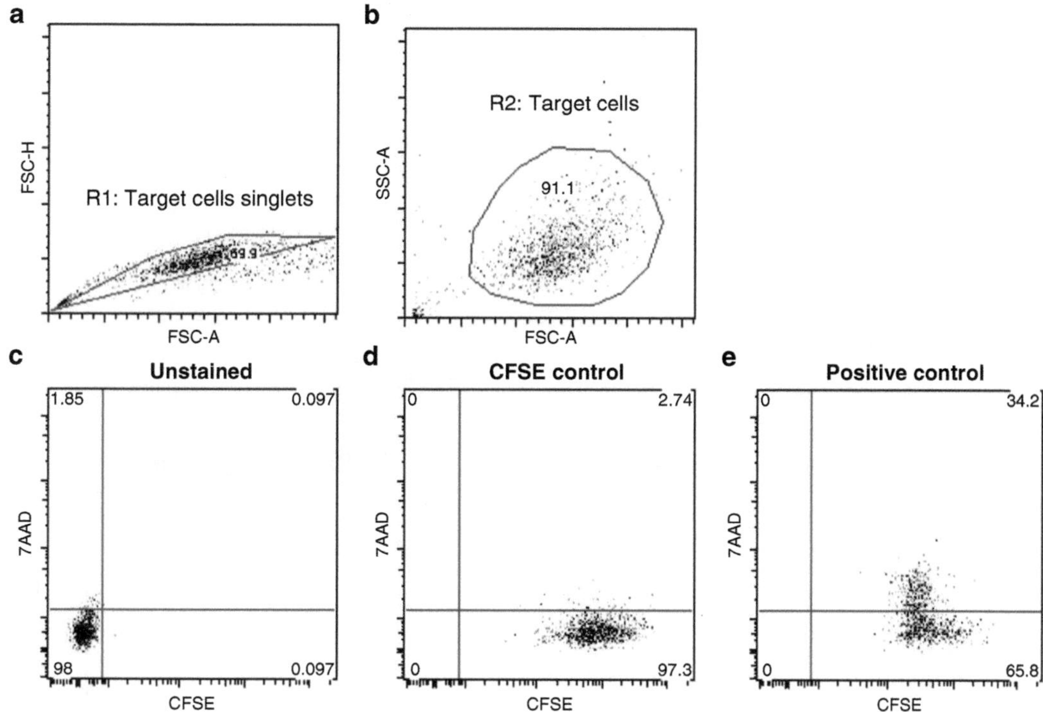

Fig. 2 Analysis strategy for target cells. Gate selection for further analysis (**a**, **b**; see text for details). Unstained target cells are used to establish autofluorescence limits (**c**). After CFSE staining, all target cells are CFSE+ (**d**). Temperature-induced dead cells (positive control) are CFSE + 7-AAD+ (**e**)

4. Set the R1 gated population in the FSC-A vs. SSA-A dot plot, and then make a new gate (R2) selecting target cells as shown in Fig. 2b.
5. Set the R2 gated population in the CFSE vs. 7-AAD dot plot, and define autofluorescence limits with quadrants as shown in Fig. 2c.
6. Set the "CFSE control" tube, acquire 1,000 events from the CFSE+ region, and verify that the fluorescence of target cells does not exceed the detection limit, as shown in Fig. 2d.
7. Acquire the "Positive control" tube, and verify that the 7-AAD + CFSE+ population correctly appears in the upper right quadrant and within the defined limits, as shown in Fig. 2e (*see* **Note 16**).
8. The effector cells must be acquired with the same parameter settings and worksheet defined with the target cells, except for the substitution of CFSE by PE detection parameter.
9. Acquire the "Unstained effector cells" control tube at the lowest flow speed for parameters and gate settings.
10. Make a gate (R3) in the FSC-A vs. FSC-H dot plot selecting the singlet events as shown in Fig. 3a.

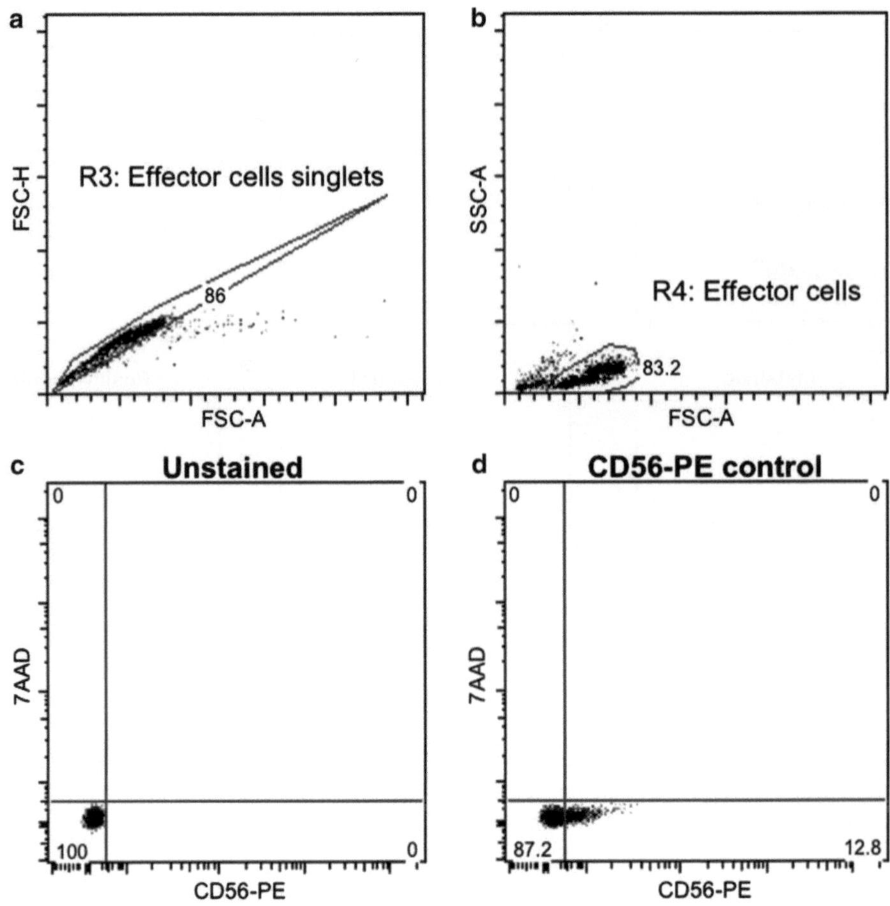

Fig. 3 Analysis strategy for effector cells. Gate selection for further analysis (**a**, **b**; see text for details). Unstained effector cells are used to establish autofluorescence limits (**c**). After CD56 staining, the percentage of NK in PBMC can be quantified (**d**)

11. Set the R3 gated population in the FSC-A vs. SSA-A dot plot, and make a new gate (R4) selecting effector cells as shown in Fig. 3b.
12. Set the R4 gated population in the CD56-PE vs. 7-AAD dot plot, and define autofluorescence limits with quadrants as shown in Fig. 3c (*see* **Note 17**).
13. Set the "CD56-PE control" tube, acquire 1,000 events from the CD56-PE+ region, and verify that the fluorescence of effector cells does not exceed the detection limit, as shown in Fig. 3d.
14. The samples containing target and effector cells must be acquired with the same parameter settings and gates defined with the target (R1, R2) and effector (R3, R4) cells, as shown in Fig. 4a.

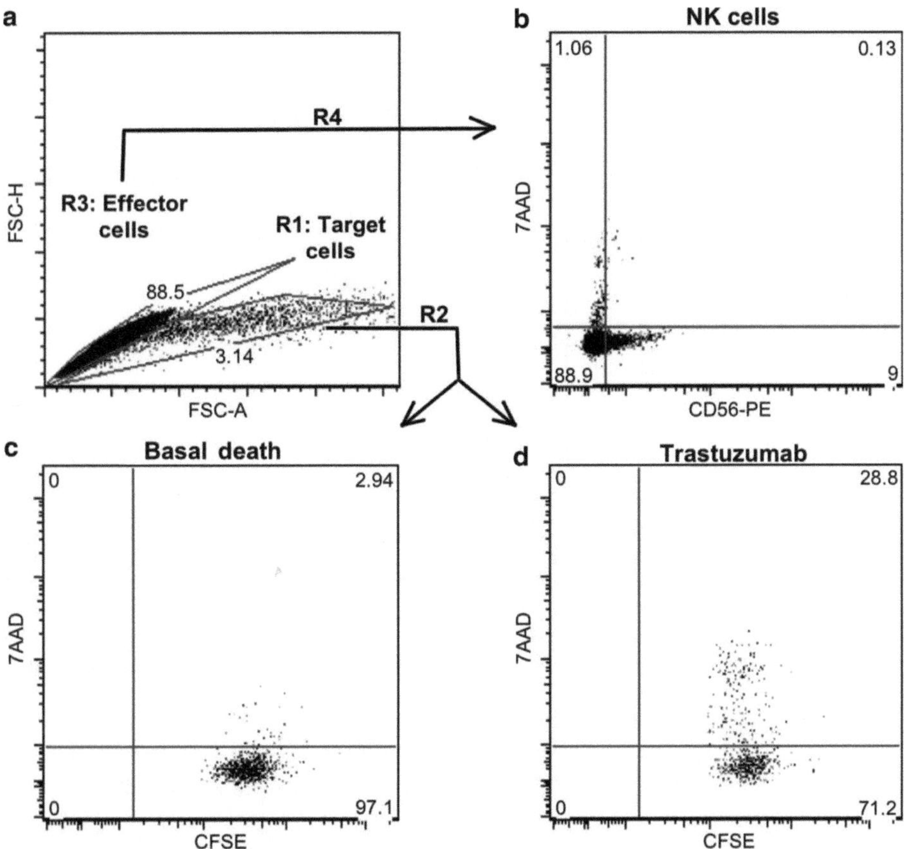

Fig. 4 ADCC assay in SK-OV-3 cells. Samples were acquired with the same parameter settings and gates defined with the controls (**a**; see text for details). NK cells can be quantified in R3 and R4 gated cells (**b**). Trastuzumab (10 μg/mL) triggers ADCC, but its excipient does not (**c**, **d**) increase the percentage of cells

15. Set the control tubes ("Basal death control" and/or excipient), and, in cells that come from R1 and R2, acquire 1,000 CFSE+ events and quantify the percentage of dead target cells (CFSE + 7-AAD+) as shown in Fig. 4c (*see* **Note 18**).
16. Set the tubes labeled "mAb concentration #" and, in cells that come from R1 and R2, acquire 1,000 CFSE+ events and quantify the percentage of dead target cells (CFSE + 7-AAD+) in the samples, as shown in Fig. 4d (*see* **Note 19**).
17. If of experimental interest, death of effector cells (CD56 + 7-AAD+) can be evaluated in CFSE− cells coming from R3 and R4 as shown in Fig. 4b.
18. Calculate the specific mAb-induced cytotoxicity by subtracting the percentage of dead target cells in appropriate control (excipient or "Basal death control") to the percentage found in "mAb concentration #". Typical results are shown in Fig. 5.

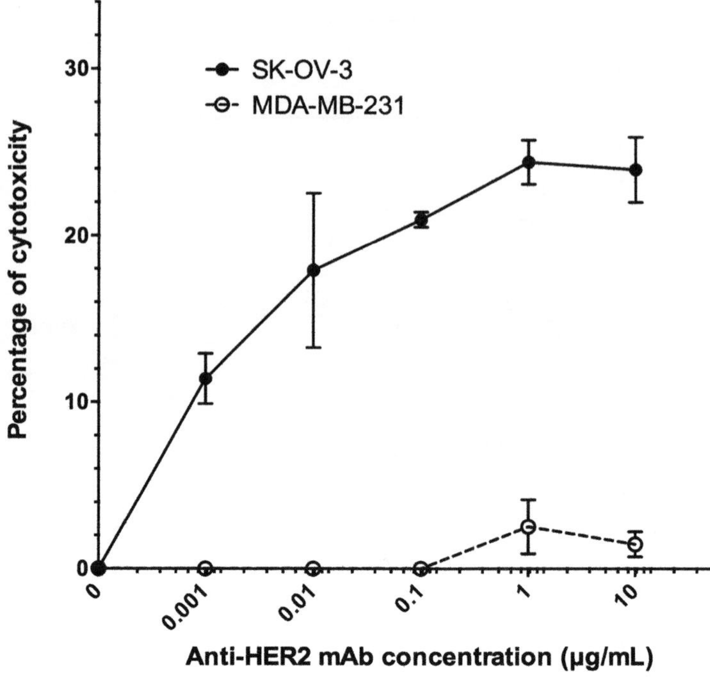

Fig. 5 Trastuzumab induces ADCC in SK-OV-3 but not in MDA-MB-231 cells. The percentage of dead target cells (CFSE + 7-AAD+) is dependent on the level of expression of HER2 as well as on the concentration of mAb

4 Notes

1. All solutions and equipment coming into contact with cells must be sterile, and proper aseptic techniques must be employed. The only exception to this could be during flow cytometry analysis.
2. We suggest performing an assay to detect expression levels of the antigen in the selected cell lines (*see* Fig. 1).
3. It has been reported that some mAbs can trigger ADCC in cell lines with low antigen expression [6]. You can also use target cells with induced downregulation of the target antigen (i.e., siRNA-transfected cells) as negative control.
4. It has been reported that PBMC preparations for clinical use are not influenced by the type of density gradient media used when comparing Ficoll-Paque and Lymphoprep™ [7], suggesting that the use of Ficoll-Paque or other density gradient media would also be useful for this protocol.
5. Blood should be obtained from healthy volunteers. If possible, repeat experiments with effector cells from different donors. We suggest that all volunteers sign a consent form approved by the Institutional Committee.

6. To maintain the Lymphoprep™/blood interface, it is helpful to hold the centrifuge tube at a 45° angle.
7. The goal of this step is the removal of platelets.
8. This concentration considers a 1:80 (target:effector) cell ratio. However, we suggest to evaluate different ratios (from 1:20 to 1:100) during standardization in order to find the condition with the best signal-to-noise ratio.
9. This staining protocol is effective in a range of $0.1–5 \times 10^7$ cells.
10. Run at least duplicates for each condition.
11. When using a mAb in a pharmaceutical presentation, a sample with the excipient should be included as a key additional control.
12. CFSE dye reacts with amine groups and should not be used with amine-containing buffers or lysine- or protein-coated slides.
13. To ensure uniform labeling, it is important that you begin with a single-cell suspension (no aggregates).
14. You should not expect changes in cell viability after CFSE.
15. There is the high variability in ADCC induction between donors [8]. Since the predominant effectors of ADCC are NK cells [4], staining with anti-CD56 allows normalization between donors.
16. The percentage of heat-induced dead cells in positive control is dependent on the target cell line used. Typically, you should expect responses over 30 %.
17. It has been reported that, in healthy subjects, NK cells account for 7–30 % of total lymphocytes [9].
18. Typically, basal death goes from 0 to 8 % of target cells.
19. Events from R1 and R2 may include some effector cells that would show as CFSE−. If this is the case, you must consider only CFSE+ cells in cytotoxicity analysis in order to avoid underestimation of the effect. To do this, open an SSC-A vs. CFSE dot plot, gate the CFSE+ cells, and set that gated population to the CFSE vs. 7-AAD dot plot.

Acknowledgments

We thank Dr. Frank Robledo-Avila and Dra. Mayra Pérez-Tapia for the critical review of this manuscript. Assays presented in this work were performed at "*Unidad de Desarrollo e Investigación en Bioprocesos* (*UDIBI*), *Instituto Politécnico Nacional.*" The work is supported by a grant from "Instituto Científico Pfizer, México" and grants IN219613 (PAPIIT-UNAM) and 225313 (Conacyt) to MAV-V.

References

1. Scott AM, Wolchok JD, Old LJ (2012) Antibody therapy of cancer. Nat Rev Cancer 12:278–287
2. Adler MJ, Dimitrov DS (2012) Therapeutic antibodies against cancer. Hematol Oncol Clin North Am 26:447–481, vii
3. Harrison A, Liu Z, Makweche S et al (2012) Methods to measure the binding of therapeutic monoclonal antibodies to the human Fc receptor FcgammaRIII (CD16) using real time kinetic analysis and flow cytometry. J Pharm Biomed Anal 63:23–28
4. Levy EM, Roberti MP, Mordoh J (2011) Natural killer cells in human cancer: from biological functions to clinical applications. J Biomed Biotechnol 2011:676198
5. Derby E, Reddy V, Kopp W et al (2001) Three-color flow cytometric assay for the study of the mechanisms of cell-mediated cytotoxicity. Immunol Lett 78:35–39
6. Collins DM, O'Donovan N, McGowan PM et al (2012) Trastuzumab induces antibody-dependent cell-mediated cytotoxicity (ADCC) in HER-2-non-amplified breast cancer cell lines. Ann Oncol 23:1788–1795
7. Yeo C, Saunders N, Locca D et al (2009) Ficoll-Paque versus Lymphoprep: a comparative study of two density gradient media for therapeutic bone marrow mononuclear cell preparations. Regen Med 4:689–696
8. Pollara J, Hart L, Brewer F et al (2011) High-throughput quantitative analysis of HIV-1 and SIV-specific ADCC-mediating antibody responses. Cytometry A 79:603–612
9. Green MR, Kennell AS, Larche MJ et al (2005) Natural killer cell activity in families of patients with systemic lupus erythematosus: demonstration of a killing defect in patients. Clin Exp Immunol 141:165–173

Chapter 17

Gene Disruption Using Zinc Finger Nuclease Technology

Sara Granja, Ibtissam Marchiq, Fátima Baltazar, and Jacques Pouysségur

Abstract

Zinc finger nucleases are reagents that induce DNA double-strand breaks at specific sites that can be repaired by nonhomologous end joining, inducing alterations in the genome. This strategy has enabled highly efficient gene disruption in numerous cell types and model organisms opening a door for new therapeutic applications. Here, we describe the disruption of CD147/basigin by this technique in a human cancer cell line.

Key words DNA double-strand breaks, Gene targeting, Knockout, Zinc finger nuclease

1 Introduction

Zinc finger nucleases (ZFNs) are a class of engineered DNA-binding proteins that have redefined genome engineering. The ZFN plasmids bind a specific DNA sequence, inducing a double-strand break (DSB). This break is repaired by the non-perfect endogenous repair mechanism called nonhomologous end joining (NHEJ). The DBS will be misrepaired by the addition and/or deletion of nucleotides. This disruption results in multiple mutations that could induce nonsense gene products in both alleles leading to a specific gene knockout. The ZFN structure consists of a DNA-binding domain—zinc finger protein (ZFP)—linked to a nuclease domain of the FOKl restriction enzyme [1].

The ZFP region contains a tandem array of Cys2–His2 fingers that each recognizes 3 bp of DNA. This domain can have from 3 to 6 fingers binding to a target of 9–18 bp, respectively. The catalytic domain composed by the enzyme FOKl requires dimerization to cut; thus, a pair of ZFNs is required. The dimerization restriction avoids cleavage at single binding sites [2]. Genomic modification with ZFN was first described more then 10 years ago, and a huge progress in this technique was observed in the last few years. This is now a well-established method in several organisms [3].

Martha Robles-Flores (ed.), *Cancer Cell Signaling: Methods and Protocols*, Methods in Molecular Biology, vol. 1165, DOI 10.1007/978-1-4939-0856-1_17,

ZFN-targeted mutagenesis has been achieved in zebrafish [4], rat embryo [5], mouse [6], sea urchin [7], frog [8], Drosophila [9], plants [10], *C. elegans* [11] and human cells [12].

Applications of ZFNs represent a complementary strategy to siRNA enabling highly efficient gene disruption in human cells, opening a door for new therapeutic applications [1]. The aim of our study is to understand the tumorigenic influence of lactate transport by monocarboxylate transporters (MCT1 and MCT4) in human cells by inhibiting their chaperone CD147/basigin. Here, we demonstrate an efficient disruption of the gene CD147/basigin in the A549 human lung adenocarcinoma cell line. A similar procedure has already been described in our laboratory in the LS174T colon adenocarcinoma cell line [12]. In this chapter, we show a protocol used to knock-out CD147/basigin in A549 cell line.

2 Materials

2.1 ZFN

1. CompoZr Knockout ZFNs (Sigma-Aldrich): ZFN plasmids targeting basigin exon 2 encoding signal peptide of basigin splice variants 1 and 2 were designed and obtained from Sigma-Aldrich. Store at −80 °C.
2. Each kit contains, besides the plasmid DNA for each ZFN, a pair of primers (forward and reverse) for screening of mutations in the DNA locus. Make aliquots of 10 μM. Store at −20 °C.

2.2 Plasmid Transformation and Isolation

1. Chemically competent *E. coli*.
2. LB medium: For 1 l of water dissolve 5 g yeast extract, 10 g tryptone and 10 g of NaCl. Adjust to pH 7.0. Autoclave and allow to cool to 55 °C. Add the antibiotic kanamycin. Store at room temperature (RT). For LB plates, add to the LB medium 20 g agar before autoclaving. Fill 100 mm diameter plates, and allow to solidify.
3. Plasmid isolation/purification Maxi Kit (QIAGEN) [13].

2.3 Cell Lines

1. A549 human adenocarcinoma cell line.
2. Culture media: Dulbecco's Modified Eagle's medium (DMEM) supplemented with 10 % inactivated serum, penicillin (10 μg/ml), and streptomycin (50 μg/ml).

2.4 Reagents

1. PBS-10 %: For 1 l of water dissolve 80 g NaCl, 2 g KCl, 16.1 g K_2HPO_4 and KH_2PO_4. Allow to dissolve, and adjust the pH to 7.4. Store at RT.
2. PBS–1 % serum and 1 mM EDTA: Dilute PBS-10 % with water to 1 %, add 1 % of serum (FCS) and 1 mM EDTA (*see* **Note 1**).
3. JetPRIME® DNA and siRNA Transfection Reagent (Polyplus-transfection SA): The kit contains a JetPRIME buffer to dilute DNA and the transfection reagent. Store at 4 °C (*see* **Note 2**).

2.5 Antibodies

1. Mouse monoclonal EMMPRIN antibody (Santa Cruz Biotechnology): Store at 4 °C.
2. Mouse monoclonal antibody EMMPRIN/CD147 (R&D Systems): Store at −20 °C.
3. R-Phycoerythrin-conjugated AffiniPure Goat Anti-Mouse IgG (Jackson ImmunoResearch): Store at 4 °C in the dark.
4. Horseradish peroxidase anti-mouse antibody (Promega): Store at 4 °C.

2.6 Equipment

1. Centrifuge and microcentrifuge.

3 Methods

3.1 Plasmid Transformation and Isolation

This protocol should be carried out on ice unless otherwise specified.

1. Transform the ZFN plasmids into chemically competent *E. coli* cells.
2. Add 50 ng/μl of each plasmid into a vial of chemically competent *E. coli* cells separately.
3. Incubate on ice for 30 min.
4. Subject the cells to a thermal shock at 42 °C for 45 s, and put them on ice for 2 min.
5. Transfer into 900 μl of pre-warmed LB medium, and place to shake horizontally (200 rpm) for 1 h at 37 °C.
6. Plate 10–100 μl from each transformation on a plate with LB agar medium containing the antibiotic for selection (*see* **Note 3**), and incubate overnight at 37 °C (*see* **Note 4**).
7. Pick isolated colonies, and subculture them in a tube containing 5 ml of LB medium with the antibiotic of selection (*see* **Note 3**). Allow growing overnight (orbital incubator, 37 °C, 225 rpm).
8. Isolate the plasmid with the QIAGEN Plasmid isolation Maxi Kit (*see* **Note 5**).

3.2 Cell Line Transfection

All procedures should be carried out under sterile conditions in a laminar airflow chamber (*see* **Notes 6** and 7).

1. Seed A549 cells with 60 % confluence (1×10^6 cells) in a 100 mm diameter dish.
2. After 24 h, transfect the cells using JetPRIME DNA transfection reagent.
3. Dilute 10 μg of DNA (5 μg of each ZFN plasmid) into 500 μl JetPRIME® buffer. Mix by vortexing (*see* **Note 8**).

4. Add 20 µl of JetPRIME reagent vortex, and spin down briefly. Incubate the mixture for 10 min.
5. Add 500 µl of the complex (drop by drop) into the cells, and mix gently.
6. Replace the transfection medium after 6 h by fresh medium.
7. Culture the cells for 1–2 weeks in a humidified atmosphere of 5 % CO_2 at 37 °C.

3.3 Fluorescence-Activated Cell Sorting

All procedures should be carried out under a laminar airflow chamber. During incubation times, place the cells and PBS–1 % serum on ice. Cell clumps should be avoided by resuspending the cells during long incubations (*see* **Note 9**).

1. Place up to 1×10^6 cells in a tube. Centrifuge for 5 min at $700 \times g$. Discharge supernatant (*see* **Note 10**).
2. Wash once with cold PBS–1 % serum. Centrifuge for 5 min at $700 \times g$. Discharge supernatant.
3. Per 1×10^6 cells add 100 µl of CD147 antibody diluted 1/100 in PBS 1× 1 % serum. Incubate for 30 min on ice (*see* **Note 11**).
4. Wash once with PBS 1× serum 1 % by centrifugation at $700 \times g$ for 5 min. Aspirate supernatant.
5. Incubate the cells with the R-Phycoerythrin-conjugated AffiniPure Goat Anti-Mouse IgG (Jackson ImmunoResearch) diluted 1/1,000 in PBS–1 % serum. Incubate for 30 min on ice (*see* **Note 12**).
6. Wash once with PBS 1× FCS 1 % by centrifugation at $700 \times g$ for 5 min. Aspirate supernatant.
7. Add 500 µl PBS 3 % serum 1 mM EDTA, and filter the samples in a fluorescence-activated cell sorting (FACS) tube.
8. Select and sort CD147-negative cells, and plate one cell per well in a 96-well plate with 100 µl DMEM supplemented with 10 % inactivated serum and penicillin (10 µg/ml). Allow the cells to grow in a humidified atmosphere of 5 % CO_2 at 37 °C (*see* **Notes 13** and **14**).
9. After growing in 96-well plates, each clone should be subcultured for the screening of CD147/basigin cells (*see* **Note 15**).

3.4 Screening of CD147/Basigin Cells

1. Freeze some cells of each clone in a vial with serum 1 % DMSO (*see* **Note 16**).
2. Subculture each clone in two different wells. Add 1 mM metformin to one well (for a preselection *see* **Note 17**), and let the other well growing to propagate the clone.
3. Extract proteins from the clones that die in the preselection with metformin.

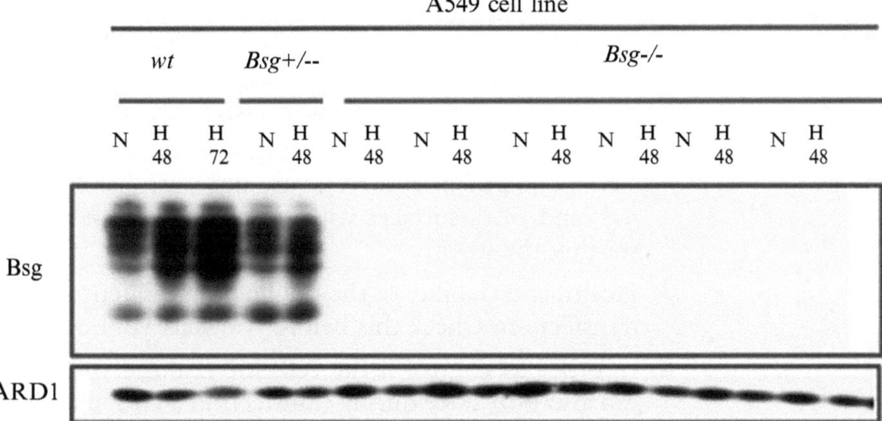

Fig. 1 Western blot analysis of Bsg in transfected cells. A549 wt cells (lanes 1–3) $Bsg^{+/-}$ cells (lanes 4 and 5) and $Bsg^{-/-}$ cells (lanes 6–17) exposed to normoxia (N) and hypoxia 1 % O_2 (H). ARD1 was used as a loading control

4. Perform a Western blot to analyze the expression of CD147/basigin in these clones as shown in Fig. 1 (*see* **Note 18**).
5. Select the clones that were completely negative by Western blot (*see* **Note 19**).
6. Extract genomic DNA from cells.
7. Confirm gene disruption by performing a PCR with the primers provided by ZFN kit (*see* **Note 20**). Send to sequence your PCR products (*see* **Note 21**).

4 Notes

1. Prepare this solution immediately before use. PBS should be cold and sterile.
2. The delivery of ZFN plasmids into the cells could be done with different transfection protocols. The choice of transfection technology can strongly influence transfection efficiency. From all the protocols we tested, we chose this transfection protocol because it is fast and easy to perform, causes minimal cytotoxicity for this cell line, and requires smaller amount of DNA to transfect.
3. The appropriate antibiotic is described in the datasheet of the zinc finger plasmids sent by the manufacturer. In this specific case, the antibiotic used for selection was kanamycin.
4. Plate two different volumes to ensure that at least one plate will have well-spaced colonies.

5. QIAGEN-tip 500 protocol was followed as described by the manufacturer. In the end, DNA was redissolved in approximately 100 μl.
6. Aseptic techniques and the proper use of laboratory equipment are essential when working with cell cultures. Always use sterile equipment and reagents, and wash hands, reagent bottles, and work surfaces with surfanios or 70 % ethanol before starting the work.
7. Health and viability of the cell line can influence the efficacy of transfection. Check this before you start your experiment.
8. In order to control if the cells internalized the plasmids, you can also introduce during transfection a GFP plasmid (0.5 μg). This will give you the information that the transfection went good and cells were properly transfected.
9. This approach cannot be used for all molecules. You need the cells viable so that the antibody should be able to recognize protein without internalization. In case this is not possible, transfection with GFP plasmid should be performed and the cells should be analyzed by FACS after 48 h of transfection. In this case, cells positive for GFP should be sorted.
10. Try to stain the maximum of cells possible to have the number of events necessary for FACS analysis (around 6×10^6 cells). Use separate tubes for staining of 1×10^6 each to make sure that all cells are uniformly stained.
11. Do not forget to have one tube as negative control for CD147/basigin expression. This tube should not be incubated with primary antibody. Incubation time can vary between 30 min and 1 h. Do not forget to resuspend the cells to avoid deposition that could lead to inefficient staining.
12. Protect all tubes from light so as not to degrade the fluorochrome.
13. There are some cell lines that are not able to grow as single cells. For that, 100 μl of conditioned media should be added to the well. The growth factors available in the conditioned media can help cells to attach and to growth.
14. As single cells (clones) are plated in each well, cells will need approximately up 3–4 weeks to grow and generate a cell population. Thus, after 2 days of sorting the cells, make up the well volume to 200 μl with fresh medium to ensure enough nutrients for cell growth.
15. Clones with different size and phenotypes are obtained. You should not forget that besides the selection of negative cells stained by the cell sorter, different cells are also originated as heterozygotic cells (with only one allele cut), homozygotic cells (with the two alleles cut), and cells that have not been

transfected. Each clone should be identified as desired (e.g., 1, 2, 3, …). Select the maximum of clones, and subculture them. This step is critical. Manipulate carefully your cells to avoid contamination.

16. Freezing media depends on the cell line. It normally consists in growth medium (RPMI, DMEM, etc.) containing 5–10 % serum.
17. Metformin is a drug from biguanide class that impairs oxidative phosphorylation. We perform this treatment as a preselection for putative negative clones. The ones that have disruption of CD147/basigin and consequently a decrease in lactate export (inhibiting glycolysis) will die after blocking respiration.
18. Western blot can vary according to the lab. In our lab and for this study, we separated 40 μg of protein on 8 % SDS polyacrylamide gels and transfer onto polyvinylidene difluoride membranes (Millipore). Membranes were incubated with the monoclonal antibody of Bsg/CD147 (R&D Systems) diluted 1/1,000 and then with a horseradish peroxidase anti-mouse antibody (Promega) diluted 1/5,000. Bands were detected with ECL (Amersham Biosciences or Millipore).
19. Select only 5–8 clones to further characterize the decreasing amount of cells to manipulate.
20. PCR reaction was carried out in a total volume of 20 μl, consisting of 200 ng of gDNA, 0.5 μM of both sense and antisense primers (sent with the ZFN kit), 200 μM of dNTPs, 1.5 mM of $MgSO_4$, 1× Taq Buffer, and 10 U of KOD hot start DNA polymerase (all reagents from Novagene). The reaction consisted of an initial denaturation at 95 °C for 3 min, followed by 31 cycles with denaturation at 95 °C for 10 s, annealing at 68 °C for 30 s, and extension at 68 °C for 30 s, followed by a final extension during 3 min at 68 °C.
21. New primers targeting a smaller fragment in the targeted area should be designed and synthesized. These new nested PCR products should be sent for sequencing to confirm the disruption of your gene in your clones.

References

1. Urnov FD, Rebar EJ, Holmes MC et al (2010) Genome editing with engineered zinc finger nucleases. Nat Rev Genet 11:636–646
2. Carroll D (2011) Genome engineering with zinc-finger nucleases. Genetics 188:773–782
3. Casci T (2011) Keeping ZFNs on target. Nat Rev Genet 12:667
4. Foley JE, Yeh J-RJ, Maeder ML et al (2009) Rapid mutation of endogenous zebrafish genes using zinc finger nucleases made by Oligomerized Pool ENgineering (OPEN). PloS ONE 4:e4348
5. Mashimo T, Takizawa A, Voigt B et al (2010) Generation of knockout rats with X-linked severe combined immunodeficiency (X-SCID) using zinc-finger nucleases. PLoS ONE 5:e8870
6. Carbery ID, Ji D, Harrington A et al (2010) Targeted genome modification in mice using zinc-finger nucleases. Genetics 186:451–459

7. Ochiai H, Fujita K, Suzuki K-I et al (2010) Targeted mutagenesis in the sea urchin embryo using zinc-finger nucleases. Genes Cells 15: 875–885
8. Young JJ, Cherone JM, Doyon Y et al (2011) Efficient targeted gene disruption in the soma and germ line of the frog Xenopus tropicalis using engineered zinc-finger nucleases. Proc Natl Acad Sci U S A 108:7052–7057
9. Beumer KJ, Trautman JK, Bozas A et al (2008) Efficient gene targeting in Drosophila by direct embryo injection with zinc-finger nucleases. Proc Natl Acad Sci U S A 105:19821–19826
10. Hicks DG, Longoria G, Pettay J et al (2004) In situ hybridization in the pathology laboratory: general principles, automation, and emerging research applications for tissue-based studies of gene expression. J Mol Histol 35:595–601
11. Morton J, Davis MW, Jorgensen EM et al (2006) Induction and repair of zinc-finger nuclease-targeted double-strand breaks in Caenorhabditis elegans somatic cells. Proc Natl Acad Sci U S A 103:16370–16375
12. Floch L, Chiche J, Marchiq I et al (2012) CD147 subunit of lactate/H+symporters MCT1 and hypoxia-inducible MCT4 is critical for energetics and growth of glycolytic tumors. Proc Natl Acad Sci U S A 109:20166
13. QIAGEN: sample and assay technologies. http://www.qiagen.com/. Accessed 23 Oct 2013

INDEX

Martha Robles-Flores (ed.), *Cancer Cell Signaling: Methods and Protocols*, Methods in Molecular Biology, vol. 1165, DOI 10.1007/978-1-4939-0856-1, © Springer Science+Business Media New York 2014

E

F

G

H

I

K

L

M

N

O

P

MIX
Papier aus verantwortungsvollen Quellen
Paper from responsible sources
FSC® C105338

If you have any concerns about our products,
you can contact us on
ProductSafety@springernature.com

In case Publisher is established outside the EU,
the EU authorized representative is:
Springer Nature Customer Service Center GmbH
Europaplatz 3, 69115 Heidelberg, Germany

Printed by Libri Plureos GmbH
in Hamburg, Germany